Elastic Energy Methods of Design Analysis

Elastic Energy
Methods of
Design Analysis

Ralph J. Harker
University of Wisconsin
Madison, Wisconsin

Elsevier
New York · Amsterdam · Oxford

Elsevier Science Publishing Co., Inc.
52 Vanderbilt Avenue, New York, New York 10017

Sole distributors outside the United States and Canada:

Elsevier Applied Science Publishers Ltd.
Crown House, Linton Road, Barking, Essex IG11 8JU, England

Library of Congress Cataloging in Publication Data

Harker, Ralph J., 1915–
    Elastic energy methods of design analysis.

    Includes index.
    1. Elastic analysis (Theory of structures)
2. Structural design.   3. Girders.   I. Title.
TA653.H37   1986   624.1'71   85-27385
ISBN 0-444-01035-1

Current printing (last digit):
10 9 8 7 6 5 4 3 2 1

Manufactured in the United States of America

To Marcia and Jean

Contents

Preface

THERE ARE FEW FACILITIES of more importance to the design engineer than the ability to analyze a structure or machine component for strength and deformation under service loading. This subject is broadly referred to as strength of materials. Behavioral characteristics are largely predicated upon linear elastic materials, with elastic strain and static equilibrium satisfied in differential elements and in entire volumetric solids.

Academically, most engineering students receive introductory courses in statics and strength of materials, and many have complementary courses in mechanical or structural design. Although these topics seem basic, there are historically formidable conceptual and philosophical difficulties to their understanding. Many of the insights required are acquired through experience and repeated exposure. Even then, there is often some difficulty with the underlying principles.

The principal methods used in this country with respect to beams include the shear diagram as the source of the bending moment diagram by integration and the derivation of the elastic curve from the moment distribution by double integration. The latter technique is also applied to indeterminate beams. At an advanced level, Castigliano's theorem is advocated as the preferred method for analysis of more complex geometries in deflection or redundancies. This theorem requires evaluation of the partial derivative of total elastic energy in a structure with respect to a load.

Although these proper classical methods have some specific inherent advantages, they collectively tend to obscure the physical significance of structural bending and to complicate analytical processes.

In the following presentation, the conservation of energy is the single underlying principle upon which the entire methodology is structured. This concept is both fundamental and understandable. Evaluation of elastic or strain energy in a system permits incorporation on an integrated basis of direct, bending, and torsionally stressed elements for deflections or redundancies. The advantages of such a unified approach are obvious.

Elastic energy solutions are not obtained, however, without a basic understanding of the physical nature of systems, of static analysis, of free-body diagrams, and of moment and torque diagrams. In fact, as this method relates intimately to the physical nature of a problem being solved, the various geometries and loading effects must be completely visualized. Additionally, there are often a number of options available with respect to the exact procedural path that is followed. A proper choice is often crucial to obtaining the least complicated solution, and this facility improves rapidly with experience.

Solutions to the more complex problems can usually be effected by two methods. One is by using fundamental elastic energy relationships, including auxiliary loadings. The other method is by resolving a structure into its elastic components, by decoupling redundant cases, and by invoking superposition. In the latter approach numerous tabulated quantities are provided to expedite the analysis, and these solutions are therefore usually simpler. For checking purposes, results of one method must agree with the other. If errors appear, differences in the corresponding coefficients of the two solutions will identify the source of the errors, and a reconciliation can be obtained.

The energy methods outlined in this book are characterized by conceptual simplicity, and techniques are further provided to minimize the computational effort required in the solutions. Conversely, the successful application of this method does require understanding, intuition, and judgment with respect to the physical subtleties of a problem, and an appreciation of the fundamental laws of mechanics that govern static behavior.

It is my hope that this book will serve as a vehicle by which the reader will become proficient, with serious study, in the concepts and techniques associated with a broadly based understanding of the significance of strain energy as a crucial component in the strength of materials domain.

Acknowledgment is made of the historical basis of this treatise in an initial book titled *Elastic Energy Theory*, by the late Professor J. A. Van den Broek of the Engineering Mechanics Department of the University of Michigan. The subject matter has been significantly altered, reorganized, and expanded, but follows the same unique philosophical pattern.

The material has evolved primarily from an elective course I have offered in Mechanical Engineering that has reinforced the students' understanding of basic statics and the elastic behavior of structures. It has also provided them with effective and practical methods for use in design analysis involving the determination of loads, deflections, and stresses. I now express my appreciation to all of these students as they have individually and collectively contributed to my learning process, as I have to theirs. Without this extensive mutual experience, this book could not have been written.

Ralph J. Harker

Madison, Wisconsin

Notation

$a, b, c..$	$=$	designated links or beams
A	$=$	area of section
	$=$	area under M diagram or under m diagram
	$=$	linear acceleration
$A\bar{x}$	$=$	area moment of M or m diagram
A_n, B_n	$=$	load factors for Three-Moment Theorem
b	$=$	cross-sectional width of rectangular beam
c	$=$	distance from neutral axis to extreme fiber
c_i	$=$	C_i/C_r = ratio of compliance in i to reference compliance
C	$=$	ℓ/EA = tensile–compressive compliance factor
C_x, C_y	$=$	horizontal and vertical compliance factors at a point
d	$=$	diameter of a round section
D	$=$	diameter
E	$=$	Young's modulus
f_i	$=$	F_i/F = ratio of preload in a bar to total auxiliary load on a system
F	$=$	auxiliary force applied to system
g_c	$=$	acceleration of gravity
g_i	$=$	load ratio, twice-redundant truss
G	$=$	shear modulus of elasticity
h	$=$	depth of rectangular beam section
I	$=$	transverse area moment of inertia
I/c	$=$	section modulus of beam area
J	$=$	polar moment of inertia of area of round section
J/r	$=$	section modulus of round section in torsion

K	=	spring rate
ℓ	=	length
m	=	distributed auxiliary-bending moment
M	=	distributed actual-bending moment
M'	=	applied auxiliary couple
p	=	unit internal or external pressure
p, q	=	principal coordinate axes
P	=	internal preload in a link without external load
q	=	constantly distributed load per unit length
	=	maximum unit load
q_i	=	Q_i/Q = ratio of actual link load to externally applied load
Q	=	actual load
r_i	=	R_i/R_r = ratio of link load to reference reaction
R	=	radius of curvature
R_i	=	external reaction
t	=	auxiliary distributed torsional couple
t_i	=	T_i/T = link load ratio, redundant link removed
T	=	applied load simulating Q, redundant link removed
	=	actual torque
	=	temperature
U_e	=	total external energy input
U_s	=	total internal strain energy
V	=	vertical reaction or shear
	=	volume
x, y, z	=	coordinate axes
$\bar{x}$	=	centroidal distance
α, β, γ	=	elastic coefficients for deflection
Δ	=	relative deflection between two points
δ	=	change in length of a link
ϵ	=	unit strain or elongation
θ, ϕ, ψ	=	angles
λ	=	coefficient of thermal expansion
	=	EI/GJ = bending–torsional stiffness ratio
μ	=	mass per unit length
ξ	=	x/ℓ = dimensionless linear coordinate
ρ	=	mass per unit volume
Ω	=	angular rotational velocity

Dimensional Units

CALCULATIONS OF LOADS, BENDING MOMENTS, stresses, deflections, and slopes require numerical data relating to the geometry of a structure and the loads imposed. These can be specified in either SI or British units. Various options are also exercised depending upon the physical size of the system; however, the analytical procedures and results developed can be of a general nature if values are deferred to the final step for quantification. Examples provided in the following chapters are usually carried to a numerical conclusion except for modulus of elasticity, loads, and terms involving the absolute dimensions of the structure. In either the SI or British systems the assumed or *normal* dimensional units are shown in the table. In combination these units will provide consistent results. For instance, lengths given numerically in a problem in centimeters will then result in a deflection in centimeters, with other units as suggested.

Table i Suggested or normal dimensional units for the quantitative determination of loads, deflections, or stresses.

QUANTITY	RECTILINEAR			POLAR		
	SYMBOL	BRITISH	SI	SYMBOL	BRITISH	SI
DEFLECTION	x, y, δ, Δ	in.	cm	θ	rad	
LENGTH	x, ℓ			R	in.	cm
AREA	A	$in.^2$	cm^2			
SECTION MODULUS	I/c	$in.^3$	cm^3	J/r	$in.^3$	cm^3
MOMENT OF INERTIA	I	$in.^4$	cm^4	J	$in.^4$	cm^4
LOAD	Q	lb	N	M, T	in.lb	N·cm
DISTRIBUTED LOAD	q	lb/in.	N/cm			
STIFFNESS	K			K	in.lb/r	N·cm/r
COMPLIANCE	C	in./lb	cm/N	C	r/in.lb	r/N·cm
ENERGY	U	in.lb	J	U	in.lb	J
STRESS	σ			τ		
MODULUS OF ELASTICITY	E	$lb/in.^2$	N/cm^2	G	$lb/in.^2$	N/cm^2
PRESSURE	p					
DENSITY	ρ	$lb/in.^3$	kg/cm^3			

1

Systems with Axial Flexibilities

WE BEGIN THIS TREATISE on elastic behavior with the usual assumption that, for most engineering materials, deformation is directly proportional to load. With this premise we are able to apply the principle of superposition, which permits us to add the effects of component loads algebraically or vectorially, as they create stresses, reactions, or displacements.

This introductory chapter concerns the displacements resulting from loads in statically determinate systems in which the elasticity is restricted to simple axial tension–compression elements. These basic planar structures or mechanisms provide an introduction to elastic energy methodology and a review of elementary static equilibrium. They include pin-connected trusses and mechanical linkages, such as simple links, springs, or cables.

Except when springs are involved, deformations are usually small, especially relative to bending or torsional effects, which are considered in later chapters. Direct stress contributions in more complex systems are occasionally significant, and if so, relationships developed in Chaps. 1 and 2 are applicable and can be incorporated in the complete solutions.

1.1   The Basic Elastic Link

The most elementary type of elastic behavior is displayed by the prismatic bar of any cross-sectional shape loaded in tension or compression along the central axis. Neglecting local stress distortions at the ends, assuming elastic material characteristics, no column-related buckling, and gradual applica-

1

tion of the load, we have uniform direct stress σ over any cross section and throughout the length (Fig. 1.1):

$$\sigma = \frac{Q}{A} \tag{1.1}$$

Unit elongation is

$$\epsilon = \frac{\delta}{\ell} = \frac{\sigma}{E} = \frac{Q}{EA} \tag{1.2}$$

and total elongation is directly proportional to the load:

$$\delta = \epsilon \ell = \left(\frac{\ell}{EA} \right) Q = CQ \tag{1.3}$$

where σ = unit direct stress
    A = cross-sectional area
    E = Young's modulus of elasticity
    ℓ = total effective length of the link

In Eq. (1.3) C is a system constant indicating the *compliance* of the bar. The larger this factor, the more flexible the link. Conversely, this property is

Figure 1.1 Elastic deflection of a simple tensile bar is directly proportional to the applied load. External work done, indicated by the shaded triangle, is converted to strain energy in the bar.

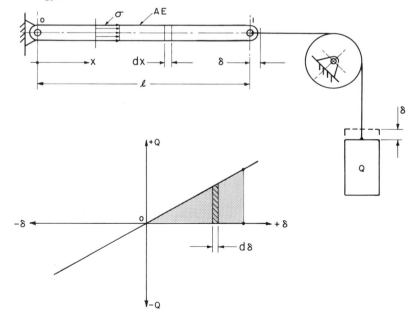

often defined by the *stiffness*, or *spring rate*, K:

$$C = \frac{\delta}{Q} = \frac{1}{K} = \text{deflection per unit load}$$

$$K = \frac{Q}{\delta} = \frac{1}{C} = \text{load per unit deflection}$$

For our purposes the compliance factor C is usually the most convenient term.

The factor K is interpreted (Fig. 1.1) as the slope of the load-deflection line through the origin.

1.2 Stored Elastic Energy

A purely elastic member does not dissipate energy when stressed. Rather it accumulates *potential energy* as it deforms, and this energy is recoverable, as in the winding of a clock spring. To quantify the *internal energy* in a bar, we take an elementary volume $A\,dx$ (Fig. 1.1) for which the element of strain energy is

$$dU_s = \frac{1}{2} Q\epsilon\,dx = \frac{1}{2} Q\left(\frac{\sigma}{E}\right) dx = \frac{1}{2} \frac{Q^2}{EA}\,dx \tag{1.4}$$

where dU_s is $\frac{1}{2}$ the product of the load and the differential elongation. This is a triangular area under the load-deflection curve, with the one-half factor accounting for the fact that the average load during the loading is $\frac{1}{2}$ of the maximum force applied.

For the total bar the entire strain energy is

$$U_s = \int_0^{\ell} \frac{1}{2} \frac{Q^2}{EA}\,dx = \frac{1}{2} \frac{Q^2}{E} \int_0^{\ell} \left(\frac{1}{A}\right) dx \tag{1.5}$$

With A under the integral sign, a variable function $A(x)$ can be accommodated. Otherwise, A can be factored out with E:

$$U_s = \frac{1}{2}\left(\frac{\ell}{EA}\right) Q^2 = \frac{1}{2} CQ^2 = \frac{1}{2} \frac{Q^2}{K} \tag{1.6}$$

Thus the elastic energy stored in a constant bar is directly proportional to its compliance, inversely proportional to its stiffness, and directly proportional to the square of the applied load, whether tensile or compressive.

1.3 External Work Applied by Loading

Having determined the nature and magnitude of the internal elastic energy in a bar, we shift our attention to the source of the stress, elongation, and energy. In Fig. 1.1 the load is applied conceptually by a weight Q connected to the link by a cable at 1.

Considering the downward displacement of Q, it appears that there has been a net change in potential energy of $Q\delta$, and this is true; however, if the weight is lowered gradually the cable tension increases from 0 to Q, as shown in the graph. Work done externally on the bar is $Q\,d\delta$ incrementally, and the total *external energy* transmitted and transformed to elastic energy is given by the triangular area

$$U_e = \tfrac{1}{2}Q\delta \tag{1.7}$$

The supplementary energy ($\tfrac{1}{2}Q\delta$) is attributed to the lowering process in which the force on the hand has decreased from Q to 0 as the deflection increased from 0 to δ. This work done by the weight on the hand accounts for the missing ($\tfrac{1}{2}Q\delta$).

If the weight were released suddenly, a free vibration would ensue with a total initial energy of ($Q\delta$). During decay to the same final equilibrium position, the extraneous energy is dissipated by damping as heat:

$$Q\delta = U_e + U_d = \tfrac{1}{2}Q\delta + U_d$$
$$U_d = \tfrac{1}{2}Q\delta$$

To summarize, a single static loading on an elastic structure (simple or complex) involves a specific quantity of potential energy. External loading is the source, and it enters the bar (Fig. 1.1) at 1, flowing to the left. It becomes stored by the mechanism of deformation throughout the entire stressed volume, much the same as a type of energy reservoir.

The process is reversible; however, for analytical purposes it is only necessary to consider the loading phase to establish the essential energy balance that provides the necessary information for deflection solutions.

1.4   *Deflection Determination*

As a simple example, we now use energy equivalence to calculate the vertical deflection of 1 due to Q (Fig. 1.2a), assuming the bar to be rigid relative to the spring. Equating the external energy at 1 to the internal stored in the spring at 2,

$$U_e = \frac{Q_1 y_1}{2} = U_s = \frac{1}{2}\left(\frac{1}{K_2}\right)Q_2^2 = \frac{1}{2}C_2 Q_2^2 \tag{1.8}$$

where Q_2 is the force exerted on K_2. From equilibrium

$$\sum M_0 = Q_1 a - Q_2 \ell = 0$$
$$Q_2 = \left(\frac{a}{\ell}\right)Q_1$$

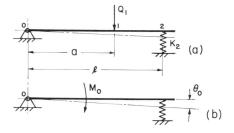

Figure 1.2 Small deflections of the pivoted rigid bar are determined using either transverse or rotational coordinates.

Substituting

$$\frac{Q_1 y_1}{2} = \frac{1}{2K_2}\left(\frac{a}{\ell}Q_1\right)^2$$

$$y_1 = \left[\left(\frac{a}{\ell}\right)^2 \frac{1}{K_2}\right]Q_1$$

This simple problem can be solved by other methods using force and geometric relations, but using conservation of energy we have only applied statics for force relationships.

1.5 *Rotational Displacement*

Strain energy is a simple scalar quantity defined by the compliance of elastic elements under axial loads for these basic systems; however, external energy can be applied translationally or angularly. In Fig. 1.2 Q_1 is equivalent to a moment of $M_0 = (Q_1 a)$ about the pivot, and we equate external to internal energy as before:

$$\frac{M_0 \theta_0}{2} = \left(\frac{1}{2K_2}\right)Q_2^2$$

From statics $Q_2 = M_0/\ell$ and

$$\theta_0 = \left(\frac{1}{K_2 \ell^2}\right)M_0 = \left(\frac{a}{K_2 \ell^2}\right)Q_1$$

1.6 *Multiple Elastic Elements*

With several links or springs, the procedure is similar to obtain a single deflection at a single load. For example, in Fig. 1.3 a two-bar linkage carries a horizontal load of Q_1, and we will determine the horizontal deflection x_1

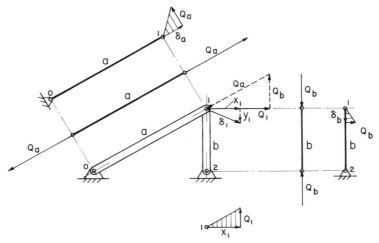

Figure 1.3 External work input at 1, stored as elastic energy in a and b, determines the horizontal displacement x_1.

of the pin. Energy balance for i elements requires that

$$U_e = \frac{Q_1 x_1}{2} = U_s = \sum \frac{C_i Q_i^2}{2}$$

$$x_1 = Q_1 \sum C_i \left(\frac{Q_i}{Q_1} \right)^2 \tag{1.9}$$

There are two factors to be evaluated in the summation for each link—the compliance and the load ratio. The former involves the link geometry and material and the latter static equilibrium.

We have a force polygon at 1 in Fig. 1.3, with summation of forces on the pin equal to 0. Q_a and Q_b are reacted by axial-link forces in tension and compression, respectively. Equation (1.9) becomes

$$x_1 = Q_1 \left[C_a \left(\frac{Q_a}{Q_1} \right)^2 + C_b \left(\frac{Q_b}{Q_1} \right)^2 \right] \tag{1.10}$$

Although Q_b is technically negative since it is compressive, when squared all terms are positive. Physically, this means that the deformation of all links in a system are *additive* to produce the total deflection of a given load.

The relative deflections of a and b are to scale in Fig. 1.3 for $C_b / C_a = 1.5$, and the hatched triangles indicate that a stores more energy than b, but collectively the two small triangles equal the area of the large triangle:

$$\frac{Q_1 x_1}{2} = \frac{Q_a \delta_a}{2} + \frac{Q_b \delta_b}{2} \tag{1.11}$$

Loads at the ground points, Q_a at 0 and Q_b at 2, do not enter the energy equation as there is no deflection possible at these support pins. Therefore, no work is done during the application of Q_1.

Obviously, the pin at 1 has moved vertically as well as horizontally. In fact, the vertical deflection is equal to the compression of b, or $C_b Q_b$. Generally, the deflection of a load will not coincide with the direction of application of a load. Using Eq. (1.9) leads to a single component deflection —the component *in the direction of the load*. This follows from the argument of the work done. Practically, however, this is usually the component of greatest importance.

1.7 *Complementary Energy*

In Fig. 1.1 and Eq. (1.6) the strain energy developed in an elastic member is indicated for a single axial load. Application of two successive tensile loads F and Q (Fig. 1.4) leads to a similar result, with the final load $(F + Q)$ and the final elongation $(F + Q)/K$ or $C(F + Q)$. Total, or final, potential energy in the spring is

$$U_s = \frac{C}{2}(F + Q)^2 = \frac{C}{2}\left(F^2 + Q^2 + 2FQ\right) \tag{1.12}$$

As expanded we note the strain energy is composed of three terms:

$(U_s)_F = CF^2/2 = (CF)F/2 = $ strain energy stored as the spring force increases from 0 to F

$\quad\quad = (U_e)_F = $ external energy provided by the application of the first load F only (Fig. 1.4a)

Figure 1.4   Under successive loads F and Q, the complementary energy developed is represented by the shaded rectangular area $F\delta_Q$.

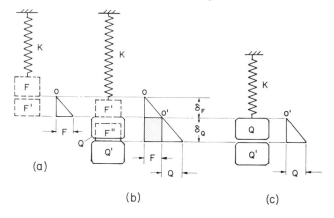

(a)

(b)

(c)

$(U_s)_Q = CQ^2/2 = (CQ)Q/2 =$ strain energy associated with Q only as the load increases from F to $(F + Q)$ (Fig. 1.4b, c)

$= (U_e)_Q =$ external energy provided by the application of the second load only, or if Q were applied singly

$(U_s)_{FQ} = CFQ = (CQ)F = complementary$ strain energy corresponding to the superposition of the second stress upon the first

$= (U_e)_{FQ} =$ external energy defined by *constant* force F acting through the displacement (CQ), or through the displacement caused by the application of the second force Q

We will term the initial force the *preload*, or *auxiliary load*. The second force represents the *actual load*, and the second displacement, δ_Q, the *actual displacement*, which is *in the direction of F* and is the objective of the loading sequence.

As seen in Fig. 1.4, the three energy terms above correspond to two triangular areas and a rectangular area, respectively. The area of the significant rectangle is CFQ or $F\delta_Q$.

To determine deflection we equate the complementary energies:

$$(U_e)_{FQ} = F\delta_Q = (U_s)_{FQ} = CFQ \tag{1.13a}$$

$$\delta_Q = CQ = \frac{Q}{K} \tag{1.13b}$$

Cancellation of F shows the solution to be independent of the magnitude of the auxiliary load and it can be taken as unity.

In Fig. 1.4 with F and Q acting vertically, there is an apparent ambiguity with respect to the total energy input by F and Q during the individual applications. There is a net change in potential energy of $F\delta_F$ during the first phase, and $Q\delta_Q$ during the second, rather than the triangular energy areas shown in Fig. 1.4a and c. This change is attributed to the *gradual* lowering of these weights, which requires resistive work during the release as the supporting force decreases from maximum to 0 and accounts for the missing triangles.

But during the lowering of Q, F is in place and *fully acting*, and the corresponding change in its potential energy is $F\delta_Q$, and we have again confirmed the rectangular area.

1.8 *Deflection in Any Direction*

Equation (1.13a) can be extended to determine the displacement of any point in a structure in any direction with single or with multiple elastic elements. For instance, with the two-bar system in Fig. 1.5, there is a horizontal load, and we will find the vertical displacement at 1. Deflections are shown exaggerated, because elastic deflections usually cause negligible

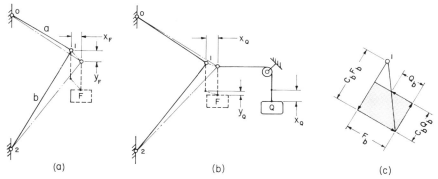

Figure 1.5 Horizontal and vertical deflections occur at 1 due to both the auxiliary load F and the actual load Q. The complementary-energy rectangle for b, shown in (c), indicates first compressive and then tensile loading.

changes in system geometry. This exaggeration may not always be true if actual springs are involved rather than links.

In this sequence, F is applied gradually in Fig. 1.5a, causing the weight F to move down and to the right. External work is added in the vertical direction of $\frac{1}{2}Fy_F$, but no force acts horizontally and no work results from x_F.

Next, Q is applied by a vertical weight, which loads the linkage horizontally by a cable, causing displacements x_Q and y_Q of the pin and of F. No work is done by the former, as there is again no horizontal component of F, but the work related to y_Q with F fully acting is the complementary external energy $(U_e)_{FQ}$. This energy is stored as counterpart internal complementary energy in the individual links, as indicated in Sec. 1.7, yielding the relations

$$Fy_Q = \sum C_i F_i Q_i \tag{1.14a}$$

$$y_Q = Q \sum C_i \left(\frac{F_i}{F}\right)\left(\frac{Q_i}{Q}\right) \tag{1.14b}$$

$$= C_a Q \sum c_i f_i q_i \tag{1.14c}$$

where y_Q = displacement of a structure at the point of application of F and in the direction of F due to the actual load Q

$C_a Q$ = reference displacement if Q were applied directly to a reference element a as a simple tensile load on that element

$c_i = C_i/C_a$ = ratio of compliance factor at any element to the compliance of the reference element C_a

$q_i = Q_i/Q$ = ratio of load on an individual element to the applied external load, or the *actual load factor*

$f_i = F_i/F$ = ratio of preload on an individual element to the auxiliary load, or the *auxiliary load factor*

With respect to f_i and q_i, these are the actual loads obtained if, for convenience, we take $F = Q = 1$, which is usually desirable.

Figure 1.5c shows the nature of the strain energy components for the link b.

1. Due to F there is a compressive displacement of 1 towards 2 of $C_b F_b$ under a load of F_b and an energy of $\frac{1}{2} C_b F_b^2$.
2. Due to Q there is a tensile displacement of 1 away from 2 of $C_b Q_b$ under the tensile load Q_b with an added energy of $\frac{1}{2} C_b Q_b^2$.
3. In the presence of F, Q creates the shaded complementary area, or the rectangle $C_b F_b Q_b$.

Since F_b and Q_b are of opposite sense, the product complementary energy is *negative*; however, a is subjected to two successive tensile loads creating a *positive* complementary energy term. The summation in Eqs. (1.14) is algebraic, and y_Q results from a difference. If these two terms should be equal and opposite, there is no vertical deflection of 1. This can be achieved in design by choice of the proper c_i ratio; that is, by using a particular ratio of cross-sectional areas for the given lengths.

We observe that Eq. (1.14c) is general in the sense that all three terms in the summation are normalized. The c_i terms represent only relative compliances. The force ratios derive only from the geometry of the linkage or the angles. Thus in these dimensionless terms, the structure could be of any physical size and carry any load. Absolute scale is obtained in the $C_a Q$ term.

1.9 Deflection by Compliance Superposition

In Fig. 1.5b the actual load Q produces two orthogonal components at 1, calculated from

$$x_Q = Q\left[C_a q_a^2 + C_b q_b^2\right] \tag{1.15a}$$

$$y_Q = Q\left[C_a f_a q_a + C_b f_b q_b\right] \tag{1.15b}$$

$$\delta_1 = \sqrt{x_Q^2 + y_Q^2} \tag{1.15c}$$

It is apparent that both components of the actual displacement are determined collectively by the axial deformations of the two links indicated by the two terms in brackets, and the contributions are relative to these terms. If one link were rigid ($C_i = 0$), pin displacement would be determined entirely by the elastic behavior of the remaining link.

For instance, if the link a were stiff in Fig. 1.5b, the Q loading would cause 1 to rotate about 0, with 1 moving up and to the right. Deflection components would be

$$x_{Q0} = Q[C_b q_b^2] \tag{1.16a}$$

$$y_{Q0} = Q[C_b f_b q_b] \tag{1.16b}$$

Conversely, with b rigid, $C_b = 0$ and 1 rotates about 2, and

$$x_{Q2} = Q[C_a q_a^2] \tag{1.17a}$$

$$y_{Q2} = Q[C_a f_a q_a] \tag{1.17b}$$

Load factors f_i and q_i derive from statics and the geometry and are independent of link stiffness. The factor $f_i q_i$ is therefore a useful design parameter.

Thus deflection behavior of a structure with any number of links can be reduced conceptually to a series of component deflections, each calculated with only one elastic member responsible for that component, with all other links in the system rigid. This enables us to better visualize the sources of calculated displacements if it is necessary to control elastic behavior in a design situation.

1.10 System Compliance

We finally define the composite compliance of a system, *primary*, in the direction of the applied force

$$C_{xx} = \left(\frac{x_Q}{Q}\right) = C_a[q_a^2 + c_b q_b^2] \tag{1.18a}$$

and a *secondary*, or cross compliance, in the y direction due to a force in the x direction of

$$C_{yx} = \left(\frac{y_Q}{Q}\right) = C_a[f_a q_a + c_b f_b q_b] \tag{1.18b}$$

On a completely dimensionless basis the actual deflection components are related to the reference deflection by

$$\frac{x_Q}{C_a Q} = [q_a^2 + c_b q_b^2] \tag{1.19a}$$

$$\frac{y_Q}{C_a Q} = [f_a q_a + c_b f_b q_b] \tag{1.19b}$$

System compliance is not restricted to normal components and can be defined in any angular direction.

1.11 *Reciprocity of Displacements at a Point*

In Fig. 1.5 we have determined the vertical deflection due to a horizontal load at 1. Conversely, we can apply the load vertically and find the resulting horizontal deflection. This solution is equivalent to interchanging F and Q; that is, the load is vertical and the auxiliary load is horizontal. This interchanges f_i and q_i in Eq. (1.15b), but the values are as before:

$$x_{Qy} = Q[C_a q_a f_a + C_b q_b f_b] \qquad (1.20)$$

The quantity in brackets is *unchanged* and we have the important result:

$$x_{Qy} = y_{Qx} \qquad (1.21)$$

Using the complementary energy argument we have verified one interpretation of Maxwell's theorem: If a force at a point in a system causes a displacement in a second direction, the same force applied at that point in the second direction will cause an identical deflection in the original direction of the force. We are not limited to orthogonal directions.

1.12 *Directional Considerations*

We have considered only vertical and horizontal effects in Fig. 1.5. Although these components are convenient, the structure can be loaded in any polar direction in the plane (Fig. 1.6a). The displacement component which results can similarly be obtained in any direction by applying the auxiliary load F in that direction (Fig. 1.6b).

As a special case, Q can be colinear with either link. If it is aligned with a, then there is no actual load in b, and $q_a = 1$ and $q_b = 0$. Thus in Eq. (1.14c) there can be no terms involving the products $f_i q_i$, with the latter 0

Figure 1.6 Actual and auxiliary loads can be directed in any angular direction in the plane. If Q is coaxial with a, a transverse displacement δ_b will occur due to the presence of link b.

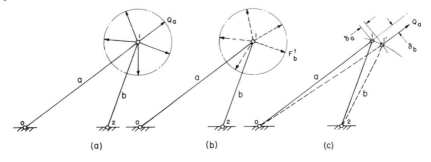

(a)   (b)   (c)

for any links except the aligned link. The aligned link then has an axial displacement of $C_a Q$ (Fig. 1.6c), but also has a component transverse to the link related to any transverse F loading (Fig. 1.6b), and to the complementary energy in this link only.

$$\delta_{FQ} = (C_a Q) f_a \tag{1.22}$$

Taking F perpendicular to b, we determine the component of the displacement of 1 caused by the rotation of the unloaded link b, δ_b. This behavior conforms to the condition of no elastic change in b due to Q and of the independence of the result of C_b, which could be rigid.

Another possibility would be for F to be aligned with a link (say b). This is then the only link subjected to auxiliary loading f_b, and we have obtained the axial deflection of b only:

$$\delta_{bQ} = C_a Q \sum c_i f_i q_i = C_b Q_b \tag{1.23}$$

1.13 *Principal Elastic Axes*

Displacement, resulting from a load at a point in an elastic structure, does not coincide with the direction of the force except in special situations. In Fig. 1.5 we can determine a component in the load direction and a second perpendicular to the load, and the vector sum of these two components represents the magnitude and direction of the *total*, or *resultant displacement*.

In Fig. 1.7 we resolve Q into horizontal and vertical components and use an xy coordinate system with Q acting at any angle with respect to the x axis. Two deflection components result from Q_x (Fig. 1.7b) and two from Q_y (Fig. 1.7c). Summing

$$x = C_x Q \cos \theta + C_{xy} Q \sin \theta \tag{1.24a}$$

$$y = C_y Q \sin \theta + C_{yx} Q \cos \theta \tag{1.24b}$$

where C_x = horizontal system compliance factor
 C_y = vertical system compliance factor
 $C_{xy} = C_{yx}$ = transverse system compliance factor, proven equal from Maxwell's theorem

The square of the resultant displacement is (Fig. 1.7d)

$$\left(\frac{\delta_Q}{Q} \right)^2 = C_x^2 \cos^2 \theta + C_y^2 \sin^2 \theta + C_{xy}^2 + (C_x + C_y) C_{xy} \sin 2\theta$$

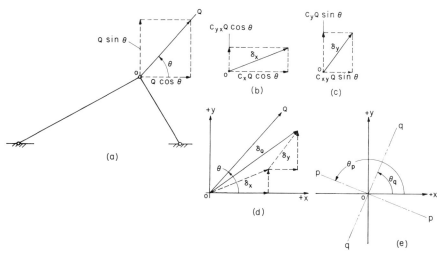

Figure 1.7 Q applied at θ has x and y components resulting in four displacement components (d). Perpendicular principal axes p and q are shown in (e).

Setting the derivative equal to 0 for maximum or minimum deflection per unit load,

$$\frac{d\delta_Q^2}{d\theta} = -2C_x^2\sin\theta\cos\theta + 2C_y^2\sin\theta\cos\theta + 2(C_x + C_y)C_{xy}\cos 2\theta = 0$$

from which

$$\tan 2\theta_p = \left(\frac{2C_{xy}}{C_x - C_y}\right) \qquad (1.25)$$

where θ_p denotes the angular direction in which Q is directed *to produce maximum deflection*. It is the direction of least stiffness and defines a *principal elastic axis*. And if Q is applied in this direction, there is no transverse component, or coupling; that is, *the force and the resulting displacement are colinear.*

Another principal axis exists perpendicular to the first, as a solution of Eq. (1.25), corresponding to minimum deflection, or maximum stiffness, at an angle θ_q.

We can further evaluate the maximum and minimum compliance factors from the previous relations:

$$C_p = \left(C_x + C_{xy}\tan\theta_p\right) \qquad (1.26a)$$

$$C_q = \left(C_x + C_{xy}\tan\theta_q\right) \qquad (1.26b)$$

We must be consistent in the use of signs and realize that C_x and C_y will generally both be positive; however, C_{xy} carries algebraic sign and can be either. If C_{xy} is positive, C_{yx} is also positive and vice versa.

Principal directional elastic axes are analogous to principal axes of the cross section of a beam, which are the uncoupled directions for bending. Both concepts are important in vibratory analysis, as they define directions of uncoupled vibratory modes when supporting a concentrated mass.

1.14 Displacement at a Second Point

The previous discussion has concentrated on elastic deformations at a single point, and a number of fundamental relationships have been developed for this most basic case. More generally, we are interested in the deflection at any point in a structure or mechanism due to a load. All the results that have been developed are applicable to this problem.

In Fig. 1.8*a* we have a typical five-bar linkage, with the ground link 0–3 a rigid element. To calculate the vertical deflection at the load we normalize Eq. (1.10) to

$$y_2 = C_a Q_2 \sum_a^d c_i q_i^2 \qquad (1.27)$$

Actual load factors are obtained from statics using closed polygons for

Figure 1.8 Deformation of a system can be caused by the contributing deformations of all links (*a*) or by just one element (*b*).

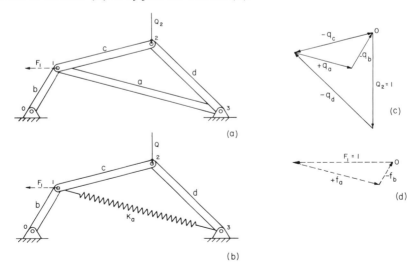

equilibrium at pins 2 and 1 (Fig. 1.8c). Compressive link loads are shown to be negative. Compliance factors must be calculated for each link.

To find the horizontal deflection of 1 we require an auxiliary loading F_1, carried to ground by links a and b, with c and d unstressed. This polygon (Fig. 1.8d) provides the auxiliary load factors for Eq. (1.14c):

$$x_1 = C_a Q_2 \sum_{a}^{d} c_i f_i q_i \qquad (1.28)$$

With only two positive terms summed, this displacement is caused by the compressive shortening of b and the tensile elongation of a. Since the result is positive, the deflection is to the left as assumed. If F_1 had been taken to the right, f_a would have been negative and f_b positive, indicating a negative x_1. As in statics, an incorrectly assumed direction is thus identified.

It is logical that the solution is independent of the stiffness of c and d because displacement at 1 is controlled by the elasticity of a and b, which links receive the applied load through c at 1.

1.15   *The Single Spring System*

If one bar in a statically determinate system is relatively flexible, for instance, a tensile helical spring a (Fig. 1.8b), we have a similar solution; however, all links, except a, are now effectively rigid ($C_i = 0$). Then elastic energy is only stored in a and Eq. (1.27) becomes

$$y_2 = \left(\frac{Q_2}{K_a} \right) q_a^2 \qquad (1.29)$$

Equation (1.28) becomes

$$x_1 = \left(\frac{Q_2}{K_a} \right) f_a q_a \qquad (1.30)$$

where the loading factors are the same as before in Fig. 1.8c and d. The load in the spring is $(q_a Q_2)$ and the extension is $(q_a Q_2)/K_a$.

Displacements are assumed to be small so the geometry shown is only slightly distorted. Otherwise, the load factors should be modified to an average position.

1.16   *Relative Deflections*

With a pinned frame (Fig. 1.9a), loaded by colinear forces Q, we have no pin grounded, and therefore, no reference position or coordinate origin. The

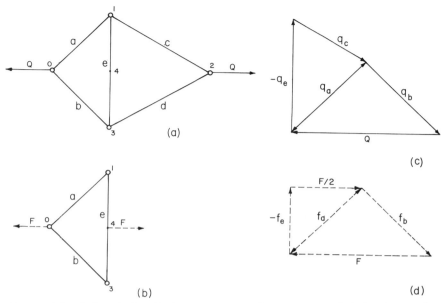

Figure 1.9   Only relative displacements occur in an ungrounded structure, calculated by considering pairs of equal and opposite loads (actual or auxiliary).

points 0 and 2 move apart, and this relative displacement can be calculated using the basic energy argument

$$\frac{Qx_{02}}{2} = \frac{\Sigma C_i Q_i^2}{2} \tag{1.31}$$

If the point 0 is grounded, 2 moves to the right the absolute distance x_{02}. If 2 is grounded 0 moves to the left a distance x_0. We see that the x_{02} deflection is essentially independent of how the structure moves with respect to ground. Using Eq. (1.31)

$$x_{02} = C_a Q \left[2 c_a q_a^2 + 2 c_c q_c^2 + c_e q_e^2 \right] \tag{1.32}$$

where load factors are determined in Fig. 1.9c. Only three are required because of the symmetrical system. Relative displacement of the two ends of any given link is simply $C_i Q_i$.

If we wish to find the relative deflection between any two points (say 0 and 4), we must use a pair of opposed auxiliary loads (Fig. 1.9b). As applied to the frame there is no reaction to ground and we have static equilibrium. The auxiliary load factors are shown in Fig. 1.9d, and the displacement is

$$x_{04} = C_a Q \sum c_i f_i q_i \tag{1.33}$$

Note the F loading develops both bending and compression in e and no loading in c or d. The auxiliary bending does not affect the complementary energy consideration because there is zero bending in e due to actual loading. The only links which have a load product are a, b, and e. x_{04} results from a combination of compression in e and elongation in a and b.

We could similarly determine the relative deflection x_{24} involving c, d, and e. The compression of e is seen to contribute to the relative deflections in both triangular branches.

Examples

1.1. A pivoted uniform bar weighing 26 lb is supported by two springs in series as shown. Find:
   (a) The load in the springs
   (b) The static vertical deflection at 2
   (c) The value of K_b to limit y_2 to 0.10 in.

Solution:
(a) Summing moments about the pivot,

$$\sum M_0 = 70Q_a \cos 45° - 40(26) = 0$$

$$Q_a = Q_b = 21.01 \text{ lb} \qquad q_a = q_b = 21.01/26 = +0.808$$

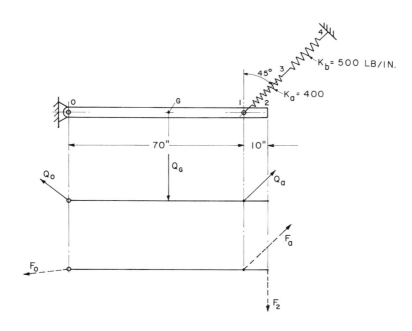

(b) Applying the auxiliary load $F_2 = 1$,

$$M_0 = 70F_a\cos 45° - 80F_2 = 0$$

$$\left(\frac{F_a}{F_2}\right) = f_a = f_b = 1.616$$

From Eq. (1.28), with $c_b = \left(\dfrac{1/500}{1/400}\right) = 0.80$

$$y_2 = \tfrac{1}{400}(26)[1.616(0.808) + 0.80(1.616)(0.808)]$$
$$= 0.153 \text{ in.}$$

(c) Load factors are unchanged, but we solve for c_b

$$0.10 = \tfrac{1}{400}(26)(1.306)[1 + c_b]$$

$$c_b = 0.178 \qquad C_b = 0.000445 \qquad K_b = 2247 \text{ lb/in.}$$

1.2. The two-bar linkage is made of steel, with $\ell_a = 540$ cm and $A_a = 6.2$ cm². Both links have the same cross section and are uniform. Calculate the system compliances with respect to a horizontal load.

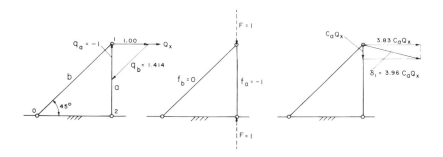

Solution: Using a tabulated format for normalized Eqs. (1.27) and (1.28) for the direct and transverse effects,

i	c	f	q	cq^2	cfq
a	1	-1	-1	1	1
b	1.414	0	1.414	2.828	0
			Σ	3.828	1

$$x_1 = C_a Q_x(3.83) \qquad C_a = \frac{540(10)^{-6}}{6.2(20)} = 4.35(10)^{-6} \text{ cm/N}$$

$$C_x = 16.7(10)^{-6} \text{ cm/N}$$

$$C_{yx} = 4.4(10)^{-6} \text{ cm/N}$$

1.3. Repeat Example 1.2 for vertical loading of the pin.

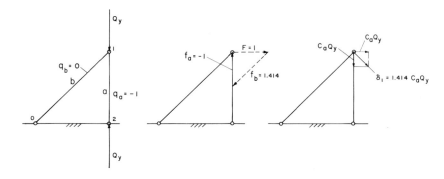

Solution: With a similar tabulation, we have

i	c	f	q	cq^2	cfq
a	1	-1	-1	1	1
b	1.414	1.414	0	0	0
			Σ	1	1

$$y_1 = C_a Q_y (1.00) \qquad C_y = 4.4(10)^{-6} \text{ cm/N}$$
$$C_{xy} = 4.4(10)^{-6} \text{ cm/N}$$

1.4. For the two bar structure in Examples 1.2 and 1.3, find:
(a) The angular directions of the principal elastic axes
(b) The corresponding maximum and minimum system compliances

Solution:
(a) From Eq. (1.25)

$$\tan 2\theta_p = \left(\frac{-2C_a}{3.83C_a - C_a} \right) = -0.704$$
$$\theta_p = -17.6° = +162.4°$$
$$\theta_q = 90 - 17.6 = 72.4°$$

With conventional xy axes (Fig. 1.7e), $C_{xy} = C_{yx} = -C_a$.
(b) From Eqs. (1.26)

$$C_p = [3.83 + (-1)(-0.317)] C_a = 4.15 C_a$$
$$C_q = [3.83 + (-1)(3.15)] C_a = 0.69 C_a$$

The system is stiffer in the q direction than the p direction by a factor of $4.15/0.69 = 6.0$.

1.5. The four-bar truss carries a vertical load Q_1. Find:
   (a) The vertical deflection of 1
   (b) The horizontal deflection of 2

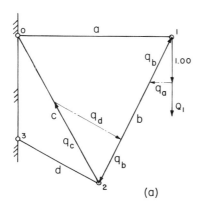
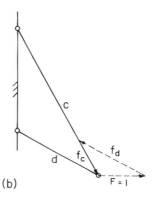

(a)   (b)

i	l	A	c	f	q	cq^2	cfq
a	2.00	12	1	0	0.50	0.25	0
b	2.20	14	0.94	0	-1.13	1.20	0
c	2.25	12	1.13	-0.96	2.08	4.89	-2.26
d	1.20	19	0.38	1.67	-1.70	1.10	-1.08
					$\sum$	7.44	-3.34

Solution:

(a) Actual load factors q_i are determined from two polygons using the common load q_b and from the summation cq^2

$$y_1 = 7.44 C_a Q_1$$

(b) The auxiliary load F loads only c and d, the negative summation indicating an actual deflection to the left of

$$x_1 = 3.34 C_a Q_1$$

2

Indeterminate Systems with Axial Flexibilities

THE ANALYTICAL CONCEPTS of the previous chapter are now expanded to include the determination of load distribution in a structure or mechanism that has one or more degrees of redundancy; that is, there are additional constraints beyond the minimum necessary for stability and static equilibrium analysis. This redundant feature can be in the form of excess links or supports to ground. Elastic energy, and in particular complementary energy, provides techniques by which we can calculate loads developed in the various elements of a structure. These in turn depend upon geometric considerations and relative stiffnesses, with the stiffer links tending to carry more load. In fact, it is the pattern of elastic behavior that governs how a system distributes the loads which it carries; however, the elastic energy analysis only supplements the conditions of static equilibrium. Elasticity augments but does not preempt statics.

In one approach to indeterminacy, we use superposition, selecting one link as redundant and determining the load in it by the energy argument. The effect of this link is then added to the loads produced in the statically determinate reduced system, decoupled by the removal of the redundant element. Energy relations from Chap. 1 can also be applied to indeterminate cases by calculating several deflection combinations using superposition of deflections to satisfy constraints.

Displacements of the indeterminate structure can be obtained after the redundant solution, removing constraints to reduce the problem to static determinacy. Auxiliary loading can then be utilized to calculate any desired deflection.

22

The indeterminate structure, unlike the determinate, can experience internal loads due to misalignment or differential thermal expansion. These are also solved by energy methods and preloads of this type superimposed on external or working loads.

This discussion is mainly confined to the single redundancy case, which is certainly the most important practically; however, equations are indicated by means of which the methods developed can be extended to the twice redundant system.

2.1 Identification of Redundancies

Figure 2.1a illustrates several ways in which the determinate truss becomes redundant. In Fig. 2.1b links c and d have been added. While it is still true that $\Sigma F_x = \Sigma F_y = 0$ at the central pin, these equations do not suffice with four forces acting. Division of the load involves the several flexibilities and elastic characteristics, and the structure is twice redundant.

The same linkage has been provided with a horizontal guide in Fig. 2.1c, which provides support with respect to vertical displacement due to Q_1. This is one redundant constraint, but if the spatial direction of the guide should coincide with the resultant direction of the deflection of 1, no constraint is

Figure 2.1 The determinate linkage (a) is once redundant if guided (c), and twice redundant with two additional links (b). In (d) there is one redundant spring, constituting a redundant link.

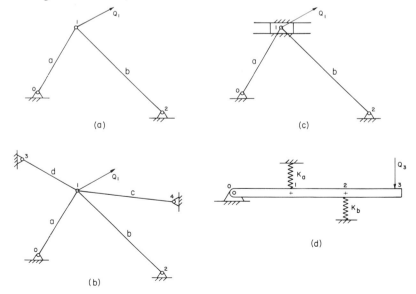

provided by the guide. Thus there is a degree of interplay between the loading and the structure that makes it difficult to define redundancy too specifically.

Figure 2.1d indicates a simple link with two springs. Since Q_3 can be carried by either, a second spring is redundant, and the division of the load obscured.

For solution purposes, we can select any of the constraints as the redundant elements. In Fig. 2.1b we can eliminate any two links and retain any two in order to reduce the system to determinacy. In Fig. 2.1c we can remove a and carry Q_1 by means of the guide and b. In Fig. 2.1d we can consider either K_a or K_b the redundant spring. The choice should be made to simplify the analytical effort if possible.

2.2 *Solution by Link Preload*

In Fig. 2.2a there is one redundancy, which we arbitrarily take as the link a. Removing a, the reduced truss, Fig. 2.2b is statically determinate. Load distribution in this situation is obtained by application of a unit load T in the direction of Q_1, and the decoupled link load factors t_b and t_c are obtained from statics. We next reintroduce the effects of a by evaluating the actual load in a due to Q_1 using complementary energy equivalence.

Taking the original truss with neither Q_1 nor T acting (Fig. 2.2c), we preload the structure by means of an auxiliary loading in a corresponding to unit tensile force. We effectively apply two equal and opposite loads

Figure 2.2 The once-redundant structure is analyzed in two phases—first by decoupling (b) and then by internal preloading (c).

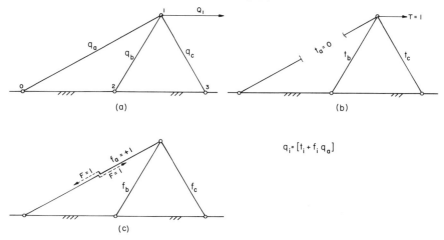

$$q_i = [t_i + f_i\, q_a]$$

($F = 1$), which can be visualized as controlled by a turnbuckle at the junction shown, inducing a compressive load f_b and a tensile load f_c. In *a* we have $f_a = +1.00$.

The pair of F loads constitutes *internal* auxiliary loading. This contrasts with the methods in Chap. 1 in which a single F loading (Fig. 1.5*a*) provides *external* auxiliary loading.

The actual loading (Fig. 2.2*a*) corresponds to link loads, which are the sum of the effects of Fig. 2.2*b* and *c*. By superposition, the load in any link is

$$Q_i = (t_iQ_1 + f_iQ_a) = Q_1[t_i + f_iq_a]$$
$$\left(\frac{Q_i}{Q_1}\right) = q_i = [t_i + f_iq_a] \tag{2.1}$$

Now to calculate the unknown factor q_a, we use the equivalence of complementary energy as in Sec. 1.7:

$$(U_e)_{FQ} = (U_s)_{FQ} = \sum C_iF_iQ_i \tag{2.2}$$

But the external work $(U_e)_{FQ}$ done by the auxiliary loading as the actual loading is applied is 0, because there can be no external manifestation of the dual or *internal F loading*. Viewed another way, if the junction in Fig. 2.2*c* moves toward 1 as Q_1 is applied, there is positive work done by F, which is acting on the lower part of link *a*; however, the F load on the upper part of *a* is displaced against the sense of the force, corresponding to negative work. The two effects are equal and opposite, with zero net work and zero change in energy.

This important relation is basic to all indeterminate solutions by the energy method, and in this instance we have

$$\sum C_iF_iQ_i = \sum C_if_iQ_1[t_i + f_iq_a] = 0 \tag{2.3a}$$

Cancelling Q_1 and C_a, we have

$$\sum c_if_it_i + q_a\sum c_if_i^2 = 0$$
$$q_a = \frac{-\sum c_if_it_i}{\sum c_if_i^2} \tag{2.3b}$$

where $q_a = (Q_a/Q_1)$ = load in reference link for unit external load
$c_i = (C_i/C_a)$ = ratio of compliance of any link to that of any reference compliance
$f_i = (F_i/F_a)$ = load in any link per unit tensile preload in the reference link
$f_a = +1$
$t_i = (T_i/T)$ = load in any link per unit external force with system decoupled by removal of reference link
$t_a = 0$
$\sum$ = summation of all links in the structure

Since the summation in Eq. (2.3a) is 0, we expect both negative and positive products of F_iQ_i, and the algebraic sum of the rectangular areas representing the complementary strain energies in the links is 0. It is convenient to tabulate the factors in Eq. (2.3b) as shown in the examples. A negative result indicates compression under load of the reference link.

Once q_a is determined, we return to Eq. (2.1) for the final load factors due to actual loading.

2.3   Redundant Solution by Reaction Preload

For a slightly modified problem (Fig. 2.3), we consider an external constraint the reference redundancy, or link, and follow a generally similar procedure. With the roller support removed at 3 to effect decoupling, the truss 1-2-3 tends to rotate clockwise as Q_1 is applied, with the loading carried by a and b. In determining the degree of assistance provided by the roller, we evaluate the reference redundancy R_3.

As before we decouple the redundant element, apply $T = 1$ in Fig. 2.3b and obtain t_a and t_b. Auxiliary loading is obtained by applying F_3 vertically, from which we find the f_i factors.

Finally, applying Q_1, there is no external work related to the preload. Reactions were induced at 0 and 2, but there was no displacement at these points. Neither was there vertical displacement at 3. The roller moved to the right, but there was no horizontal force, and therefore no work in this sense.

Figure 2.3   Support redundancy requires decoupling and preloading at the support.

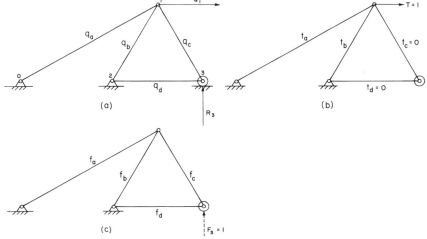

We have satisfied the conditions which led to Eq. (2.3b), and

$$\left(\frac{R_3}{Q_1}\right) = q_3 = \frac{-\Sigma(cft)_i}{\Sigma(cf^2)_i} \tag{2.4}$$

Superimposing the effect of the roller force on the link loads,

$$\left(\frac{Q_i}{Q_1}\right) = q_i = [t_i + f_i q_3] \tag{2.5}$$

This solution, which parallels Sec. 2.2, is identical if we visualize the reaction at 3 to be equivalent to a vertical link below the roller, infinitely short and stiff ($C_3 = 0$). Then as the reference redundant link, it supplies the preload and is the decoupling element. Being stiff, it involves no energy and does not enter the summation.

2.4 Redundant Solution by Displacement Restoration

We can further view the redundant linkage as a load-deflection problem. For example, take the case of Sec. 2.3 (Fig. 2.3), with the procedure indicated in Fig. 2.4.

The first step is to remove the roller support (Fig. 2.4a) and to determine the deflection δ_{3Q} caused by Q_1 for the statically determinate system. This is indicated in Fig. 2.4b with an auxiliary loading applied at 3.

Figure 2.4 Redundancy is solved by superimposing support displacement on displacement due to load.

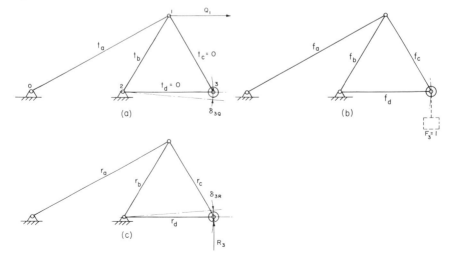

From Eq. (1.28)

$$\delta_{3Q} = C_a Q_1 \sum_a^b (cft)_i \qquad (2.6)$$

where t_i represents the load-related factors in the *decoupled* condition.

The summation only includes a and b as there are no t_i loads in c and d. Point 3 is really an extension kinematically of link b.

Elastic stiffness of the structure, with respect to applied load at 3, is now calculated by applying a vertical force R_3 (Fig. 2.4c). Using the relation for deflection at a load in the direction of the load,

$$\delta_{3R} = C_a R_3 \sum (cr^2)_i \qquad (2.7)$$

where all links are involved and $r_i = R_i/R_3$, resembling q_i, but not caused by Q_1. Note also that the r_i factors are identical to the f_i factors, except for sign reversal, and in r_i^2 this makes no difference.

Since the actual deflection of 3 is 0, this is achieved by the superposition of the two opposed displacements. Equating absolute values

$$\delta_{3Q} = C_a Q_1 \sum (cft)_i = \delta_{3R} = C_a R_3 \sum (cr^2)_i$$

$$\frac{R_3}{Q_1} = q_3 = \frac{\sum (cft)_i}{\sum (cr^2)_i} \qquad (2.8)$$

Equations (2.3b), (2.4), and (2.8) are basically equivalent. There is no negative sign in (2.8) because the equal displacements were assumed in opposite directions, and R_3 was taken in its true sense.

The displacement–restoration principle can also be used at links. For instance in Fig. 2.2a, we could remove the pin at 0 and find the separation due to Q_1. Then applying a load at the lower hold of a, we determine the force–deflection relationship and use Eq. (2.8) in which q_3 becomes q_a.

Final link loads in Fig. 2.4 are found by superimposing the Q_1 and R_3 effects (Fig. 2.4a) and c or by superimposing the t_i and r_i terms.

2.5 *Deflections in the Indeterminate System*

Referring to a simple redundant case (Fig. 2.5a), assume a vertical load Q_1, for which the primary energy relations in Sec. 1.6 suffice:

$$y_1 = C_a Q_1 \sum (cq^2)_i \qquad (2.9)$$

where the redundancy solution must precede the deflection solution to establish the link loads; however, we have simplifying alternatives.

In Fig. 2.5b, c, and d, we decouple removing one link, and all three remaining linkages are stable and determinate. If we then apply an auxiliary load at 1 vertically, we can find f_i factors for any combination of remaining links, and from complementary energy considerations, we can see that Q_1

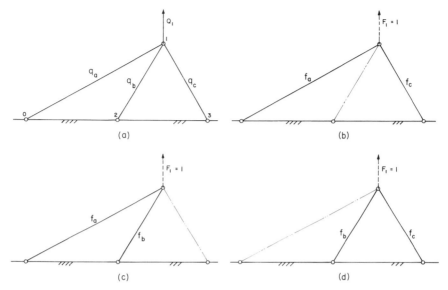

Figure 2.5 Vertical displacement at 1 can be analyzed using alternative auxiliary-loading combinations after the redundancy solution.

deflects

$$y_1 = C_a Q_1 \sum (cfq)_i \tag{2.10}$$

Although different f_i values result in Fig. 2.5b–d, the results will be identical in Eq. (2.10), and only two terms are involved rather than three.

If we wanted the horizontal displacement of 1 due to vertical Q_1, we would be required to use an auxiliary F loading horizontally. This could be done with the complete system (Fig. 2.5a), but this leads to another indeterminate problem. Instead we decouple by removing any link and apply a horizontal F load at 1 to the statically determinate case and apply

$$x_1 = C_a Q_1 \sum (cfq)_i \tag{2.11}$$

2.6 Justification for Decoupled Preload

In Eqs. (2.10) and (2.11) we have obtained f_i loadings from a decoupled or reduced system to solve for displacements in a completely coupled system. We seem to be analyzing a pseudolinkage having different elastic characteristics from the true linkage. In fact, there are three different systems which can be used (Fig. 2.5), each of which will carry preload under conditions of static stability. Equivalence of complementary energy for a complete system [Eq. (1.13a)] requires summation for all links which have preload f_i and actual load q_i. Links which are infinitely stiff ($C_i = 0$) or those which have zero preload, or those which have zero actual load drop from the summation; however, products for the remaining links still must satisfy the equations for the conservation of energy.

Arguing the physical situation, in Fig. 2.5a the displacement of 1 must conform simultaneously to the geometric and elastic behavior of all connected links, as it represents a single point. Thus any two links must also concur in both the horizontal and vertical components, and information factored into Eq. (2.11) by *any two links* can define the elastic compliance of the system at 1. Conformity of all connected links at 1 is established initially in the redundancy solution.

We conclude that *in preloading an indeterminate elastic system for the purpose of determining deflections, any or all redundancies can be removed to reduce it to static determinacy.*

This feature greatly expedites analysis by the energy method and should be invoked whenever possible in the calculation of displacements.

2.7 *Loads Caused by Misalignment*

It is impossible to stress a determinate structure at assembly. In Fig. 1.3 a slightly long or short link has a negligible effect on the geometry and the mating links at the pin can be assembled by shifting without producing duress. But in Fig. 2.6a we have expended this freedom after two links are assembled. If the remaining link does not fit exactly, we must force the entire system if we are to assemble.

This causes negligible geometric changes, but causes all links to carry initial load and to have initial axial deformation. We develop link loads without, and prior to, any external load on the structure. These loads will obviously be larger for a given error for a stiffer system.

Assume link a to be short a distance δ_0 in Fig. 2.6a. This will be a positive error in the sense that it produces tensile stresses in a at assembly. We apply force P_0 to cause the deflection δ_0 elastically and permit assembly. This displacement results from stresses in the complete structure, and can be evaluated as a deflection of a determinate linkage at the load and in the direction of the load. From this basic relation

$$\delta_0 = C_a P_0 \sum \left(cf^2 \right)_i \tag{2.12a}$$

$$P_0 = \left(\frac{\delta_0}{C_a} \right)\left(\frac{1}{\sum \left(cf^2 \right)_i} \right) \tag{2.12b}$$

$$P_i = f_i P_0 \tag{2.12c}$$

where $\delta_0 =$ cumulative misalignment error for system
  $P_0 =$ preload force in link adjacent to measured error
  $P_i =$ preload in any link
  $f_i =$ relative load factor with respect to the reference link

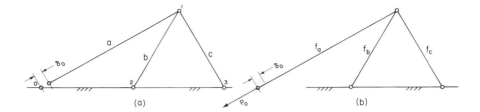

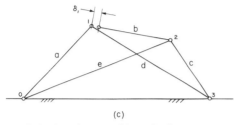

Figure 2.6 Loads caused by misalignment are determined from system compliance with respect to the displacement error.

No auxiliary loading is required. Observation of the $(cf^2)_i$ terms will indicate the relative contribution of each link to the total assembly force, considering the individual terms in Eq. (2.12b). P_0 is directly proportional to δ_0 and inversely proportional to the absolute reference compliance factor C_a.

If external loads are also applied, an independent indeterminate solution is required. Then all link loads are superimposed algebraically, and

$$Q_i = P_i + Q_1[t_i + f_i q_a] \tag{2.13}$$

2.8   *Thermally Induced Loads*

Temperature changes can similarly induce loads in an indeterminate structure by altering the effective free length of the component links. Several possibilities exist:

1. The entire structure, including those connecting the ground points, is of material having a common thermal coefficient of expansion and exposed to a uniform change in temperature.
2. A system as in (1), but with temperature differentials present within the structure.
3. A structure in which different links have different coefficients of expansion, with the entire system subjected to a common temperature change.

4. A system as in (3), but with temperature differentials within the structure.

Except in (1) we will have internal loads, stresses, and displacements developed in the entire redundant structure. It is assumed that assembly takes place at a common temperature.

If we isolate a single link with a temperature differential (2), its effective change in free length is

$$\delta_i = -\lambda \ell_i (\Delta T) \tag{2.14}$$

where λ is the common coefficient of thermal expansion and the negative sign is necessary to correspond to a positive or tensile-link stress if the link temperature is below the temperature of the rest of the system.

The approach is analogous to that in Sec. 2.7, and if link a in Fig. 2.6a is cooled by ΔT after assembly, the equivalent error δ_0 develops. Thermally induced loads depend upon the system stiffness, with relations in Eqs. (2.12).

In Fig. 2.6a and b the simplest case is shown with a length difference between 0 and 1 (the former a ground point). More generally, this difference could be specified between any two points, or for a link anywhere in the linkage (Fig. 2.6c). This condition is also possible in the misalignment cases in Sec. 2.7, but we now have loads developed after assembly. We will refer to both as preloads P_i (distinguished from actual loads Q_i).

In Fig. 2.6c shortening or lengthening of b causes loads in all links because of this imposed *relative displacement* δ_1, resulting in P_b in b. This force pulls 1 towards 2, loading a and d. It also causes the left end of b to move toward 1, loading b, c, and e.

Total strain energy in all five bars is given by $\frac{1}{2} \Sigma C_i P_i^2$, which is in turn equal to the external energy as P_b moves to the right at 1 and an equal P_b moves to the left with the end of b. Without specifying which portion is attributable to each, the total relative motion closes the gap, and

$$\delta_1 = C_a P_b \Sigma (cf^2)_i \tag{2.15}$$

which parallels Eqs. (2.12).

We also conclude that relative deflection solutions in a once-redundant structure do not require the indeterminate procedure outlined in Sec. 2.2.

2.9   Twice-Redundant Systems

Occasionally, it may be necessary to analyze a structure with two redundant axial flexibilities, (Fig. 2.1b). The method follows from previous techniques in this chapter, but requires extension to two reference links (say a and b).

As in the simpler case we first remove both these bars and apply a load of $T = 1$ in the direction of the external load to obtain the t_i factors for the remaining links. We then apply unit tensile preloads first in a and then in b, determining f_i and g_i factors, respectively. Loads in the reference links a and b, derived from complementary energy equivalence, are

$$q_a = \left(\frac{AD - BE}{E^2 - CD} \right) \quad \text{and} \quad q_b = \left(\frac{BC - AE}{E^2 - CD} \right) \tag{2.16}$$

where q_a and q_b = external load factors for the reference links = load in each link for unit external load

f_i = load in any link with unit tensile preload in the reference link a with b removed ($f_a = +1, f_b = 0$)

g_i = load in any link with unit tensile preload in the reference link b with a removed ($g_b = +1, g_a = 0$)

t_i = load in any link per unit external force with system decoupled by removal of both a and b ($t_a = t_b = 0$)

$A = \Sigma(cft)_i \quad C = \Sigma(cf^2)_i \quad E = \Sigma(cfg)_i$
$B = \Sigma(cgt)_i \quad D = \Sigma(cg^2)_i$

By superposition of the three effects

$$\frac{Q_i}{Q_1} = q_i = [t_i + q_a f_i + q_b g_i] \tag{2.17}$$

Examples

2.1. The two springs are fixed at each end and there is no preload. If $K_a = 870$ and $K_b = 650$ lb/in., find:
(a) The force in each spring
(b) The displacement of 1

Q₁= 660 LB

Solution:
(a) Taking a as the reference element, it is removed for the uncoupled condition, for which $t_b = +1$. Applying an auxiliary tensile load in a,

after reinsertion, $f_a = f_b = +1$. Using Eq. (2.3b), with $c_b = K_a/K_b = 1.34$

$$q_a = \frac{-(1.34)(1)^2}{(1)(1)^2 + 1.34(1)^2} = -0.57 \quad \text{(compression)}$$

From Eq. (2.1)

$$q_b = 1 + (-0.57)(1) = +0.43 \quad \text{(tension)}$$

$$Q_a = (-0.57)(660) = 376 \text{ lb}$$

$$Q_b = (+0.43)(660) = 284 \text{ lb}$$

(b) Having the load and the stiffness of each spring,

$$x_1 = C_a Q_a = \tfrac{1}{870}(376) = 0.43 \text{ in.} = C_b Q_b = \tfrac{1}{650}(284)$$

2.2. The square frame consists of links which all have equal cross-sectional area. Find:
   (a) The load distribution
   (b) The change in the 0–2 dimension
   (c) The change in the 1–3 dimension

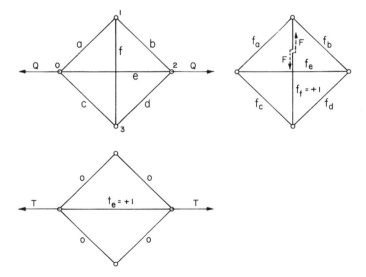

Solution:
(a) Taking f as reference and redundant, its removal causes T loading only in e. Compliances are directly proportional to the lengths, and unit tensile preload is taken in f. Tabulating results for the links:

i	c	f	t	cft	cf^2	fq_f	q
a	0.707	− 0.707	0	0	0.354	+ 0.207	+ 0.207
b	0.707	− 0.707	0	0	0.354	+ 0.207	+ 0.207
c	0.707	− 0.707	0	0	0.354	+ 0.207	+ 0.207
d	0.707	− 0.707	0	0	0.354	+ 0.207	+ 0.207
e	1	1	1	1	1	− 0.293	+ 0.707
f	1	1	0	0	1	− 0.293	− 0.293
				Σ	1	3.414	

Using Eq. (2.3b), $q_f = -1/3.414 = -0.293$.
For the other q_i factors, use Eq. (2.1)

$$q_a = (0 + (-0.707)(-0.293)) = +0.207$$

(b) $\delta_{02} = \delta_e = C_e(q_e Q) = +0.707(C_e Q)$
(c) $\delta_{13} = \delta_f = C_f(q_f Q) = -0.293(C_f Q)$

We note in the table from q_e that 70.7 percent of Q is carried by the central link, and the branches carry the remaining 29.3 percent.

2.3. In Example 2.2 link a is too short as assembly by 0.07 cm. The area of all links is 4.4 cm², and they are made of an aluminum alloy. The length of the four outer links is 30 cm. Calculate the preload induced in the links at assembly.

Solution: As shown in the sketch, a closure force at 0 is unknown. Assuming it to be unity and taking a as the reference link, we tabulate the related load

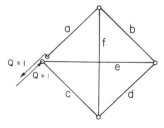

i	c	q	cq^2	$P(N)$
a	1	+ 1	1	6900
b	1	+ 1	1	6900
c	1	+ 1	1	6900
d	1	+ 1	1	6900
e	1.414	− 1.414	2.828	− 9760
f	1.414	− 1.414	2.828	− 9760
		Σ	9.657	

factors and apply Eqs. (1.9):

$$C_a = \frac{\ell}{EA} = \frac{30}{(4.4)(6.5)(10)^6} = 1.05(10)^{-6} \text{ cm/N}$$

From Eq. (2.12b)

$$P_a = \frac{0.07(10)^6}{(1.05)(9.657)} = 6900 \text{ N}$$

2.4. A stiff bar of negligible weight is supported by three springs and carries a single load Q. Deflections are sufficiently small to neglect angularity effects. The springs are all at zero load initially and can take either tensile or compressive forces. Find:
(a) The force in K_1
(b) The forces in all springs if $K_1 = K_2 = K_3$ and Q is applied at 1

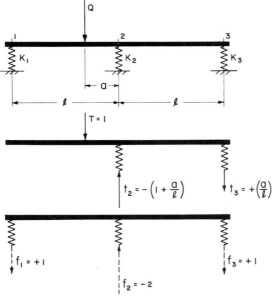

Solution:
(a) From Sec. 2.2 take K_1 as reference for the single redundancy. Decoupled t loading and preloads f are shown. From Eq. (2.3b)

$$q_1 = \frac{-\left[\dfrac{2(1 + a/\ell)}{k_2} + \dfrac{(a/\ell)}{k_3}\right]}{1 + 4/k_2 + 1/k_3}$$

(b) All K factors cancel and $a = \ell$:

$$q_1 = -\left(\tfrac{2}{3} + \tfrac{1}{6}\right) = -\tfrac{5}{6}$$

$$q_2 = t_2 + f_2 q_1 = -2 + \tfrac{5}{3} = -\tfrac{1}{3}$$

$$q_3 = t_3 + f_3 q_1 = 1 - \tfrac{5}{6} = +\tfrac{1}{6}$$

$\Sigma q = 1.00$, and the spring at 3 is in tension.

2.5. In the Fig. 2.5 linkage, assume $c_b = 3$ and $c_c = 2$, and the base angle at 0 is 30°, and at 2 and 3 is 60°. Solve for the vertical deflection due to Q_1 using:
(a) The complete system (Fig. 2.5a)
(b) Auxiliary loading in a and c (Fig. 2.5b)
(c) Auxiliary loading in a and b (Fig. 2.5c)
(d) Auxiliary loading in b and c (Fig. 2.5d)

Solution:

(a) The redundancy solution must precede any displacement analysis. Taking a as reference, we have tabular values as follows:

i	c	f	t	cft	cf^2	q
a	1	1	0	0	1	0.235
b	3	-1.155	0.577	-2	4	0.306
c	2	0.577	0.577	0.667	0.667	0.713
			Σ	-1.333	5.667	

Considering elastic energy in all links caused by the actual loading, or $\frac{1}{2}\Sigma(CQ^2)_i$, we tabulate

i	c	q	cq^2
a	1	0.235	0.055
b	3	0.306	0.280
c	2	0.713	1.017
		Σ	1.353

$$y_1 = C_a Q_1 \, (1.353)$$

In the vertical direction the system is 1.353 times as compliant as the link a individually in its axial direction.

(b) Removing link b in Fig. 2.5b, the auxiliary load F_1 preloads only a and c, from which the tabulation is

i	c	f	q	cfq
a	1	0.500	0.235	0.118
b	3	0	0.306	0
c	2	0.866	0.713	1.235
			Σ	1.353

$$y_1 = C_a Q_1 \, (1.353)$$

(c) Removing link c in Fig. 2.5c, the auxiliary load F_1 preloads only a and b, with tabulated results from complementary energy,

i	c	f	q	cfq
a	1	-1	0.235	-0.235
b	3	1.732	0.306	1.588
c	2	0	0.713	0
			Σ	1.353

$$y_1 = C_a Q_1 \ (1.353)$$

(d) Removing link a in Fig. 2.5d, the auxiliary load F_1 preloads only b and c:

i	c	f	q	cfq
a	1	0	0.235	0
b	3	0.577	0.306	0.529
c	2	0.577	0.713	0.824
			Σ	1.353

$$y_1 = C_a Q_1 \ (1.353)$$

We have verified the equivalence of four alternatives for applying energy balance for a single deflection. Completely different numerical combinations yield identical results. If the deflection were not in the direction of Q_1, we are limited to the latter three.

2.6. Determine the force in c (Fig. 2.5a) by displacement restoration (Sec. 2.4).

Solution: Applying Eqs. (2.8) we require two summations. For the numerator t loadings are due to Q_1 with 3 disconnected and F applied at 3 towards 1, causing compression in c:

i	c	f	t	cft
a	1	-1.732	-1	1.732
b	3	$+2$	$+1.732$	10.392
c	2	-1	0	0
			Σ	12.124

The denominator requires unit external load at 3 in the direction of c, tensile to null the displacement:

i	c	r	cr^2
a	1	1.732	3
b	3	-2	12
c	2	$+1$	2
		Σ	17

$$\frac{R_3}{Q_1} = q_3 = \frac{12.124}{17} = 0.713$$

as found in Example 2.5.

3

Fundamentals of
Beam Deflection

THIS CHAPTER IS CONCERNED with the planar flexural displacement of the straight beam subjected to a simple transverse load or couple at any point along the axis of the beam. The deflection parameters are transverse displacement and slope of the tangent to the elastic curve, also at any axial location. In the linear elastic system, deflection effects due to multiple loads are superimposed algebraically.

Attention is now centered on energies and deflections related only to bending deformation. Direct stress, considered previously in truss elements, is usually negligible in beams but can be included if necessary. Similarly, deflection contributions resulting from transverse shear are small and normally not considered.

In the energy methods developed, the usual objective is a simple coordinate deflection at a specified point. Although expressions for the complete or continuous elastic curve can be developed from strain energy relations, this is not the usual procedure, in contrast to the conventional double integration process for converting the moment distribution to the complete elastic curve.

The shear diagram has only slight relevance in the energy technique and limited pertinence for stress purposes. We therefore simplify matters by dispensing with this concept completely. Static equilibrium suffices to determine the bending moment in a beam.

The analysis of structures in bending is more involved than those with direct stress only (Chaps. 1 and 2). In axial elements there is only one load, and in links of constant section only one direct stress; however, in bending, the moment loading varies continuously throughout the length of the beam and stress also varies on the cross section. Strain energy distribution is thus a more complex function and no longer uniform.

It is assumed in these procedures that the usually accepted conditions for beam theory apply. The beams are relatively slender, are not abnormally deep or wide, do not buckle, and the flexural stresses are within the purely elastic domain.

3.1 Bending Moment in a Beam

For the planar situation with a simply supported beam and a single load Q_1 (Fig. 3.1a), we can determine the end reactions by applying static equilibrium equations:

$$\sum M_0 = Q_1 a - R_2 = 0 \tag{3.1a}$$

$$\sum F_y = R_0 + R_2 - Q_1 = 0 \tag{3.1b}$$

Instead of summing forces (Fig. 3.1b), we can use the moment equation

$$\sum M_2 = R_0 \ell - Q_{1b} = 0 \tag{3.1c}$$

Then

$$R_0 = \left(\frac{b}{\ell}\right)Q_1 \quad \text{and} \quad R_2 = \left(\frac{a}{\ell}\right)Q_1$$

Equilibrium equations also apply to each portion of the beam if separated in any span (Fig. 3.1b and c), with the transverse cross section subjected to a vertical force (or *shear*) and a couple (or *bending moment*), the latter of far greater importance in our present discussions. The internal moment M_x is the same quantity whether analyzing the free body to the right or left of the section:

$$M_x = R_0 x_0 - Q_1(a - x_0) \tag{3.2a}$$

$$= R_2 x_2 \tag{3.2b}$$

Sense of these two external moments on the respective portions is opposite in space, or clockwise and counterclockwise; however, both correspond to compression in the upper fibers and tension in the lower. Thus these couples are of the same sense with respect to the nature of the stresses produced. By usual convention the moments here are positive.

We also note that the shear and moment acting on one section of the beam are reacted by the other, constituting *internal loading*. This is also shown in Fig. 3.1d with an isolated short beam element. The ability of a beam to sustain loads is related to its capacity to sustain the imposed

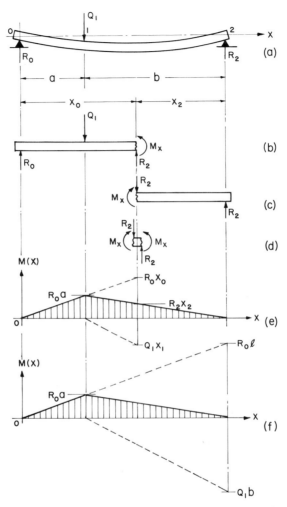

Figure 3.1 Free-body equilibrium of the beam to the left or right of any section determines the internal bending moment.

bending moments, which cause stresses and deformations in the fibers of the beam.

3.2 *The Bending Moment Diagram*

The magnitude of the internal bending moment varies continuously along the length of a loaded beam and is termed $M(x)$. A plot of this variation or function becomes the *bending moment diagram*. From Eqs. (3.2) we can determine the bending moment at any point, isolating either side of the beam. As we vary the break point, or abscissa, we are able to develop $M(x)$, shown by the shaded triangle (Fig. 3.1e) (maximum at Q_1 and 0 at the simply supported ends).

We can start to develop the diagram by taking the $R_0 x_0$ product from left to right. The slope continues to Q_1, but beyond a there are two moment effects because of Q_1 shown by the dashed lines; however, we can more easily isolate the right portion of the beam (Fig. 3.1c) and take the bending moment as $R_2 x_2$ from 2 to 1. Other options include substituting the dashed components in Fig. 3.1e and f for the total moment.

For any procedure, we must consider the *left or right sections independent of each other*, as in Fig. 3.1b and c, in order to apply free-body equilibrium. As we shall see, important simplifications result from the proper choice, usually by proceeding from both ends (0 and 2). Then the only expressions required for $M(x)$ are $R_0 x_0$ from 0 to 1, and $R_2 x_2$ from 2 to 1. The coordinate x_2 is from 2, positive to the left.

3.3 The Neutral Axis

With linear elastic behavior in a beam, direct stresses result from either axial or moment loading (Fig. 3.2), and planes remain planes during deformation. The former causes simple elongation, with the load Q acting at the centroid of the cross-sectional area. The latter causes rotational displacement with tension on one side of the beam and compression on the other, and the moment M is a couple acting in the plane of bending. If both are present, stresses are the algebraic summation (Fig. 3.2).

Bending stresses are detailed in Fig. 3.3a, illustrated by a trapezoidal beam section with a vertical axis of symmetry. The beam can bend about

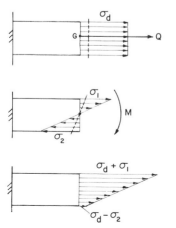

Figure 3.2 Direct and bending stress distribution combine to produce a linear variation depthwise.

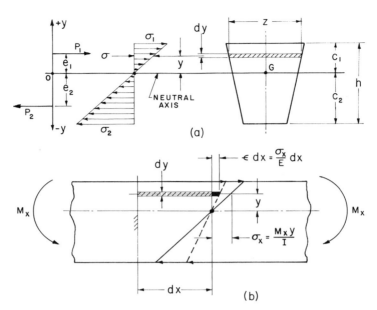

Figure 3.3 Bending stresses are 0 at the neutral axis (Fig. 3.3a). Strain energy in a beam element in the length dx is shown in Fig. 3.3b.

this principal axis, or as shown, about a perpendicular principal axis (analogous to Sec. 1.13). The principal axis is the neutral axis at $y = 0$, located at the center of gravity of the area, and is termed a *central principal axis*. The reason for the centroidal position is now demonstrated.

The bending moment in a beam creates linearly distributed stresses (Fig. 3.3a), regardless of the cross section. P_1 is the resultant of the tensile stress forces and P_2 of the compressive stresses:

$$P_1 = \int_0^{c_1} \sigma z \, dy = \int_0^{c_1} \left(\frac{y}{c_1} \sigma_1 \right) z \, dy = \frac{\sigma_1}{c_1} \int_0^{c_1} zy \, dy \qquad (3.3a)$$

where σ_1 = stress at the extreme tensile fiber
$\quad c_1$ = distance from the neutral axis to the extreme tensile fiber
$\quad z \, dy$ = elemental area parallel to y axis

Similarly,

$$P_2 = \frac{\sigma_2}{c_2} \int_0^{c_2} zy \, dy \qquad (3.3b)$$

Since the bending moment is a couple, $M = P_1(e_1 + e_2) = P_2(e_1 + e_2)$ and $P_1 = P_2$. Therefore,

$$\int_0^{c_1} zy \, dy = \int_0^{c_2} zy \, dy \qquad (3.4)$$

With these integrals representing the first moment of the upper and lower sections of the area about a common axis, this equality locates the axis *at the centroid of the area*. The resultant equal tensile and compressive forces P_1 and P_2 act at distances e_1 and e_2 from the neutral axis. In general, these distances are not equal, and the tensile moment P_1e_1 is not equal to the compressive moment P_2e_2. The bending moment is a couple, however, which is the sum of the two.

3.4 Moment of Inertia

The total resistive bending moment developed by the bending stresses is the sum of the incremental components. In Fig. 3.3a

$$dM = y(\sigma z \, dy)$$

$$M = \int_0^{c_1} y\left(\frac{y}{c_1}\sigma_1\right)z \, dy + \int_0^{c_2} y\left(\frac{y}{c_2}\sigma_2\right)z \, dy$$

$$M = \left(\frac{\sigma_1}{c_1}\right)\int_0^{c_1} zy^2 \, dy + \left(\frac{\sigma_2}{c_2}\right)\int_0^{c_2} zy^2 \, dy = \frac{\sigma_1}{c_1}\int_{-c_2}^{c_1} y^2 z \, dy \qquad (3.5)$$

The final quantity under the integral sign represents the second moment of the area about the neutral axis and is termed the *moment of inertia I* of the section. As an area characteristic there is no inertia involved, but the designation is derived from a similar expression for mass moment of inertia in dynamics.

We can evaluate the maximum bending stresses from Eqs. (3.5):

$$M = \left(\frac{\sigma_1}{c_1}\right)I = \left(\frac{\sigma_2}{c_2}\right)I$$

$$\sigma_1 = \frac{M}{(I/c_1)} \qquad \sigma_2 = \frac{M}{(I/c_2)} \qquad (3.6)$$

The minimum denominator is the *section modulus* of the cross section.

3.5 Elastic Energy in Bending

A beam carrying transverse loads, which induce bending stresses, contains internal potential energy as *strain* or *elastic* energy, distributed throughout the entire volume of the elastic material of which the beam is composed. Considering a short section of a beam (Fig. 3.1d) and particularly Fig. 3.3b, we again have planes remaining planes after bending and stress increasing linearly with the distance from the neutral axis. In the elemental beam length dx, the incremental elongation is

$$\epsilon \, dx = \left(\frac{\sigma}{E}\right) dx \tag{3.7}$$

where ϵ = horizontal unit elongation at a distance y from the neutral axis
σ = unit tensile or compressive stress in the element
E = Young's modulus of elasticity

There is a differential tensile or compressive force on the shaded element, obtained from the product of the stress and the area:

$$dQ = \sigma(z \, dy) \tag{3.8}$$

This elementary volume is essentially a prismatic bar subjected to tension or compression, with cross-sectional area $(z \, dy)$ and length dx, with characteristics as in Chap. 1. Total energy stored in this miniature spring derives from the force and elongation. We obtain this quantity for the dx interval by integrating from top to bottom:

$$dU_s = \frac{1}{2}\int_{-c_2}^{c_1} [\sigma(z \, dy)]\left[\frac{\sigma}{E}\, dx\right] = \frac{1}{2E}\int_{-c_2}^{c_1} (\sigma^2 z \, dy)\, dx \tag{3.9a}$$

and substituting (My/I) for the stress σ,

$$dU_s = \left(\frac{M_x^2}{2EI}\right) dx \tag{3.9b}$$

In order to quantify the entire internal strain energy in the beam, we must sum across the entire length, and we have an important fundamental relationship for beam analysis:

$$U_s = \int_0^\ell \frac{M(x)^2}{2EI(x)}\, dx = \int_0^\ell \frac{M^2}{2EI}\, dx \tag{3.10}$$

The bending moment will usually vary along the length of the beam and is a function abbreviated to M. Section moment of inertia I can also vary functionally, but it is usually a constant or a series of constants.

3.6 *Displacement of a Load*

To find beam deflection at a load, as in Sec. 1.4 we equate external work done on the system by the gradual application of a load to the internal strain energy in the beam using Eq. (3.10):

$$U_e = \frac{Q_1 y_1}{2} = U_s = \int_0^\ell \frac{M^2}{2EI}\, dx \tag{3.11}$$

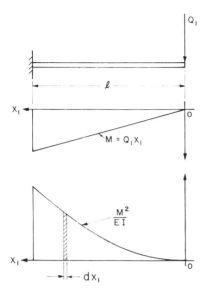

Figure 3.4 The end-loaded cantilever develops a triangular moment diagram and a parabolic distribution of elastic energy.

For the simple constant cantilever (Fig. 3.4), the triangular moment distribution yields

$$\frac{Q_1 y_1}{2} = \frac{1}{2EI} \int_0^\ell M^2 \, dx_1 = \frac{Q_1^2}{2EI} \int_0^\ell x_1^2 \, dx_1$$

$$y_1 = \frac{Q_1 \ell^3}{3EI} \tag{3.12}$$

and we recognize this famous equation.

Visualizing the distribution of the elastic energy relative to the length of the beam, M^2/EI is parabolic (Fig. 3.4) increasing rapidly toward the base. Maximum contribution to the definite integral and to the end deflection occurs in this region. Similarly, on a transverse section there is negligible stress in the vicinity of the neutral axis and maximum at the extreme fibers. Therefore with local internal energy storage varying as the square of the stress [Eq. (3.9)], the end deflection is caused largely by elastic deformations in the upper and lower fibers of the beam. The largest contribution comes from these located at the base or root of the cantilever.

3.7 *Rotational Displacement of a Couple*

If the constant cantilever beam has an end moment (Fig. 3.5), the argument is identical relative to the internal bending energy, but we must express the

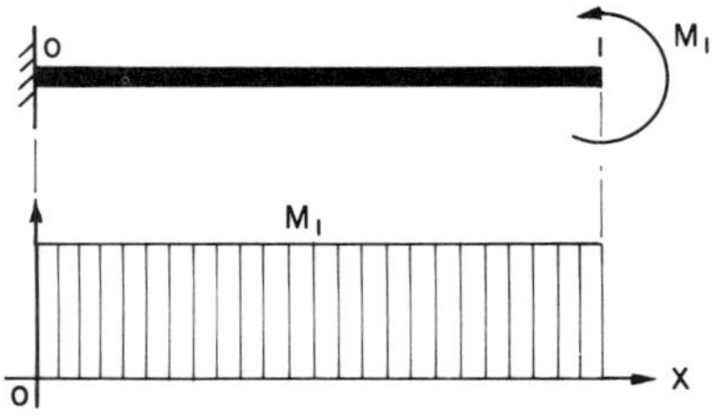

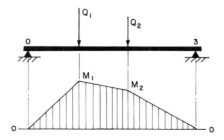

Figure 3.5 The simple cantilever subjected to an end couple has positive, constant bending moment.

input in rotational coordinates. For the constant bending moment,

$$U_e = \frac{M_1\theta_1}{2} = U_s = \int_0^\ell \frac{M^2}{2EI}\,dx \qquad (3.13a)$$

$$\frac{M_1\theta_1}{2} = \frac{M_1^2}{2EI}\int_0^\ell dx \qquad (3.13b)$$

$$\theta_1 = \frac{M_1\ell}{EI} \qquad (3.13c)$$

The slope of the free end in radians is directly proportional to the couple and to the length and inversely proportional to the stiffness factor EI.

3.8 Symmetrical Loads

Equations (3.11) and (3.13b) are simple and basic but limited to solutions with a single coordinate displacement. For example with several loads (Fig. 3.6), we can determine the reactions and $M(x)$. The total strain energy can be evaluated from Eq. (3.10); however, the external energy involves two different loads with two different deflections.

For conservation of energy, we can write

$$\frac{Q_1 y_1}{2} + \frac{Q_2 y_2}{2} = \int_0^\ell \frac{M^2}{2EI}\,dx \qquad (3.14)$$

This equation is valid but does not enable us to solve for y_1 or y_2.

Figure 3.6 Bending moment diagram for the simply supported beam with two loads.

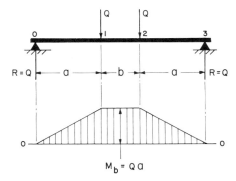

Figure 3.7 Symmetrical loading of the simple beam produces two equal deflections.

In special cases, Eq. (3.10) can be applied to several external loads successfully. With symmetry (Fig. 3.7), the deflections are equal. For the symmetrical moment distribution we have

$$2\left(\frac{Qy_1}{2}\right) = 2\int_0^{\ell/2} \frac{M^2}{2EI}\,dx$$

$$y_1 = \frac{1}{QEI}\int_0^a (Qx_0)^2\,dx_0 + \int_0^{b/2}(Qa)^2\,dx$$

$$= \frac{Q}{EI}\left[\frac{a^3}{3} + \frac{a^2 b}{2}\right] = \frac{Qa^3}{3EI}\left[1 + \frac{3}{2}\left(\frac{b}{a}\right)\right] \qquad (3.15)$$

with the second term in brackets indicating the effect of the center span b on the vertical displacements of the loads. A similar solution can be made with two symmetrical equal and opposed twin couples.

3.9   Span Loading

We further examine the significance of Eq. (3.10) in Fig. 3.8. The loading bracket stores no elastic energy if rigid, and we have a moment distribution for the elastic beam given by

$$\frac{Q_2 y_2}{2} = \int_0^\ell \frac{M^2}{2EI}\,dx \qquad (3.16)$$

We evaluate the moment diagram and the integral numerically to solve for y_2. It should be noted that we have not determined y_1 or y_3, but in the context of total system energy we find the single displacement at the load.

If the loading bracket is equivalent to a constant beam with a stiffness $(EI)_b$ (Fig. 3.8a), applying loads Q_1 and Q_3 to the main beam, there will be strain energy in the upper beam. Its bending moment is shown in Fig. 3.8d.

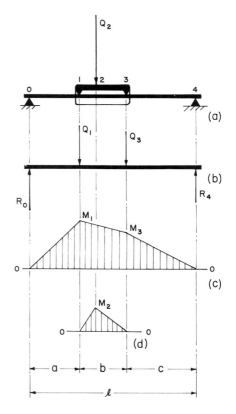

Figure 3.8 A single load is applied to a beam through a bracket bridging two contact points 1 and 3.

The energy equation becomes

$$\frac{Q_2 y_2}{2} = \int_0^{\ell} \frac{M^2}{2EI}\, dx + \int_0^b \frac{(M_b)^2}{2(EI)_b}\, dx \tag{3.17}$$

where the second integral gives the additional deflection of Q_2 caused by the superposition of the flexibility of the upper beam.

3.10 *Beams with Stepped Sections*

Changes in cross section can have a marked effect on the flexural behavior of beams, and the moment of inertia varies considerably with modest changes in dimension. With a round or square geometry, I varies as d^4 or as b^4, respectively, so a 10 percent increase corresponds to a factor of 1.46, or nearly a 50 percent gain in beam stiffness.

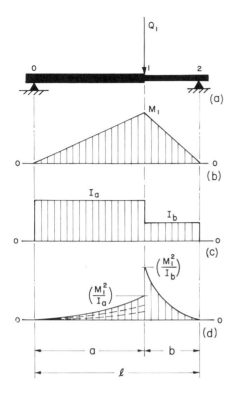

Figure 3.9 A beam with two different sections is loaded at the junction of the spans.

Figure 3.9 shows a simple beam with one change in section and a load at the transition. The resulting bending-moment diagram (Fig. 3.9*b*) depends upon the nature of the loading and supports and is *independent of beam stiffness characteristics*. Continuity of deflection and slope exists at 1, and the transverse deflection is found by determining the elastic energy in the entire beam:

$$\frac{Q_1 y_1}{2} = \int_0^a \frac{(R_0 x_0)^2}{2EI_a} \, dx_0 + \int_0^b \frac{(R_2 x_2)}{2EI_b} \, dx_2$$

$$\frac{y_1}{Q_1} = \left(\frac{R_0}{Q_1}\right)^2 \left(\frac{a^3}{3EI_a}\right) + \left(\frac{R_2}{Q_1}\right)^2 \left(\frac{b^3}{3EI_b}\right)$$

(3.18)

where the terms represent the contribution of each span to the total compliance. These relative effects are seen graphically in Fig. 3.9*d*. The two integrals in Eqs. (3.18) are proportional to the areas under the respective parabolic curves. In the *a* interval the ordinates are reduced with $I_a > I_b$.

Substituting for the reactions and reducing to dimensionless parameters,

$$y_1 = \frac{Q_1 b^3}{3EI_b} \left[\left(\frac{a}{\ell} \right)^2 \left\{ \left(\frac{a}{b} \right) \left(\frac{I_b}{I_a} \right) + 1 \right\} \right] \tag{3.19}$$

The first factor is a reference deflection corresponding to a simple cantilever of length b and load Q_1, fixed to ground at 2. If $I_a = I_b = I$, the expression becomes

$$y_1 = \frac{Q_1 \ell^3}{3EI} \left(\frac{a}{\ell} \right)^2 \left(\frac{b}{\ell} \right)^2 \tag{3.20}$$

which is a basic formula for the deflection at a load for any point on a uniform beam simply supported at both ends.

A further alternative occurs if I_a is relatively rigid. Then the beam is straight from 0 to 1, with flexure from 1 to 2. As shown by the dashed curves (Fig. 3.9d), the area under these parabolas decreases as I_a becomes larger, with the ordinates and the area 0 in the limit. For this condition, Eq. (3.19) reduces to

$$y_1 = \frac{Q_1 b^3}{3EI_b} \left(\frac{a}{\ell} \right)^2 \tag{3.21}$$

We see that any number of changes in section can be incorporated in the basic energy relation. There is no complication other than integrating with the proper limits for each span having a different value of I.

It is possible for different materials to be welded axially to produce a composite beam lengthwise. Although rare, this also requires only a break in the integration procedure at the juncture.

3.11 Beams with Continuously Varying Sections

A more general elastic beam with a concentrated load has a cross section which varies continuously along the length. For any given geometry, the coordinates of the $I(x)$ distribution can be computed. Then a plot of $M(x)^2/I(x)$ provides a curve, the area under which establishes the integral value in Eq. (3.11). The graphical integration can employ one of several common numerical procedures. Occasionally, the $I(x)$ distribution is a known function or can be so approximated. Then integration is usually possible, resulting in a direct solution for deflection.

For instance, in Fig. 3.10 we have a cantilever of rectangular cross section. The width is constant, but the depth varies parabolically from 0 at the free end. This, incidentally, is a classic illustration of a beam with constant bending stress at the top and bottom surfaces, and we have for

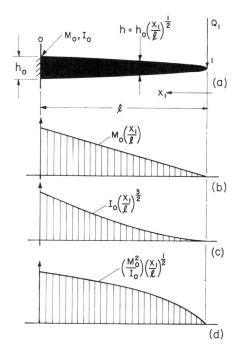

Figure 3.10 The cantilever beam with rectangular cross section has constant width and parabolically varying depth.

the section

$$h = h_0\sqrt{\frac{x_1}{\ell}} \tag{3.22a}$$

$$I = \frac{1}{12}bh^3 = \frac{1}{12}bh_0^3\left(\frac{x_1}{\ell}\right)^{3/2} = I_0\left(\frac{x_1}{\ell}\right)^{3/2} \tag{3.22b}$$

$$I/c = \frac{1}{6}bh^2 = \frac{1}{6}bh_0^2\left(\frac{x_1}{\ell}\right) \tag{3.22c}$$

For the cantilever, both bending moment and section modulus are directly proportional to the distance from the end, and the constant ratio results in a constant stress:

$$\sigma(x_1) = \frac{M_0(x_1/\ell)}{(I/c)_0(x_1/\ell)} = \frac{Q_1\ell}{(I/c)_0} \tag{3.23}$$

The deflection relation is

$$\frac{Q_1 y_1}{2} = \int_0^\ell \frac{(Q_1 x_1)^2}{EI_0(x_1/\ell)^{3/2}}\, dx_1$$

$$y_1 = 2\left[\frac{Q_1\ell^3}{3EI_0}\right] \tag{3.24}$$

Since the quantity in brackets represents the deflection of a cantilever beam of constant cross section, the removal of material to produce the parabolic shape causes the beam to be *exactly* $\frac{1}{2}$ *as stiff* as the corresponding prismatic cantilever.

3.12 Beam Deflection by Component Cantilevers

We can also visualize the simple beam as composed of two cantilever elements (Fig. 3.11), with the cantilever deflections as the basis for the deflection at the load. If the bases are connected at the load (1), there is a common tangent at this point corresponding to the slope developed by Q. Loads on the two cantilevers are the end reactions R_0 and R_2. We have the slope in terms of the difference of the cantilever deflections:

$$\theta_1 = \left(\frac{y_2 - y_0}{\ell}\right) = \frac{1}{\ell}\left[\frac{aQb^3}{\ell 3EI_2} - \frac{bQa^3}{\ell 3EI_0}\right] = \frac{Qab}{3E\ell^2}\left[\frac{b^2}{I_2} - \frac{a^2}{I_0}\right] \quad (3.25)$$

The deflection is

$$y_1 = y_0 + a\theta_1 = y_0 + \frac{a}{\ell}(y_2 - y_0) = \left(\frac{b}{\ell}\right)y_0 + \left(\frac{a}{\ell}\right)y_2$$

$$= \frac{Q\ell^3}{3E}\left(\frac{a}{\ell}\right)^2\left(\frac{b}{\ell}\right)^2\left[\left(\frac{a}{\ell}\right)\frac{1}{I_0} + \left(\frac{b}{\ell}\right)\frac{1}{I_2}\right] \quad (3.26)$$

Although this illustration consists of two simple cantilevers, we can obviously extend the method to a beam with several changes in section within each span. Each must then be analyzed independently to obtain the end deflections due to end loads. Finally, Eqs. (3.25) and (3.26) are applied.

Figure 3.11 Elastic curve of beam corresponds to two inverted cantilevers of lengths a and b, and can be so analyzed for deflection.

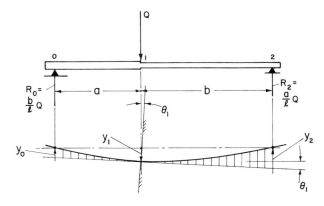

As shown, we have only determined the slope and deflection *at the point of application of the load*.

3.13 Combined Bending and Axial Elements

Although deflections associated with bending are usually an order of magnitude greater than those due to direct effects, discussed in Chaps. 1 and 2, we now consider these components collectively. In Fig. 3.12 a cable supports the outer end of a pinned beam, with bending in the beam and tension in the cable. Additionally, the beam is subjected to a compressive deformation.

For conservation of energy due to the application of Q_1,

$$\frac{Q_1 y_1}{2} = \int_0^2 \frac{M^2}{2EI_a} dx + \sum \frac{C_i Q_i^2}{2} \tag{3.27a}$$

$$y_1 = \frac{1}{Q_1} \left[\int_0^2 \frac{M^2}{EI_a} dx + C_a Q_a^2 + C_b Q_b^2 \right] \tag{3.27b}$$

Figure 3.12 A cable-supported beam has a deflection at Q_1 caused by beam flexure and axial deformations of *a* and *b*.

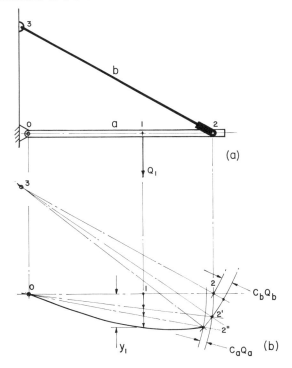

The moment diagram will be similar to Fig. 3.9*b*, and the compressive beam force $(-Q_a)$ acts axially creating no moment. Compression in *a* causes the negative load factor, but when squared the result is positive. Motion of 2 toward 0 causes a rotation of 2 about 3 and a clockwise rotation of the beam.

Other flexibilities can be included in Eqs. (3.27). For example, the load Q_1 could be applied through a spring element, adding a $C_1 Q_1^2$ term.

Examples

3.1. The bevel gear, considered rigid, has a thrust component of 1600 N that produces bearing loads and bending of the steel shaft. Calculate
(a) The linear deflection in the direction of Q
(b) The angular deflection of the gear
(c) The deflection in (a) including the effects of radial-bearing flexibility, if both bearings have a radial compliance of $45(10)^{-6}$ cm/N

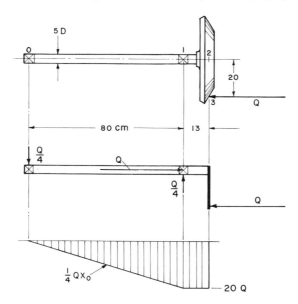

Solution:
(a) From the basic energy relation [Eq. (3.11)],

$$\frac{Q x_3}{2} = \int_0^{80} \frac{\left(\frac{Q}{4} x_0\right)^2}{2 EI} dx_0 + \int_0^{13} \frac{(20Q)^2}{2 EI} dx_2$$

$$x_3 = \frac{Q}{EI}\left(\frac{1}{16}\frac{(80)^3}{3} + (20)^2(13)\right)$$

$$I = \frac{(5)^4}{64} = 30.7 \text{ cm}^4 \qquad E = 20(10)^6 \text{ N/cm}^2$$

$$x_3 = \frac{1600}{20(10)^6(30.7)}(10,670 + 5,200)$$

$$= 0.028 + 0.013 = 0.041 \text{ cm}$$

Note the major contribution of the 80 cm span to the total deflection.

(b) $$\theta_2 = \frac{0.041}{20} = 0.0021 \text{ rad}$$

(c) Including direct energy terms from Chap. 1,

$$\frac{Qx_3}{2} = \int_0^{\ell} \frac{M^2}{2EI} \, dx + \sum \frac{C_i Q_i^2}{2}$$

$$x_3 = 0.041 + 2(45)(10)^{-6}(1600)\left(\tfrac{1}{4}\right)^2$$

$$= 0.041 + 0.009 = 0.050 \text{ cm}$$

3.2. The rotor is essentially rigid in the center section and has a concentrated transverse load. Determine the general expression for the deflection at the load.

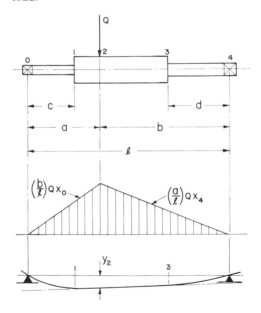

Solution:

$$\frac{Qy_2}{2} = \int_0^1 \frac{\left(\frac{b}{\ell} Q x_0\right)^2}{2EI_c} \, dx_0 + \int_4^3 \frac{\left(\frac{a}{\ell} Q x_4\right)^2}{2EI_d} \, dx_4$$

$$y_2 = Q\left[\left(\frac{b}{\ell}\right)^2 \frac{c^3}{3EI_c} + \left(\frac{a}{\ell}\right)^2 \frac{d^3}{3EI_d}\right]$$

Note the second factors involving the cube of the lengths c and d are basic cantilever deflections, which derive from the cantilever nature of the outboard sections; however, the a and b dimensions account for the straight center section which is tilted relative to the axis of the rotor.

3.3. The mesh of the two gears results in a total change in center-to-center distance in the plane of the shafts of 0.005 in. due to flexure of the two shafts. What is the separating force in the plane at the teeth that produces this deflection?

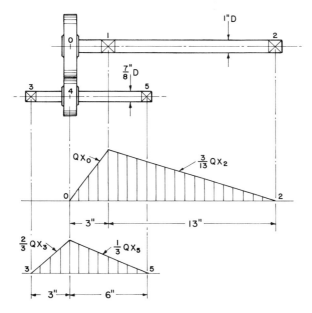

Solution:

$$\frac{Qy_{04}}{2} = \int_0^3 \frac{(Qx_0)^2}{2EI_0}\,dx_0 + \int_0^{13} \frac{\frac{3}{13}(Qx_2)^2}{2EI_0}\,dx_2$$

$$+ \int_0^3 \frac{\left(\frac{2}{3}Qx_3\right)^2}{2EI_3}\,dx_3 + \int_0^6 \frac{\left(\frac{1}{3}Qx_5\right)^2}{2EI_3}\,dx_5$$

$$y_{04} = \frac{Q}{E}\,\frac{(3)^3 + \left(\frac{3}{13}\right)^2(13)^3}{(3)(0.049)} + \frac{\left(\frac{2}{3}\right)^2(3)^3 + \left(\frac{1}{3}\right)^2(6)^3}{(3)(0.0288)}$$

$$0.005 = \frac{Q}{3(30)(10)}\,6\,(2940 + 1250)$$

$$Q = 107\text{ lb}$$

3.4. Two simple beams of different EI values are end connected by helical springs of different constants. Find the general expression for the vertical deflection at 4.

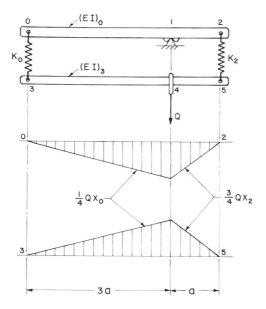

Solution: We must include both flexural energy in the beams and strain energy in the springs. The latter will probably be considerably greater than the former, and the springs can be modeled as simple tensile links. Bending moment diagrams for the two beams are identical:

$$\frac{Qy_4}{2} = \int_0^{3a} \frac{\left(\frac{Q}{4}x_0\right)^2}{2(EI)_0}\, dx_0 + \int_0^{a} \frac{\left(\frac{3}{4}Qx_2\right)^2}{2(EI)_0}\, dx_2$$

$$+ \int_0^{3a} \frac{\left(\frac{Q}{4}x_3\right)^2}{2(EI)_3}\, dx_3 + \int_0^{a} \frac{\left(\frac{3}{4}Qx_5\right)^2}{2(EI)_3}\, dx_5$$

$$+ \frac{1}{2K_0}\left(\frac{Q}{4}\right)^2 + \frac{1}{2K_2}\left(\frac{3}{4}\right)Q^2$$

$$y_4 = Q\left[\frac{\left(\frac{1}{16}\right)(3a)^3 + \left(\frac{9}{16}\right)(a)^3}{3(EI)_0} + \frac{\left(\frac{1}{16}\right)(3a)^3 + \left(\frac{9}{16}\right)(a)^3}{3(EI)_3}\right.$$

$$\left. + \frac{1}{16K_0} + \frac{9}{16K_2}\right]$$

$$= \frac{Q}{16}\left[(12a^3)\left(\frac{1}{(EI)_0} + \frac{1}{(EI)_3}\right) + \frac{1}{K_0} + \frac{9}{K_2}\right]$$

All spring elements are effectively in series.

4

Beams with Concentrated Loads

IN CHAP. 3 THE ELASTIC-ENERGY PROCEDURE was extended to the simple deflection combination of a linear displacement due to a force or an angular deflection due to a couple. Although rather restrictive, this type of problem is important practically. The methods of solution developed provide the principles which are required to attack the more complex bending cases which will be discussed in this and later chapters. We are still concerned with deflection of the determinate straight beam, but we limit our attention to point loading, with distributed loading to follow in Chap. 5.

As with any elastic structure, a single load at a given point produces displacements at all other points in the structure. There are also slopes caused by a load at the load and at all other coordinate points. Similarly, a moment load causes transverse deflections and slopes throughout a system. These coupled effects are determined using complementary-energy concepts introduced in Chap. 1.

4.1 Transverse Deflections Due to a Force

In Fig. 4.1 a simple beam carries a load Q_2 that defines a continuous elastic deflection curve. Any coordinate point on this curve can be determined by applying an *auxiliary load* to induce an *auxiliary-moment distribution* (Fig. 4.1b). We use dashed lines to identify the auxiliary conditions.

Considering that the auxiliary loading precedes the actual as shown, the potential energy input during the loading sequence to determine the deflec-

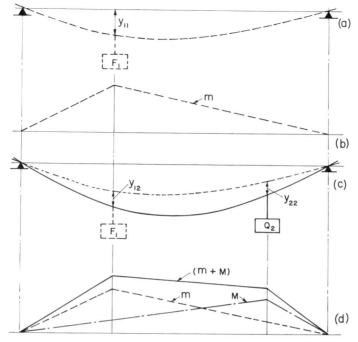

Figure 4.1 Actual loading Q_2, applied after auxiliary load F_1 results in superposition of deflection and bending moment distributions.

tion at 1 with a load at 2 is composed of three effects due to the gradually applied static loading:

$$U_e = \left(\frac{F_1 y_{11}}{2} \right) + \left(\frac{Q_2 y_{22}}{2} \right) + F_1 y_{12} \qquad (4.1)$$

where F_1 = auxiliary load at point of desired deflection
  Q_2 = actual system load at 2
  y_{11} = deflection at 1 due to F_1
  y_{22} = deflection at 2 due to Q_2
  y_{12} = deflection at 1 due to Q_2

and the component terms represent, respectively:

1. External work done on the system by the simple application of F_1 as in Eq. (3.11).
2. External work done on the system during the displacement y_{22} of Q_2.
3. A change in potential energy that accompanies the lowering of the weight F_1 as Q_2 is applied.

In (1) and (2) the applied forces increase gradually from 0 to maximum, but F_1 is *fully acting* during the application of Q_2.

Having established the total external energy in Eq. (4.1), we turn to the internal strain energy induced by the dual loading. From Eq. (3.10) the final stored energy is

$$U_s = \int_0^\ell \frac{(m + M)^2}{2EI} \, dx \qquad (4.2)$$

where
m = auxiliary-moment distribution due to F_1 only
M = actual-moment distribution due to Q_2 only
$(m + M)$ = total final bending-moment distribution

Expansion of Eq. (4.2) clarifies the nature of the strain energy:

$$U_s = \int_0^\ell \frac{m^2}{2EI} \, dx + \int_0^\ell \frac{M^2}{2EI} \, dx + \int_0^\ell \frac{2mM}{2EI} \, dx \qquad (4.3)$$

The first two integrals correspond to the first two energy terms in Eq. (4.1), respectively

$$\frac{F_1 y_{11}}{2} = \int_0^\ell \frac{m^2}{2EI} \, dx \qquad (4.4a)$$

$$\frac{Q_2 y_{22}}{2} = \int_0^\ell \frac{M^2}{2EI} \, dx \qquad (4.4b)$$

Therefore, the third terms in Eqs. (4.1) and (4.3) must be equal to each other:

$$F_1 y_{12} = \int_0^\ell \frac{mM}{EI} \, dx \qquad (4.4c)$$

This important equation relates the *auxiliary external work to the related change of elastic energy* in the entire volume of the beam and defines the *complementary energy* as in Secs. 1.7 and 1.8.

4.2 *Distribution of Internal Elastic Energy*

Equation (4.3) indicates that a beam subjected to two successive loadings has three component internal energy distributions. Although not essential to the solution of problems by energy techniques, it is instructive to consider the nature of these functions. Thus in Fig. 4.1 we have F_1 applied at $x_0/\ell = 0.3$ and Q_2 applied at $x_0/\ell = 0.8$, and we evaluate the terms in the relation

$$m^2 + M^2 + 2mM = (m + M)^2 \qquad (4.5)$$

These functions in Fig. 4.2 are normalized with respect to the maximum ordinate $(m + M)^2$ at $x_0/\ell = 0.3$, and we assume $F_1 = Q_2$ to obtain relative numerical results.

In a differential length of beam dx internal elastic energy is proportional to these ordinates, and

1. m^2 defines that related to the application of the auxiliary load F_1.
2. M^2 defines that related to the application of the actual load Q_2.
3. $2mM$ defines complementary bending energy related to the application of Q_2 in the presence of F_1.
4. $(m + M)^2$ defines the total final bending energy after the application of both F_1 and Q_2.

In Fig. 4.2 we note maximum m^2 at the point of maximum auxiliary-bending moment (at F_1), with parabolic decrease to 0 at the ends. Similarly, M^2 is maximum at Q_2, with both of these primary curves shown dashed. The complementary energy distribution ($2mM$) is maximum at the center C, and is an inverted parabola from A to B, with symmetry terminated at

Figure 4.2 Three-component elastic-energy distributions contribute to the total energy in a beam.

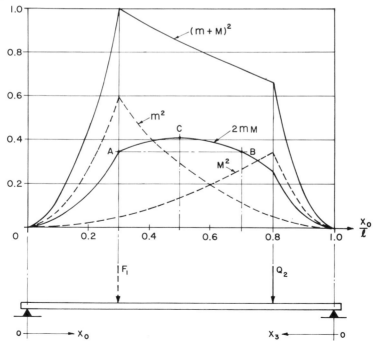

the discontinuity at A. Total energy distribution $(m + M)^2$ is the sum of ordinates of the other curves.

Total internal energy quantities, or the integral values, are proportional to the respective areas under the curves. Thus (1), (2), and (3) sum to (4), and in this example the complementary energy is roughly $\frac{1}{2}$ of the final energy.

This example has involved the basic case of two positive triangular moment diagrams. But with opposite loads (Fig. 4.3), the sense of the forces, bending moments, and bending stresses produced by F_1 and Q_2 is reversed, resulting in the distributions shown. The m^2 and M^2 curves are identical to those in Fig. 4.2, but the $(2mM)$ values are negative, although of the same absolute values. Total energy distribution is reduced, as it now represents an algebraic difference. It is also 0 at D when $(m + M) = 0$.

In the strain energy distributions, auxiliary, actual, and total energies must always be positive because they are squared terms, and correspond to positive potential energy storage; however, the complementary energy ordinates and integral value can be positive or negative as shown. It is

Figure 4.3 Similar to Fig. 4.2, but with F_1 and Q_2 of opposite sense. The negative complementary distribution $(2mM)$ reduces the total strain energy.

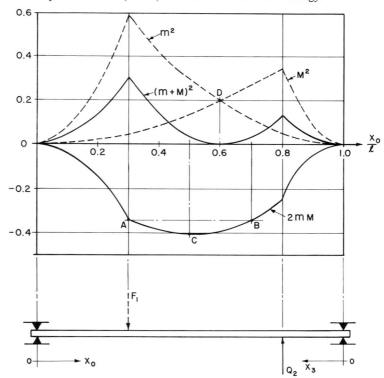

essential in the energy solutions that $(2mM)$ *carry algebraic sign.* As indicated in Sec. 3.1, the sense of the bending stresses produced by the moments is the criterion for algebraic sign. If stresses at a surface fiber due to m and M are both tensile or both compressive, the product is positive and vice versa.

The physical significance of a negative complementary energy integral relates to the coupled external energy, and in Fig. 4.3 F_1 will increase rather than decrease as Q_2 is applied. With a displacement *against the force*, the work done is negative.

4.3   *The Volumetric Internal Energy Analogy*

In considering the various philosophical aspects of the complementary energy in bending, a three-dimensional model has been suggested,[5] that will appeal to the graphically minded.

Viewing the simple beam in Fig. 4.4 in three dimensions, we can plot the auxiliary diagram vertically and the actual moment horizontally, generating rectangles in all planes perpendicular to the longitudinal axis of the beam. All rectangles are cornered on the x axis. The total enclosed volume is then proportional to the complementary energy for a constant section beam, since

$$V = \int_0^\ell (mM)\, dx \quad \text{and} \quad (U_s)_{FQ} = \int_0^\ell \frac{mM}{EI}\, dx \qquad (4.6)$$

Volumes in this illustration consist of a pyramid at each end with

$$V_{01} = \frac{A_1 \ell_{01}}{3} = \frac{m_1 M_1 \ell_{01}}{3} \qquad (4.7a)$$

$$V_{23} = \frac{A_2 \ell_{23}}{3} = \frac{m_2 M_2 \ell_{23}}{3} \qquad (4.7b)$$

Figure 4.4   Complementary elastic energy is proportional to a three-dimensional volume.

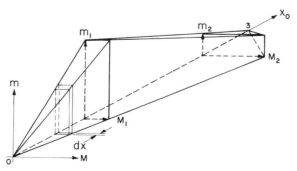

The center sectional geometry defines a *prismoid*, with six planar surfaces and parallel bases, with a volume of

$$V_{12} = \frac{\ell_{12}}{3} \left[(A_1 + A_2) + (A_{12} + A_{21}) \right] \tag{4.7c}$$

where $A_1 = m_1 M_1$
$A_2 = m_2 M_2$
$A_{12} = \frac{1}{2} m_1 M_2$
$A_{21} = \frac{1}{2} m_2 M_1$

Or, for the entire beam the complementary elastic energy is

$$(U_s)_{FQ} = \frac{1}{EI} \left[\frac{A_1 \ell_{02}}{3} + \frac{A_2 \ell_{13}}{3} + \frac{A_{12} \ell_{12}}{3} + \frac{A_{21} \ell_{12}}{3} \right] \tag{4.8}$$

The first two terms relate to a pair of back-to-back pyramids with rectangular bases. The second two terms relate to two *equivalent* pyramids with triangular bases.

Equation (4.8) can be used for the deflection solution of the fundamental beam with a single load. The analogy can also be extended to include spans with various inertias.

4.4 *External Energy Characteristics*

Successive applications of F_1 and Q_2 in Fig. 4.1 result in three components of work done on the system of external means [Eq. (4.1)]. These loading phases are shown in Fig. 4.5a and b, with the energy represented by the areas under the respective curves:

$$U_e = \int_0^A F_1 \, dy_1 + \int_C^D Q_2 \, dy_2 + \int_A^B F_1 \, dy_1 \tag{4.9}$$

The complementary factor corresponds to the rectangle $F_1 y_{12}$, as developed in Chaps. 1 and 2.

If both F_1 and Q_2 are applied gradually but *simultaneously*, the forces would increase simultaneously and linearly from 0 to B and D, respectively (Fig. 4.5c and d). Although this appears to be a different situation, results are identical to the successive procedure. Noting the shaded triangles,

$$(U_e)_F = \frac{F_1(y_{11} + y_{12})}{2} + \frac{F_1 y_{12}}{2} = F_1 \left(\frac{y_{11}}{2} + y_{12} \right) \tag{4.10a}$$

$$(U_e)_Q = \frac{Q_2(y_{21} + y_{22})}{2} - \frac{Q_2 y_{21}}{2} = \frac{Q_2 y_{22}}{2} \tag{4.10b}$$

with the complementary quantity again shown to be $F_1 y_{12}$. This area is exactly twice the area of the triangle $0AB$ in Fig. 4.5c.

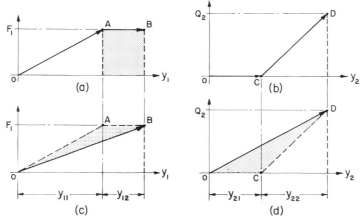

Figure 4.5 Either stepwise or simultaneous application of F_1 and Q_2 results in equivalent external energies.

4.5 Deflections Due to Load

To illustrate a simple solution using Eq. (4.4c), we take a cantilever with an end load (Fig. 4.6a) and analyze the resulting elastic curve. Applying the auxiliary load at 1, there are two triangular moment diagrams, and

$$F_1 y_{12} = \int_0^\ell \frac{mM}{EI} \, dx = \int_0^a \frac{(F_1 x_1)(Q_2 x_2)}{EI} \, dx_1$$

$$y_{12} = \frac{Q_2}{EI} \int_0^a x_1 (x_1 + b) \, dx_1$$

Before integration we have reduced the function to a single variable. By choosing x_1, the lower limit is 0. Since m is 0 in the b interval, there is no complementary energy in this portion of the beam:

$$y_{12} = \frac{Q_2}{EI} \left[\frac{a^3}{3} + \frac{ba^2}{2} \right] = \frac{Q_2 a^3}{3EI} \left[1 + \frac{3}{2} \left(\frac{b}{a} \right) \right] \qquad (4.11)$$

The first term represents the deflection of 1 as a simple cantilever with Q applied at 1. The second term accounts for the additional deflection at 1 due to the outboard moment (bQ_2).

If we require the slope at 1 due to Q_2, an auxiliary couple M' is applied (Fig. 4.6b), and

$$M' \theta_{12} = \int_0^\ell \frac{mM}{EI} \, dx = \int_b^\ell \frac{M'(Q_2 x_2)}{EI} \, dx_2$$

$$\theta_{12} = \frac{Q_2 \ell^2}{2EI} \left[1 - \left(\frac{b}{\ell} \right)^2 \right] \qquad (4.12a)$$

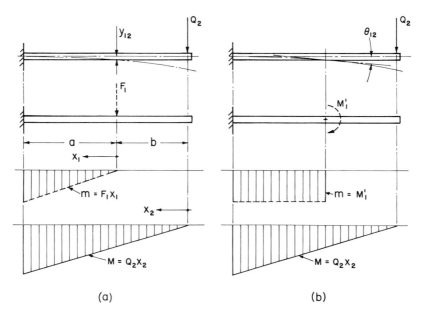

Figure 4.6 Auxiliary and actual bending moment distributions for the cantilever with end load are used to solve for the deflection at any point (a) or the slope at any point (b).

Or, converting to x_1,

$$\theta_{12} = \frac{Q_2}{EI} \int_0^a (x_1 + b)\, dx_1$$

$$= \frac{Q_2 a^2}{2EI}\left[1 + 2\left(\frac{b}{a}\right)\right] \tag{4.12b}$$

Again, the second term accounts for the contribution of the b distance.

4.6 Deflections Due to a Couple

An alternate form of loading at a point is a couple. Although less common than the force load, the rotational is both a possibility and a fundamental case. Applying an end couple to a cantilever (Fig. 4.7a), we have a transverse deflection at 1 given by

$$F_1 y_{12} = \int_0^\ell \frac{mM}{EI}\, dx = \frac{M_1}{EI}\int_0^a (F_1 x_1)\, dx_1$$

$$y_{12} = \frac{M_1}{EI}\left(\frac{a^2}{2}\right) \tag{4.13}$$

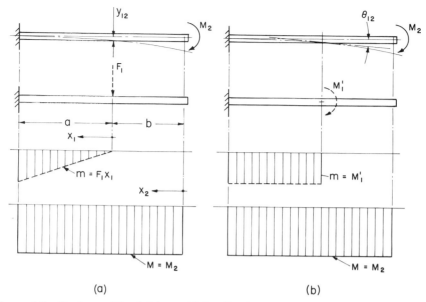

Figure 4.7 Similar to Fig. 4.6, but with loading by an end couple.

The end moment also develops a slope distribution found by applying an auxiliary couple (Fig. 4.7*b*). From energy equivalence

$$M_1'\theta_{12} = \int_0^\ell \frac{mM}{EI}\,dx = \int_0^a \frac{M'M_2}{EI}\,dx_1$$

$$\theta_{12} = \frac{M_2 a}{EI} \tag{4.14}$$

We observe the end force produces a somewhat complex elastic curve in the cantilever; however, the end couple causes parabolic and linear distributions for deflection and slope, respectively.

In Eqs. (4.11)–(4.14) the point 1 can be coincident with 2; then $a = \ell$ and $b = 0$.

4.7 *Reciprocity of Influence Factors*

If we reverse the situation leading to Eq. (4.11), we apply Q_1 and determine y_{21} (Fig. 4.8). Then the deflection of the extending beam is obtained from

$$F_2 y_{21} = \int_0^\ell \frac{mM}{EI}\,dx = \int_0^a \frac{(F_2 x_2)(Q_1 x_1)}{EI}\,dx_1$$

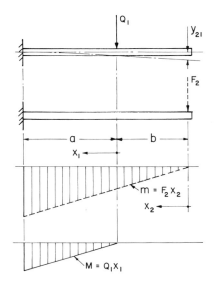

Figure 4.8 Auxiliary and actual bending moment diagrams for finding the deflection with the load applied at any point.

Again, we have $\int_0^a (x_2 x_1)\, dx_1$ for an identical definite integral:

$$y_{21} = \frac{Q_1 a^3}{3EI}\left[1 + \frac{3}{2}\left(\frac{b}{a}\right)\right] \tag{4.15}$$

If $Q_1 = Q_2$ or if each is a unit load ($y_{12} = y_{21}$), Maxwell's theorem is easily proven in this example by the interchange of the m and M diagrams.

Deflection expressions are made more general by converting to *influence* or *compliance* factors, indicating the deflection for unit load, or the ratio of deflection to load. For any two points on a given beam, there are a number of such ratios. We have:

1. Direct factors at the loads.
2. Coupled factors for two points.
3. Force or couple loads.
4. Linear or slope deflection.

For any two points in an elastic system, complete compliance relationships require specification of 16 factors (Table 4.1). As shown, reciprocity reduces the numerical values to 10.

In this notation

α = deflection per unit load
β = angular deflection per unit load, *or* linear deflection per unit couple
γ = angular deflection per unit couple

Table 4.1 Coupled elastic influence coefficients relating deflection at any point on a cantilever to the end (a) and complete factors for the simple beam relative to any two points (b).

Sign convention: $+Q$, $+y \rightarrow$; $+M$, $+\theta \circlearrowright$

Coefficient	Relates	(a)	(b)
α_{11}	$\dfrac{y_1}{Q_1}$	$\dfrac{A}{3}\left(\dfrac{a}{\ell}\right)^3$	$\dfrac{A}{3}\left(\dfrac{a}{\ell}\right)^2\left(\dfrac{b}{\ell}\right)^2$
α_{12} $\big(\alpha_{21}\big)$	$\dfrac{y_2}{Q_1}$, $\dfrac{y_1}{Q_2}$	$\dfrac{A}{3}\left(\dfrac{a}{\ell}\right)^2\left[1+\dfrac{1}{2}\left(\dfrac{b}{\ell}\right)\right]$	$\dfrac{A}{6}\left(\dfrac{a}{\ell}\right)\left(\dfrac{d}{\ell}\right)\left[1-\left(\dfrac{a}{\ell}\right)^2-\left(\dfrac{d}{\ell}\right)^2\right]$
α_{22}	$\dfrac{y_2}{Q_2}$	$\dfrac{A}{3}$	$\dfrac{A}{3}\left(\dfrac{e}{\ell}\right)^2\left(\dfrac{d}{\ell}\right)^2$
β_{11}	$\dfrac{y_1}{M_1}$, $\dfrac{\theta_1}{Q_1}$	$\dfrac{B}{2}\left(\dfrac{a}{\ell}\right)^2$	$\dfrac{B}{3}\left(\dfrac{a}{\ell}\right)\left(\dfrac{b}{\ell}\right)\left[1-2\left(\dfrac{a}{\ell}\right)\right]$
β_{12}	$\dfrac{y_2}{M_1}$, $\dfrac{\theta_1}{Q_2}$	$B\left(\dfrac{a}{\ell}\right)\left[1-\dfrac{1}{2}\left(\dfrac{a}{\ell}\right)\right]$	$\dfrac{B}{6}\left(\dfrac{d}{\ell}\right)\left[2\left(\dfrac{e}{\ell}\right)-3\left(\dfrac{a}{\ell}\right)^2-\left(\dfrac{e}{\ell}\right)^2\right]$
β_{21}	$\dfrac{y_1}{M_2}$, $\dfrac{\theta_2}{Q_1}$	$\dfrac{B}{2}\left(\dfrac{a}{\ell}\right)^2$	$\dfrac{B}{6}\left(\dfrac{a}{\ell}\right)\left[-1+\left(\dfrac{a}{\ell}\right)^2+3\left(\dfrac{d}{\ell}\right)^2\right]$
β_{22}	$\dfrac{y_2}{M_2}$, $\dfrac{\theta_2}{Q_2}$	$\dfrac{B}{2}$	$\dfrac{B}{3}\left(\dfrac{e}{\ell}\right)\left(\dfrac{d}{\ell}\right)\left[1-2\left(\dfrac{e}{\ell}\right)\right]$
γ_{11}	$\dfrac{\theta_1}{M_1}$	$C\left(\dfrac{a}{\ell}\right)$	$\dfrac{C}{3}\left[1-3\left(\dfrac{a}{\ell}\right)+3\left(\dfrac{a}{\ell}\right)^2\right]$
γ_{12} $\big(\gamma_{21}\big)$	$\dfrac{\theta_1}{M_2}$, $\dfrac{\theta_2}{M_1}$	$C\left(\dfrac{a}{\ell}\right)$	$\dfrac{C}{6}\left[2+3\left(\dfrac{a}{\ell}\right)^2-6\left(\dfrac{e}{\ell}\right)+3\left(\dfrac{e}{\ell}\right)^2\right]$
γ_{22}	$\dfrac{\theta_2}{M_2}$	C	$\dfrac{C}{3}\left[1-3\left(\dfrac{e}{\ell}\right)+3\left(\dfrac{e}{\ell}\right)^2\right]$

$$A=\frac{\ell^3}{EI} \qquad B=\frac{\ell^2}{EI} \qquad C=\frac{\ell}{EI}$$

Table 4.2 Similar to Table 4.1, but for an intermediate and an outboard point (a) and for two opposed outboard points (b).

Sign convention: $+Q$, $+y \rightarrow$; $+M$, $+\theta \circlearrowright$

Coefficient	Relates	(a)	(b)
α_{11}	$\dfrac{y_1}{Q_1}$	$\dfrac{A}{3}\left(\dfrac{a}{\ell}\right)^2\left(\dfrac{b}{\ell}\right)^2$	$\dfrac{A}{3}\left(\dfrac{a}{\ell}\right)^2\left[1+\left(\dfrac{a}{\ell}\right)\right]$
α_{12} $\big(\alpha_{21}\big)$	$\dfrac{y_2}{Q_1}$, $\dfrac{y_1}{Q_2}$	$\dfrac{A}{6}\left(\dfrac{a}{\ell}\right)\left(\dfrac{d}{\ell}\right)\left[-1+\left(\dfrac{a}{\ell}\right)^2\right]$	$\dfrac{A}{6}\left(\dfrac{a}{\ell}\right)\left(\dfrac{b}{\ell}\right)$
α_{22}	$\dfrac{y_2}{Q_2}$	$\dfrac{A}{3}\left(\dfrac{d}{\ell}\right)^2\left[1+\left(\dfrac{d}{\ell}\right)\right]$	$\dfrac{A}{3}\left(\dfrac{b}{\ell}\right)^2\left[1+\left(\dfrac{b}{\ell}\right)\right]$
β_{11}	$\dfrac{y_1}{M_1}$, $\dfrac{\theta_1}{Q_1}$	$\dfrac{B}{3}\left(\dfrac{a}{\ell}\right)\left(\dfrac{b}{\ell}\right)\left[1-2\left(\dfrac{a}{\ell}\right)\right]$	$-\dfrac{B}{3}\left(\dfrac{a}{\ell}\right)\left[1+\dfrac{3}{2}\left(\dfrac{a}{\ell}\right)\right]$
β_{12}	$\dfrac{y_2}{M_1}$, $\dfrac{\theta_1}{Q_2}$	$\dfrac{B}{6}\left(\dfrac{d}{\ell}\right)\left[-1+3\left(\dfrac{a}{\ell}\right)^2\right]$	$-\dfrac{B}{6}\left(\dfrac{b}{\ell}\right)$
β_{21}	$\dfrac{y_1}{M_2}$, $\dfrac{\theta_2}{Q_1}$	$\dfrac{B}{6}\left(\dfrac{a}{\ell}\right)\left[-1+3\left(\dfrac{d}{\ell}\right)^2\right]$	$\dfrac{B}{6}\left(\dfrac{a}{\ell}\right)$
β_{22}	$\dfrac{y_2}{M_2}$, $\dfrac{\theta_2}{Q_2}$	$\dfrac{B}{3}\left(\dfrac{d}{\ell}\right)\left[1+\dfrac{3}{2}\left(\dfrac{d}{\ell}\right)\right]$	$\dfrac{B}{3}\left(\dfrac{b}{\ell}\right)\left[1+\dfrac{3}{2}\left(\dfrac{b}{\ell}\right)\right]$
γ_{11}	$\dfrac{\theta_1}{M_1}$	$\dfrac{C}{3}\left[1-3\left(\dfrac{a}{\ell}\right)+3\left(\dfrac{a}{\ell}\right)^2\right]$	$\dfrac{C}{3}\left[1+3\left(\dfrac{a}{\ell}\right)\right]$
γ_{12} $\big(\gamma_{21}\big)$	$\dfrac{\theta_1}{M_2}$, $\dfrac{\theta_2}{M_1}$	$\dfrac{C}{6}\left[-1+3\left(\dfrac{a}{\ell}\right)^2\right]$	$-\dfrac{C}{6}$
γ_{22}	$\dfrac{\theta_2}{M_2}$	$\dfrac{C}{3}\left[1+3\left(\dfrac{d}{\ell}\right)\right]$	$\dfrac{C}{3}\left[1+3\left(\dfrac{b}{\ell}\right)\right]$

$$A=\frac{\ell^3}{EI} \qquad B=\frac{\ell^2}{EI} \qquad C=\frac{\ell}{EI}$$

Table 4.3 Similar to Table 4.1, but for two outboard points (a). In (b) a uniform beam between supports is connected to a rigid section.

Table 4.4 Similar to Table 4.1, for a beam with a rigid and a constant section (a) and a cantilever with an extended rigid section (b).

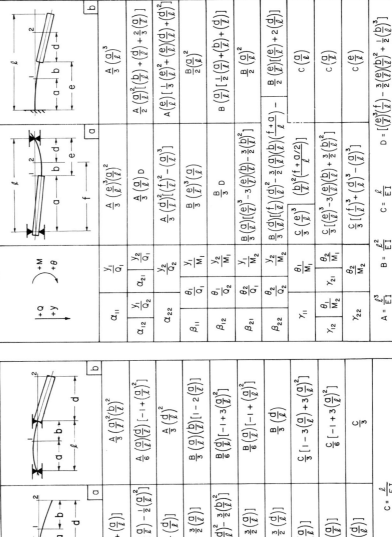

Table 4.1 indicates that $\beta_{12} \neq \beta_{21}$. This is not a violation of Maxwell's theorem, but represents a limitation beyond which reciprocity cannot be extended. In effect, we must realize that β_{12} and β_{21} are independent physical properties, and

$$\frac{\theta_1}{Q_2} \neq \frac{\theta_2}{Q_1} \quad \text{and} \quad \frac{y_2}{M_1} \neq \frac{y_1}{M_2}$$

Tables 4.1–4.4 provide complete equations for influence factors for eight fundamental types of beams. All are dimensionless with respect to beam geometry. Absolute beam size enters only the A, B, and C terms.

4.8 Characteristic Deflections of Basic Beams

Equation (4.4c) is especially adaptable to calculation of a specific deflection, and these solutions are usually direct and uncomplicated; however, complete deflection functions are also obtained using variable parameters rather than numerical. This approach has been used to develop Tables 4.1–4.4, and, in a sense, these influence factors at a coordinate point describe the

Table 4.5 Polynomial equations for the complete elastic curve of a constant cantilever with force or couple loading.

$x_0 \lessgtr a$ $x_1 \lessgtr b$	Q	M
y_1	$\frac{A}{3}\left(\frac{a}{\ell}\right)^3$	$\frac{C}{2}\left(\frac{a}{\ell}\right)^2$
$y(x_0)$	$\frac{A}{2}\left[\left(\frac{a}{\ell}\right)\left(\frac{x_0}{\ell}\right)^2 - \frac{1}{3}\left(\frac{x_0}{\ell}\right)^3\right]$	$\frac{C}{2}\left(\frac{x_0}{\ell}\right)^2$
$y(x_1)$	$\frac{A}{3}\left[\left(\frac{a}{\ell}\right)^3 + \frac{3}{2}\left(\frac{a}{\ell}\right)^2\left(\frac{x_1}{\ell}\right)\right]$	$\frac{C}{2}\left(\frac{a}{\ell}\right)\left[\left(\frac{a}{\ell}\right) + 2\left(\frac{x_1}{\ell}\right)\right]$
θ_1	$\frac{B}{2}\left(\frac{a}{\ell}\right)^2$	$D\left(\frac{a}{\ell}\right)$
$\theta(x_0)$	$B\left[\left(\frac{a}{\ell}\right)\left(\frac{x_0}{\ell}\right) - \frac{1}{2}\left(\frac{x_0}{\ell}\right)^2\right]$	$D\left(\frac{x_0}{\ell}\right)$
$\theta(x_1)$	$\frac{B}{2}\left(\frac{a}{\ell}\right)^2$	$D\left(\frac{a}{\ell}\right)$

$$A = \frac{Q\ell^3}{EI} \qquad B = \frac{Q\ell^2}{EI} \qquad C = \frac{M\ell^2}{EI} \qquad D = \frac{M\ell}{EI}$$

complete deflection curve. The main utility of the tables, however, is related to a single deflection resulting from a given load. This typifies the majority of engineering design problems.

More general results are achieved by the energy method in Table 4.5 for the cantilever, in which both force and couple loadings are included. Considerable manipulation is often required, however, to reduce the expressions to concise and usable form. We observe in Table 4.5 that the different

Figure 4.9 Elastic curve behavior for the constant cantilever showing deflection and slope variation as the load shifts along the length from Table 4.5.

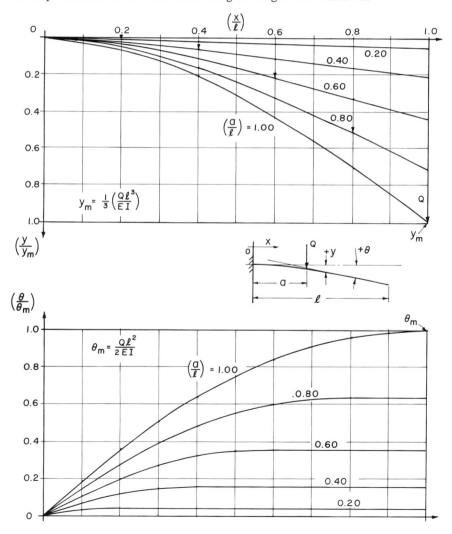

spans (a and b) are described by different functions of the dimensionless horizontal coordinates (x_0/ℓ) and (x_1/ℓ). Absolute deflections involve the reference beam factors A, B, C, and D. The distribution terms involve constant, linear, squared, and cubic terms, with curvatures having parabolic components.

The cantilever curves in Fig. 4.9 allow us to study the deflections qualitatively and quantitatively as the load shifts from the base to the free

Figure 4.10 Cantilever elastic-curve behavior as couple load shifts from fixed to free end.

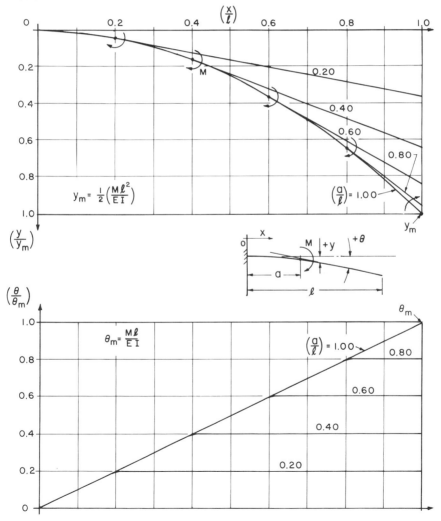

Table 4.6 Deflection equations for a constant beam with a load at any point between the simple supports.

	$a = b$	y_1
y_1	$\dfrac{C}{48}$	$\dfrac{C}{3}(E)^2$
$y(x_0)$	$\dfrac{C}{16}\left[\left(\dfrac{x_0}{\ell}\right) - \dfrac{4}{3}\left(\dfrac{x_0}{\ell}\right)^3\right]$	$\dfrac{C}{6}\left[A_1\left(\dfrac{x_0}{\ell}\right) - A_2\left(\dfrac{x_0}{\ell}\right)^3\right]$
$y(x_2)$	$\dfrac{C}{16}\left[\left(\dfrac{x_2}{\ell}\right) - \dfrac{4}{3}\left(\dfrac{x_2}{\ell}\right)^3\right]$	$\dfrac{C}{6}\left[B_1\left(\dfrac{x_2}{\ell}\right) - B_2\left(\dfrac{x_2}{\ell}\right)^3\right]$
θ_1	0	$\dfrac{DE}{3}\left[1 - 2\left(\dfrac{a}{\ell}\right)\right]$
$\theta(x_0)$	$\dfrac{D}{16}\left[1 - 4\left(\dfrac{x_0}{\ell}\right)^2\right]$	$\dfrac{D}{6}\left[A_1 - 3A_2\left(\dfrac{x_0}{\ell}\right)^2\right]$
$\theta(x_2)$	$\dfrac{D}{16}\left[-1 + 4\left(\dfrac{x_2}{\ell}\right)^2\right]$	$\dfrac{D}{6}\left[-B_1 + 3B_2\left(\dfrac{x_2}{\ell}\right)^2\right]$
$C = \dfrac{Q\ell^3}{EI}$, $D = \dfrac{Q\ell^2}{EI}$, $E = \left(\dfrac{a}{\ell}\right)\left(\dfrac{b}{\ell}\right)$	$x_0 \leqq a$	$x_2 \leqq b$
A_1	$\left(\dfrac{b}{\ell}\right)\left[1 - \left(\dfrac{b}{\ell}\right)^2\right]$	B_1
A_2	$\left(\dfrac{b}{\ell}\right)$	B_2

Last column values: $B_1 = \left(\dfrac{a}{\ell}\right)\left[1 - \left(\dfrac{a}{\ell}\right)^2\right]$, $B_2 = \left(\dfrac{a}{\ell}\right)$

Table 4.7 Similar to Table 4.6, but for couple loading.

	$a = b$	y_1
y_1	0	$\dfrac{C}{3}\left(\dfrac{a}{\ell}\right)\left[1 - 3\left(\dfrac{a}{\ell}\right) + 2\left(\dfrac{a}{\ell}\right)^2\right]$
$y(x_0)$	$\dfrac{C}{24}\left[-\left(\dfrac{x_0}{\ell}\right) + 4\left(\dfrac{x_0}{\ell}\right)^3\right]$	$\dfrac{C}{6}\left[A_1\left(\dfrac{x_0}{\ell}\right) + \left(\dfrac{x_0}{\ell}\right)^3\right]$
$y(x_2)$	$\dfrac{C}{24}\left[\left(\dfrac{x_2}{\ell}\right) - 4\left(\dfrac{x_2}{\ell}\right)^3\right]$	$\dfrac{C}{6}\left[-B_1\left(\dfrac{x_2}{\ell}\right) - \left(\dfrac{x_2}{\ell}\right)^3\right]$
θ_1	$\dfrac{D}{12}$	$\dfrac{D}{3}\left[1 - 3\left(\dfrac{a}{\ell}\right) + 3\left(\dfrac{a}{\ell}\right)^2\right]$
$\theta(x_0)$	$\dfrac{D}{24}\left[-1 + 12\left(\dfrac{x_0}{\ell}\right)^2\right]$	$\dfrac{D}{6}\left[A_1 + 3\left(\dfrac{x_0}{\ell}\right)^2\right]$
$\theta(x_2)$	$\dfrac{D}{24}\left[-1 + 12\left(\dfrac{x_2}{\ell}\right)^2\right]$	$\dfrac{D}{6}\left[B_1 + 3\left(\dfrac{x_2}{\ell}\right)^2\right]$
$C = \dfrac{M\ell^2}{EI}$, $D = \dfrac{M\ell}{EI}$	$x_0 \leqq a$	$x_2 \leqq b$
A_1	$2 - 6\left(\dfrac{a}{\ell}\right) + 3\left(\dfrac{a}{\ell}\right)^2$	B_1

Last column value: $B_1 = 2 - 6\left(\dfrac{b}{\ell}\right) + 3\left(\dfrac{b}{\ell}\right)^2$

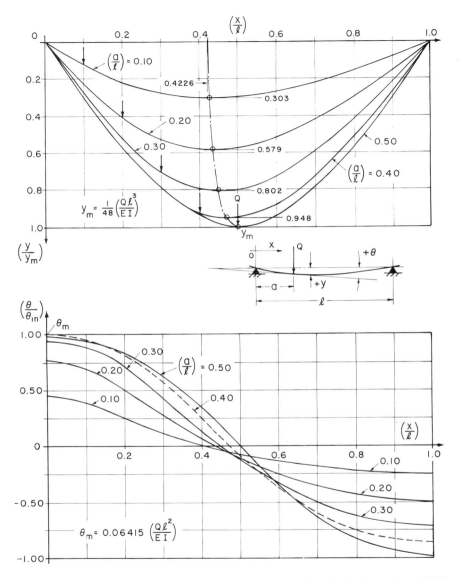

Figure 4.11 Deflection and slope characteristics for the simple beam from Table 4.6. Maximum deflection cannot occur farther than 0.08ℓ from the center, regardless of location of the load.

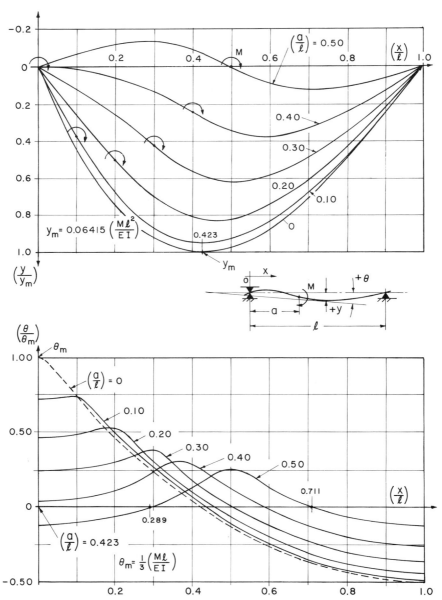

Figure 4.12 Similar to Fig. 4.11, but from Table 4.7. End couple also produces maximum deflection 0.08ℓ from the center of the beam.

$x_0 \leq a$	Q	M
y_2	$\frac{A}{3}\left[1+\left(\frac{a}{b}\right)\right]$	$\frac{C}{2}\left[1+\frac{2}{3}\left(\frac{a}{b}\right)\right]$
$y(x_0)$	$\frac{A}{6}\left(\frac{a}{b}\right)^2\left[-\left(\frac{x_0}{a}\right)+\left(\frac{x_0}{a}\right)^3\right]$	$\frac{C}{6}\left(\frac{a}{b}\right)^2\left[-\left(\frac{x_0}{a}\right)+\left(\frac{x_0}{a}\right)^3\right]$
$y(x_1)$	$\frac{A}{6}\left[2\left(\frac{a}{b}\right)\left(\frac{x_1}{b}\right)+3\left(\frac{x_1}{b}\right)^2-\left(\frac{x_1}{b}\right)^3\right]$	$\frac{C}{2}\left[\frac{2}{3}\left(\frac{a}{b}\right)\left(\frac{x_1}{b}\right)+\left(\frac{x_1}{b}\right)^2\right]$
θ_0	$-\frac{B}{6}\left(\frac{a}{b}\right)$	$-\frac{D}{6}\left(\frac{a}{b}\right)$
θ_1	$\frac{B}{3}\left(\frac{a}{b}\right)$	$\frac{D}{3}\left(\frac{a}{b}\right)$
θ_2	$\frac{B}{6}\left[2\left(\frac{a}{b}\right)+3\right]$	$D\left[1+\frac{1}{3}\left(\frac{a}{b}\right)\right]$
$\theta(x_0)$	$\frac{B}{6}\left(\frac{a}{b}\right)\left[-1+3\left(\frac{x_0}{a}\right)^2\right]$	$\frac{D}{6}\left(\frac{a}{b}\right)\left[-1+3\left(\frac{x_0}{a}\right)^2\right]$
$\theta(x_1)$	$\frac{B}{3}\left[\left(\frac{a}{b}\right)+3\left(\frac{x_1}{b}\right)-\frac{3}{2}\left(\frac{x_1}{b}\right)^2\right]$	$\frac{D}{3}\left[\left(\frac{a}{b}\right)+3\left(\frac{x_1}{b}\right)\right]$

$$A=\frac{Qb^3}{EI} \qquad B=\frac{Qb^2}{EI} \qquad C=\frac{Mb^2}{EI} \qquad D=\frac{Mb}{EI}$$

Table 4.8 Deflection equations for the simple beam with an outboard force or couple.

end. In the latter case, $(a/\ell)=b$. For instance, if $(a/\ell)=0.80$, the deflection at the end is approximately 0.70 of the maximum possible deflection y_m. Similarly, deflection at this load is about 0.50 of the reference value, and we are able to visualize the reductions in deflection corresponding to moving the load back from the free end.

Slope variations are also shown in Fig. 4.9. Since the beam is straight beyond the load point, these curves are horizontal or constant to the right of the load Q.

Figure 4.10 shows the effect of shifting the loading couple relative to the beam. Tables 4.6 and 4.7 and Figs. 4.11 and 4.12 provide the same types of information for the simple beam on end supports.

For the constant beam with an outboard load (Table 4.8), the parameter of interest is the location of the support relative to the load (a/ℓ). If $(a/\ell)=1$ (Fig. 4.13), Q is at the right end support and there is no

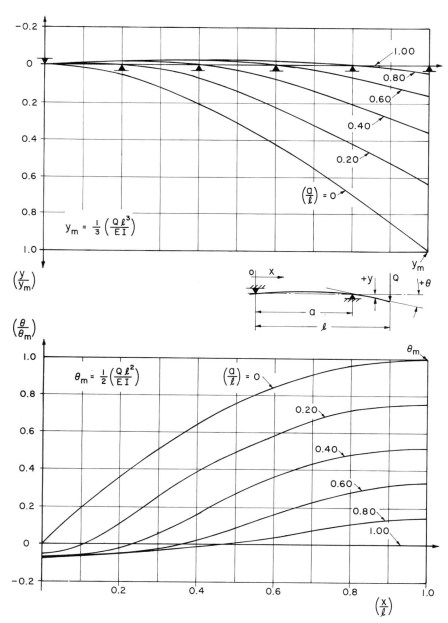

Figure 4.13 Elastic behavior due to force from Table 4.8, showing the effect of shifting the support with respect to the length.

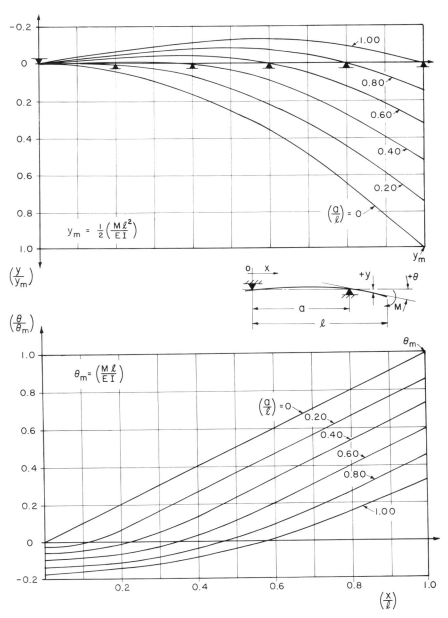

Figure 4.14 Similar to Fig. 4.13, but for couple loading at the outboard end.

deflection of the beam. As (a/ℓ) approaches 0, the beam approaches the cantilever, which we recognize from the reference deflection, both for the applied load and couple (Fig. 4.14). As the slope curves cross the axis, they identify the condition of maximum upward deflection between the supports.

4.9 *The Tapered Cantilever*

If a cantilever is a frustum of a pyramid or cone (Table 4.9), we have an I characteristic varying continuously from minimum at the free end to maximum at the base. This basic type of problem was introduced in Sec. 3.11. Theoretically, these tapered geometries can converge to an apex, but it is physically impossible to apply a load at such a point.

To illustrate the derivations in Table 4.9, we take the cantilever of constant thickness and varying width (Table 4.9a) and determine y_1 due to Q_1. With notation in Fig. 4.15.

$$\frac{Q_1 y_1}{2} = \int_0^\ell \frac{M^2}{2EI}\,dx = \frac{Q_1^2}{2E}\int_0^\ell \frac{x_1^2}{I_0(x_2/\ell_2)}\,dx_1 \qquad (4.16a)$$

Since the moment of inertia is $\frac{1}{12}bh^3$, I varies linearly as the width and is directly proportional to the distance from the extended apex. The end

Table 4.9 End deflection of tapered cantilevered beams subjected to either end force or end couple. Reference deflections C are for the stiffer equivalent constant beam having the cross section of the base.

	a	b	c
$\dfrac{y_1}{y_{QC}}$	$\dfrac{3}{(1-r)^3}\left[\dfrac{(1-r)(1-3r)}{2} - r^2\log r\right]$	$\dfrac{3}{(1-r)^3}\left[\log\dfrac{1}{r} - \dfrac{(1-r)(3-r)}{2}\right]$	$\dfrac{1}{(1-r)^3}\left[\dfrac{1}{r} - r^2 - 3(1-r)\right]$
$\dfrac{y_M}{y_{MC}} = \dfrac{\theta_Q}{\theta_{QC}}$	$\dfrac{2}{(1-r)}\left[1 + \left(\dfrac{r}{1-r}\right)\log r\right]$	$\left[\dfrac{1}{r}\right]$	$\dfrac{1}{(1-r)^2}\left[\dfrac{2r}{3} + \dfrac{1}{3r^2} - 1\right]$
$\dfrac{\theta_1}{\theta_{MC}}$	$\dfrac{1}{(1-r)}\left[\log\dfrac{1}{r}\right]$	$\dfrac{1}{2(1-r)}\left[\dfrac{1}{r^2} - 1\right]$	$\dfrac{1}{3(1-r)}\left[\dfrac{1}{r^3} - 1\right]$
	$r = \left(\dfrac{b_1}{b_0}\right)$ $I_0 = \left(\dfrac{b_0 h^3}{12}\right)$	$r = \left(\dfrac{h_1}{h_0}\right)$ $I_0 = \left(\dfrac{b h_0^3}{12}\right)$	$r = \left(\dfrac{d_1}{d_0}\right)$ $I_0 = \left(\dfrac{\pi d_0^4}{64}\right)$
	$y_{QC} = \dfrac{Q_1 \ell^3}{3EI_0}$	$y_{MC} = \dfrac{M_1 \ell^2}{2EI_0}$ $\theta_{QC} = \dfrac{Q_1 \ell^2}{2EI_0}$	$\theta_{MC} = \dfrac{M_1 \ell}{EI_0}$

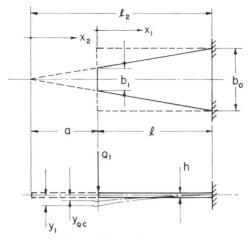

Figure 4.15 The tapered cantilevered beam has a reference triangular altitude of ℓ_2 and a reference constant beam, shown dashed.

deflection becomes

$$y_1 = \frac{Q_1 \ell_2}{EI_0} \int_0^\ell \frac{x_1^2}{(a + x_1)} \, dx_1 \tag{4.16b}$$

After integration and substitution of limits, we have

$$y_1 = \frac{Q_1 \ell_2^3}{EI_0} \left[\frac{(1 - r)(1 - 3r)}{2} - r^2 \log r \right] \tag{4.16c}$$

where

$$r = \left(\frac{a}{\ell_2} \right) = \left(\frac{b_1}{b_0} \right) \quad \text{and} \quad \left(\frac{\ell}{\ell_2} \right) = (1 - r)$$

Finally, dividing by $(Q_1 \ell^3 / 3EI_0)$, we obtain the ratio of the tapered cantilever deflection to the reference constant cantilever with an end load y_{QC}, we have the first equation in Table 4.9a.

Checking for a constant beam $r = 1$, the ratio is indeterminate; however, if $r = 0.99$, $y_1/y_{QC} = 1.0024$, sufficiently close to unity to confirm the agreement. For the completely triangular beam $r = 0$, and $y/y_{QC} = 1.5$ (Fig. 4.16). This beam, loaded at the apex, is more compliant than the constant beam by the factor of 1.5 as we have removed $\frac{1}{2}$ of the rectangular material but maintained $\frac{2}{3}$ of the stiffness. This is because most of the material in the triangle tends to be concentrated near the base to resist the bending moment that tends to maximum at the base.

The direct and coupled ratios in Fig. 4.16 indicate the increase in deflection resulting from increased taper as we deviate from the straight beam at $r = 1$. The change is most pronounced for the cone (Fig. 4.16c), for which I varies as the fourth power of the diameter.

Relations developed for the basic cantilever (Table 4.9) are applicable to other combinations. For instance, two similar tapered beams base connected and center loaded correspond to two simple tapered cantilevers in series. In other combinations (say tapered and straight beams in the same element), we can isolate the tapered beam determining the *relative* deflection and slope between the base and the tip.

Figure 4.16 The taper factor r increases the flexibility of the three types of beams in Table 4.10. Both direct and coupled elastic factors are shown.

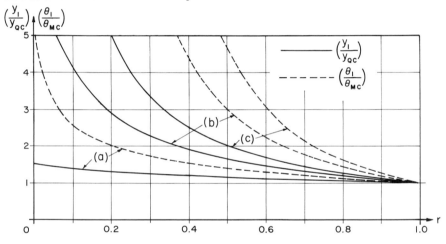

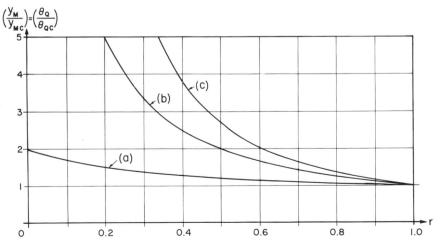

A relatively wide tapered beam will tend to depart from classical bending behavior, and deflections in these cases will deviate somewhat from the nominal calculated values.

4.10 *Arbitrarily Distributed Stiffness*

In this chapter, with concentrated loads, the bending moment diagrams consist of straight lines, either—constant, triangular, or trapezoidal. These diagrams derive from equilibrium and are, therefore, independent of beam stiffness. The beams are assumed to be planar with the centroids of the cross-sectional areas coincident with the plane, and with all neutral (principal) axes of these areas perpendicular to or in the plane of the beam.

If the beam has a complex geometry (Fig. 4.17) so it is difficult or impossible to obtain an algebraic expression for the functional variation of

Figure 4.17 The variable beam of arbitrary cross section requires the numerical integration of the internal energy functional distribution (e).

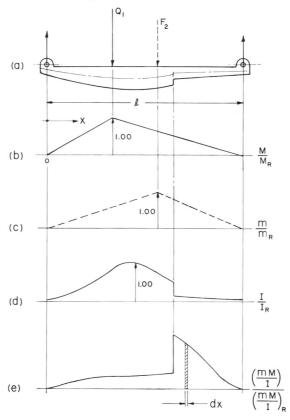

the inertia variation lengthwise, our definite integral internal energy terms are still applicable but require numerical integration.

We must then determine the centroid and moment of inertia of beam sections at intervals to establish the $I(x)$ curve (Fig. 4.17). These intervals need not be equally spaced along the beam, although this is usually the most convenient procedure. We should include the ordinates in the transverse planes of discontinuity.

With energy distribution established (Fig. 4.17e), we have the data from which to evaluate the integral from the relation

$$F_2 y_{21} = \int_0^\ell \frac{mM}{EI} \, dx = \frac{1}{E} \int_0^\ell \frac{m(x) M(x)}{I(x)} \, dx \qquad (4.17)$$

Numerical integration is somewhat awkward with respect to the numerical values and units involved; however, we can *normalize* the entire integrand by using *reference* (maximum) values for all variables (Fig. 4.17). The equation for the deflection at the second point is then

$$y_{21} = Q_1 C_R \left[\int_0^1 f(\xi) \, d\xi \right] = \alpha_{21} Q_1 \qquad (4.18)$$

where

$$C_R = \frac{(m_R)(M_R)\ell}{F_2 Q_1 EI_R} = \frac{(C_m \ell)(C_M \ell)\ell}{EI_R} = \frac{C_m C_M \ell^3}{EI_R}$$

$$= \text{reference compliance factor}$$

$$f(\xi) = \frac{\left(\dfrac{m}{m_R}\right)\left(\dfrac{M}{M_R}\right)}{\left(\dfrac{I}{I_R}\right)} = \text{dimensionless integrand function}$$

$$\xi = \frac{x}{\ell} = \text{dimensionless axial coordinate}$$

For the simple beam (Table 4.1b),

$$C_m = \frac{m_R}{F_2 \ell} = \left(\frac{e}{\ell}\right)\left(\frac{d}{\ell}\right) \qquad C_M = \frac{M_R}{Q_1 \ell} = \left(\frac{a}{\ell}\right)\left(\frac{b}{\ell}\right)$$

We note the C_m and C_M factors are dimensionless, depending only upon the axial positions of 1 and 2. The only absolute parameters occur in the $(Q_1 \ell^3/EI_R)$ term which relates to the *absolute* deflection. All other terms derive from the *relative* geometric configuration and constitute generalized results. A similar normalizing procedure can be adapted to accommodate slopes and couple loading, leading to dimensionless numerical integration.

4.11 *Comments on Accuracy*

The assumption of classical linear stress distribution throughout the width, height, and length of a beam is obviously simplistic. It is well known that geometric discontinuities create local stress distortions. These include sudden changes in section and holes. In Fig. 4.17a there will be dead or unstressed zones immediately to the left of the shoulder before the bending stress pattern resumes. It follows that our basic energy relations do not account for these effects, and that approximations exist in beams of unusual geometry and in relatively short beams.

Internal energy integrations, however, are volumetric in nature, and we are summing the entire beam as an energy reservoir. Thus while local deviations must occur, these typically correspond only to small or localized volumes and the effect on the total volumetric behavior is often negligible. Also favoring accuracy is an averaging feature in the stress nonlinearities. In some regions the stresses are greater than nominal, and in other regions less. Resulting under- and overevaluation of the local energies will provide a degree of compensation.

Another factor not usually considered is the deflection associated with local deformations at the point of application at a load or a support. These are related to *Hertz* contact stress patterns and may be termed Hertz deflections. Again, with only small volumes subjected to high localized stresses the effect upon total strain energy is often negligible.

Examples

4.1. A uniform beam on simple supports is loaded at the one-quarter point. Determine:
   (a) The deflection at the load
   (b) The deflection at the center
   (c) The slope at the support nearest the load

Solution:
(a) From Eq. (3.11)

$$\frac{Qy_1}{2} = \int_0^\ell \frac{M^2}{2EI}\, dx = \int_0^a \frac{(3/4)\,Qx_0^2}{2EI}\, dx_0 + \int_0^{3a} \frac{(1/4)\,Qx_3^2}{2EI}\, dx_3$$

$$y_1 = \frac{3}{4}\left(\frac{Qa^3}{EI}\right)$$

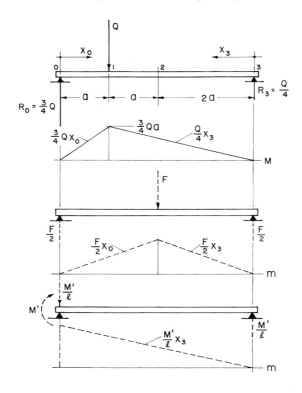

(b) From Eq. (4.4c)

$$Fy_2 = \int_0^\ell \frac{Mm}{EI}\, dx = \frac{(F/2)}{EI}\left[\int_0^a (3/4)\, Qx_0^2\, dx_0\right.$$

$$+ \int_0^{2a}(Q/4)\, x_3^2\, dx_3 + \left.\int_a^{2a} x_0\,(Q/4)\, x_3\, dx_0\right]$$

$$x_0 + x_3 = 4a \qquad x_3 = 4a - x_0$$

$$y_2 = \frac{Q}{2EI}\left[\frac{3}{4}\left(\frac{a^3}{3}\right) + \frac{1}{4}\left(\frac{8a^3}{3}\right) + \frac{1}{4}\int_a^{2a}\left(4ax_0 - x_0^2\right)\, dx_0\right]$$

$$= \frac{11}{12}\left(\frac{Qa^3}{EI}\right)$$

(c) $M'\theta_0 = \int_0^\ell \frac{Mm}{EI}\, dx = \frac{M'Q}{EI}\left[\int_0^{3a}\left(\frac{1}{4a}\right)\left(\frac{1}{4}\right) x_3^2\, dx_3\right.$

$$+ \left.\int_0^a \left(\frac{1}{4a}\right) x_3\left(\frac{3}{4}\right) x_0\, dx_0\right]$$

$$\theta_0 = \frac{Q}{EI}\left[\frac{1}{16}\left(\frac{27a^2}{3}\right) + \frac{3}{16}\int_0^a\left(4ax_0 - x_0^2\right)\, dx_0\right]$$

$$= \frac{7}{8}\left(\frac{Qa^2}{EI}\right)$$

4.2. Verify the deflection results in Example 4.1, using the influence factors in Table 4.1*b*.

Solution:

(a) $\qquad \alpha_{11} = \dfrac{\ell^3}{3EI}\left(\dfrac{1}{4}\right)^2\left(\dfrac{3}{4}\right)^2 = \dfrac{3\ell^3}{256}$

$\qquad\qquad y_1 = \dfrac{3Qa^3}{4EI}$

(b) $\qquad \alpha_{21} = \dfrac{\ell^3}{6EI}\left(\dfrac{1}{4}\right)\left(\dfrac{1}{2}\right)\left[1 - \left(\dfrac{1}{4}\right)^2 - \left(\dfrac{1}{2}\right)^2\right] = \left(\dfrac{11}{768}\right)\dfrac{\ell^3}{EI}$

$\qquad\qquad y_2 = \left(\dfrac{11}{12}\right)\dfrac{Qa^3}{EI}$

(c) Reversing the beam,

$$\beta_{21} = \dfrac{\ell^2}{6EI}\left(\dfrac{3}{4}\right)\left[-1 + \left(\dfrac{3}{4}\right)^2 + 0\right] = -\left(\dfrac{7}{128}\right)\dfrac{\ell^2}{EI}$$

For the original beam

$$\theta_0 = \left(\dfrac{7}{8}\right)\dfrac{Qa^2}{EI}$$

4.3. Verify the results of Example 4.1 using Table 4.6.

Solution:

(a) $\qquad y_1 = \dfrac{Q\ell^3}{3EI}\left(\dfrac{1}{4}\right)^2\left(\dfrac{3}{4}\right)^2 = \left(\dfrac{3}{256}\right)\dfrac{Q^3}{EI} = \left(\dfrac{3}{4}\right)\dfrac{Qa^3}{EI}$

(b) $\qquad B_1 = \dfrac{1}{4}\left(1 - \dfrac{1}{16}\right) = \dfrac{15}{64}$

$\qquad y(x_2) = \dfrac{Q\ell^3}{6EI}\left[\dfrac{15}{64}\left(\dfrac{1}{2}\right) - \dfrac{1}{4}\left(\dfrac{1}{2}\right)^3\right] = \left(\dfrac{11}{768}\right)\dfrac{Q^3}{EI} = \left(\dfrac{11}{12}\right)\dfrac{Qa^3}{EI}$

(c) $\qquad A_1 = \dfrac{3}{4}\left[1 - \left(\dfrac{3}{4}\right)^2\right] = \dfrac{21}{64}$

$\qquad \theta(x_0) = \dfrac{Q\ell^2}{6EI}\left[\dfrac{21}{64} - 0\right] = \left(\dfrac{7}{128}\right)\dfrac{Q\ell^2}{EI} = \left(\dfrac{7}{8}\right)\dfrac{Qa^2}{EI}$

4.4. Verify approximately the results of Example 4.1 from the curves in Fig. 4.11.

Solution:

(a) $\qquad\qquad \dfrac{a}{\ell} = 0.25 \qquad \dfrac{x}{\ell} = 0.25$

Interpolating, $\dfrac{y}{y_m} \approx 0.54$

$$y_1 = 0.54\left[\dfrac{1}{48}\dfrac{Q\ell^3}{EI}\right] = 0.013\dfrac{Q\ell^3}{EI} = 0.72\dfrac{Qa^3}{EI}$$

(b) $\qquad \dfrac{a}{\ell} = 0.25 \qquad \dfrac{x}{\ell} = 0.50$

Interpolating, $\dfrac{y}{y_m} \approx 0.70$

$$y_2 = 0.70\left[\dfrac{1}{48}\dfrac{Q\ell^3}{EI}\right] = 0.0146\dfrac{Q\ell^3}{EI} = 0.93\dfrac{Qa^3}{EI}$$

(c) $\qquad \dfrac{a}{\ell} = 0.25 \qquad \dfrac{x}{\ell} = 0$

Interpolating, $\dfrac{\theta}{\theta_m} \approx 0.90$

$$\theta_0 = 0.90(0.06415)\dfrac{Q\ell^2}{EI} = 0.058\dfrac{Q\ell^2}{EI} = 0.92\dfrac{Qa^2}{EI}$$

4.5. Verify the results of Example 4.1b using volumetric integration [Eq. (4.8)].

Solution:

$$m_1 = \tfrac{1}{8}F\ell \qquad m_2 = \tfrac{1}{4}F\ell \qquad M_1 = \tfrac{3}{16}Q\ell \qquad M_2 = \tfrac{1}{8}Q\ell$$

$$A_1 = \tfrac{3}{128}FQ\ell^2 \qquad A_2 = \tfrac{1}{32}FQ\ell^2 \qquad A_{12} = \tfrac{1}{128}FQ\ell^2 \qquad A_{21} = \tfrac{3}{128}FQ\ell^2$$

$$Fy_2 = (U_s)_{FQ} = \dfrac{FQ\ell^2}{3EI}\left[\dfrac{3}{128}\left(\dfrac{\ell}{2}\right) + \dfrac{1}{32}\left(\dfrac{3}{4}\right)\ell + \dfrac{1}{128}\left(\dfrac{\ell}{4}\right) + \dfrac{3}{128}\left(\dfrac{\ell}{4}\right)\right]$$

$$y_2 = \left(\dfrac{11}{768}\right)\dfrac{Q\ell^3}{EI} = \left(\dfrac{11}{12}\right)\dfrac{Qa^3}{EI}$$

4.6. Verify Example 4.1b using geometric integration.

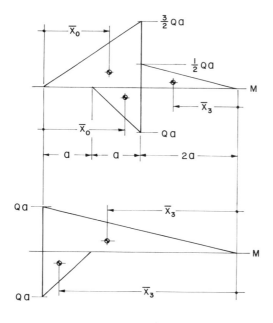

Solution: Geometric (area–moment) techniques for evaluation of definite integrals, such as Eq. (4.4c), are outlined in Chap. 5. This approach is introduced briefly in this sample problem and in Example 4.7:

$$Fy_2 = \frac{F}{2EI}\left[\int_0^{2a} x_0 M\,dx_0 + \int_0^{2a} x_3 M\,dx_3\right]$$

$$y_2 = \frac{Qa}{2EI}\left[\frac{3}{2}\left(\frac{2a}{2}\right)\frac{2}{3}(2a) - \frac{a}{2}\left(a + \frac{2}{3}a\right) + \frac{1}{2}\left(\frac{2a}{2}\right)\frac{2}{3}(2a)\right]$$

$$= \frac{Qa^3}{2EI}\left[2 - \frac{1}{2} - \frac{1}{3} + \frac{2}{3}\right] = \left(\frac{11}{12}\right)\frac{Qa^3}{2EI}$$

4.7. Also verify Example 4.1c results using geometric integration.

Solution:

$$M'\theta_0 = \frac{M'}{\ell EI}\int_0^\ell x_3 M\,dx_3$$

$$0 = \frac{1}{\ell EI}\left[\frac{Qa(4a)}{2}\frac{2}{3}(4a) - \frac{Qa^2}{2}\left(3a + \frac{2}{3}a\right)\right]$$

$$= \left(\frac{7}{8}\right)\frac{Qa^2}{EI}$$

4.8. A uniform bar with simple supports is loaded by a couple. Determine the ratio b/ℓ to produce minimum slope at 2.

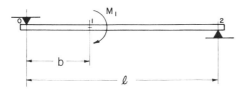

Solution: From Fig. 4.12 for zero slope at $x/\ell = 0$, $a/\ell = 0.423$. Therefore 0 is the minimum slope and occurs in a reverse sense at $b/\ell = (1 - 0.423) = 0.577$.

For minimum θ_2 with $b/\ell \leq 0.50$, at $x/\ell = 1$ this is the center location for M_1. The slope is

$$0.125\left(\frac{1}{3}\frac{M\ell}{EI}\right) = \left(\frac{1}{24}\right)\frac{M\ell}{EI}$$

4.9. A shaft, consisting of two different diameters, is simply supported as shown. Find:

(a) The deflection of Q

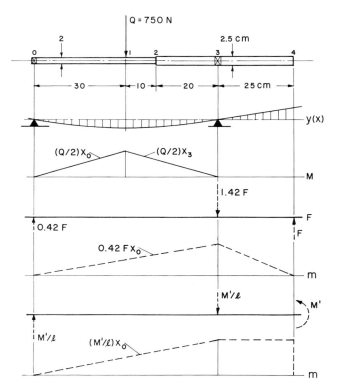

(b) The contribution of the diameter enlargement to this deflection

Solution:

(a)
$$\frac{Qy_1}{2} = \int_0^\ell \frac{M^2}{2EI}\, dx = \int_0^{30} \frac{(Q/2)\, x_0^2}{2EI_0}\, dx_0 + \int_0^{20} \frac{(Q/2)\, x_3^2}{2EI_3}\, dx_3$$
$$+ \int_{20}^{30} \frac{(Q/2)\, x_3^2}{2EI_0}\, dx_3$$

$$y_1 = \frac{Q}{12E}\left[\frac{(30)^3}{I_0} + \frac{(20)^3}{I_3} + \frac{(30^3 - 20^3)}{I_0}\right]$$

$$I_0 = \frac{\pi}{64}d^4 = \frac{\pi}{4} = 0.785 \text{ cm}^4 \qquad I_3 = 1.917 \text{ cm}^4$$

$$y_1 = \frac{Q}{12E}(34,400 + 4,200 + 24,200) = 0.196 \text{ cm}$$

(b) As a simple beam with a diameter of 2 cm and a center load,

$$y_{1C} = \frac{Q\ell^3}{48EI} = \frac{750(60)^3}{48(20)(0.785)} = 0.215 \text{ cm}$$

The larger diameter has reduced the deflection by $(0.215 - 0.196) = 0.019$ cm, or by approximately 9 percent.

4.10. In Example 4.9 transverse (radial) compliances of the bearings are $C_0 = 300(10)^{-6}$ and $C_3 = 120(10)^{-6}$ cm/N, respectively. Find the deflection of Q.

Solution: Incorporating the strain energy at the bearings from Eq. (1.6),

$$\frac{Qy_1}{2} = \int_0^\ell \frac{M^2}{2EI(x)}\,dx + \sum \frac{CQ^2}{2}$$

$$y_1 = 261(10)^{-6}Q + 300(10)^{-6}\frac{Q^2}{2} + 120(10)^{-6}\frac{Q^2}{2}$$

$$= 0.196 + 0.056 + 0.023 = 0.274 \text{ cm}$$

where the first term is the flexural deflection from Example 4.9.

4.11. In Example 4.9 determine the transverse deflection 4 (the outboard end).

Solution:

$$Fy_4 = \int_0^\ell \frac{mM}{EI}\,dx = \int_0^{30} \frac{(0.42\,Fx_0)(Q/2)\,x_0}{EI_0}\,dx_0$$

$$+ \int_{20}^{30} \frac{(0.42\,Fx_0)(Q/2)\,x_3}{EI_0}\,dx_3$$

$$+ \int_0^{20} \frac{(0.42\,Fx_0)(Q/2)\,x_3}{EI_3}\,dx_3$$

$$y_4 = \frac{0.42Q}{2E}\left[\frac{x_0^3}{3I_0}\bigg|_0^{30} + \int_{20}^{30}\frac{(60x_3 - x_3^2)}{I_0}\,dx_3 + \int_0^{20}\frac{(60x_3 - x_3^2)}{I_3}\,dx_3 \right]$$

$$= 7.875(10)^{-6}(11,460 + 11,040 + 4870) = 0.216 \text{ cm} \quad (\text{up})$$

4.12. In Example 4.9 determine the slope of the deflection curve at 4 (the outboard end).

Solution: Application of M' at 4 produces a triangular auxiliary moment diagram similar to the F diagram from 0 to 3 in Example 4.11; therefore, the definite integrals are identical with $(M'/60)$ replacing $0.42\,F$:

$$\theta_4 = \frac{Q}{2(60)\,E}(11,460 + 11,040 + 4,870) = 0.0086 \text{ rad}$$

Note in these examples no integration is required from 3 to 4 as the M diagram is 0 in this interval.

4.13. Complete the calculation for deflection of a point on the variable section beam of Fig. 4.17 using normalized numerical integration.

Solution: To apply Eq. (4.18) we tabulate the m/m_R, M/M_R, and h/h_R ratios with maximum values of unity at C, A, and B, respectively. Intervals are 0.10 of the length, with a discontinuity at D.

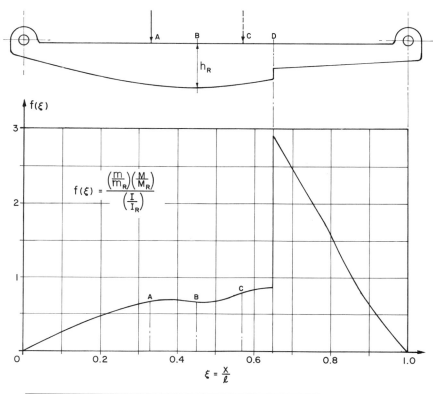

$\xi = x/l$	m/m_R	M/M_R	h/h_R	$(h/h_R)^3$	$f(\xi)$
0	0	0	—	—	0
0.10	0.18	0.30	0.57	0.19	0.28
0.20	0.35	0.60	0.77	0.45	0.47
0.30	0.52	0.90	0.91	0.76	0.62
(A) 0.33		1.00			
0.40	0.70	0.90	0.99	0.96	0.64
(B) 0.45			1.00	1.00	
0.50	0.88	0.75	0.99	0.96	0.68
(C) 0.57	1.00				
0.60	0.93	0.60	0.89	0.70	0.80
(D) 0.65	0.80	0.52	0.80	0.51	0.85
(D) 0.65	0.80	0.52	0.53	0.15	2.90
0.70	0.70	0.45	0.51	0.13	2.42
0.80	0.47	0.30	0.45	0.09	1.57
0.90	0.23	0.15	0.39	0.06	0.60
1.00	0	0	—	—	0

Numerical values of the areas under the two sections can be obtained using the trapezoidal rule, Simpson's rule, or by two equivalent triangles. The latter method is the most rapid, and is relatively accurate, with values of

$$A = \frac{(1.1)(0.65)}{2} + \frac{(2.75)(0.35)}{2} = 0.36 + 0.48 = 0.84$$

Flexural deformation of the larger section accounts for $0.36/0.84$ or 43 percent of the deflection of C. The smaller section accounts for the remaining 57 percent. Also the 0.84 factor represents the height of a rectangle that, with the same unit base, has the same area as that which lies under the curve.

We evaluate the external factors involving m_R and M_R from

$$C_m = (0.57)(0.43) = 0.245 \qquad C_M = (0.33)(0.67) = 0.221$$

$$C_m C_M = 0.0542$$

$$y_{BA} = \frac{0.0542 Q_A \ell^3}{EI_B}[0.84] = 0.046\frac{Q_A \ell^3}{EI_B}$$

Comparing this deflection with that of a constant beam having a section of the maximum I_B,

$$y_{BA} = \frac{1}{48}\frac{Q_A \ell^3}{EI_B} = 0.021\frac{Q_A \ell^2}{EI_B}$$

The compliance α_{21} is increased by a factor of $0.046/0.021 = 2.2$ by the material removed.

4.14. An aluminum alloy plate is tapered in width as shown, and loaded transversely at its maximum width. Find:

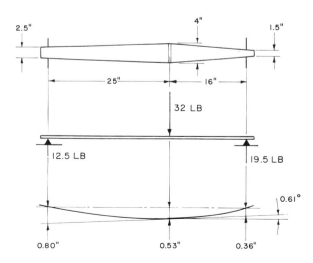

(a) The slope at the load
(b) The deflection at the load
(c) The maximum bending stress

Solution: Using the cantilever component method of Sec. 3.12 in conjunction with Table 4.9a,

$$I_0 = \tfrac{1}{12}(4)(0.30)^3 = 0.009 \text{ in.}^4$$

For the left section

$$y_{QC} = \frac{(12.5)(25)^3}{3(10)(10)^6(0.009)} = 0.72'' \qquad r = \frac{2.5}{4} = 0.625$$

$$y_1 = 0.72 \left\{ \frac{3}{(0.375)^3} \left[\frac{(0.375)(-0.875)}{2} - (0.625)^2 \log 0.625 \right] \right\}$$

$$= 0.72(1.11) = 0.80''$$

For the right section

$$y_{QC} = \frac{(19.5)(16)^3}{3(10)(10)^6(0.009)} = 0.30'' \qquad r = \frac{1.5}{4} = 0.375$$

$$y_1 = 0.30 \left\{ \frac{3}{(0.625)^3} \left[\frac{(0.625)(-0.125)}{2} - (0.375)^2 \log 0.375 \right] \right\}$$

$$= 0.30(1.21) = 0.36''$$

From Eq. (3.25)

(a) $\theta_1 = \dfrac{0.36 - 0.80}{41} = -0.0107 \text{ rad} = -0.61° \quad (\text{ccw})$

From Eq. (3.26)

(b) $y_1 = 0.80 + 25(-0.0107) = 0.53''$

(c) $\sigma_1 = \dfrac{M}{I_0/c} = \dfrac{(12.5)(25)}{(0.009/0.15)} = 5200 \text{ psi}$

5

Distributed Loads

PROGRESSING FROM BEAMS LOADED at a single point, we now consider distributed loads. These are typically caused by weights of the structural parts, by fluids, or by inertia loading due to acceleration. As in Chap. 4 loads generate deflections throughout the elastic system, but we will generally solve for only one coordinate deflection of interest in a given determination. It is possible to derive functions for the complete curve; however, this is usually considerably more cumbersome than the discrete calculation in the energy system. Again, only statically determinate cases are considered.

With assumed linear elasticity and several loads, we can *superimpose* the stresses, moments, reactions, deflections, and slopes due to each algebraically. Distributed loads in this sense involve the addition of an infinite number of component loads, but superposition is possible as an exact summation by means of integration.

Deflections are quantified by determining the bending moment functions induced by the *actual distributed* loads and by *auxiliary concentrated* loads, with solutions derived from the evaluation of the total complementary internal strain energies in flexure.

5.1 *Uniform Loading*

A constant distributed loading q, force per unit length, acts on a basic cantilever (Fig. 5.1a) applying a total load of $q\ell$. This resultant force at the midpoint of the beam produces the equilibrium condition in Fig. 5.1a, with

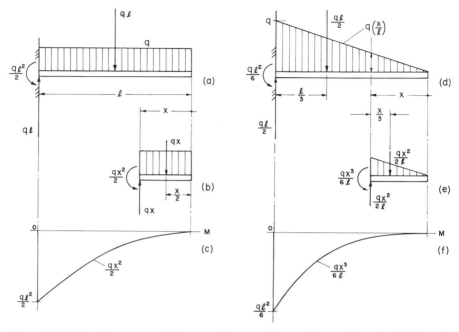

Figure 5.1 Bending moment diagrams for a cantilever with a uniformly distributed load (*a*) and with a constantly increasing distributed load (*b*).

a bending moment at the base of $q\ell(\ell/2) = (q\ell^2/2)$ and an equal and opposite supporting shear. It is incorrect, however, to use this resultant load for distributed bending moment purposes. Rather, we see in Fig. 5.1*c* that the internal couple is $qx^2/2$, or a *squared* parabolic function.

5.2 *Uniformly Increasing Loading*

Triangularly distributed loading (Fig. 5.1*d*) can be defined as increasing at a certain rate (load per unit length per unit length) or in terms of the maximum ordinate q (load per unit length) as in Fig. 5.1*a*. We will choose the latter option, therefore requiring only the single parameter q. Resultant load then equals the shear at the base and the area under the loading diagram and acts at the centroid of the load distribution triangle. The bending moment diagram (Fig. 5.1*f*) becomes

$$M(x) = q\left(\frac{x}{\ell}\right)\left(\frac{x}{2}\right)\left(\frac{x}{3}\right) = \frac{qx^3}{6\ell} \qquad (5.1)$$

and is a *cubic* parabolic variation. Bending moment at the base is $(q\ell^2)/6$ and is directly proportional to the *square* of the span.

5.3 *Elementary Deflection Solutions*

For the simple cantilever (Fig. 5.2), we determine the end deflection using an auxiliary load F_1 at the end. The complementary energy relation is, from Eq. (4.3)

$$F_1 y_1 = \int_0^\ell \frac{mM}{EI}\, dx = \frac{1}{EI} \int_0^\ell (F_1 x_1)\left(\frac{q x_1^2}{2}\right) dx_1$$

$$y_1 = \frac{q}{2EI}\left(\frac{x_1^4}{4}\right)_0^\ell = \frac{q\ell^4}{8EI} \tag{5.2}$$

With a simple definite integral, we have evaluated the end deflection. Although external auxiliary reactions exist at the wall, there is neither linear nor angular displacement associated with these loadings as q is applied. By definition, the fixed end enforces zero displacement. Therefore, of the three loadings in Fig. 5.2 related to the equilibrium of the F_1 loading, only the $F_1 y_1$ product occurs.

Figure 5.2 Determination of end deflection and end slope for a cantilever requires application of auxiliary force F and auxiliary couple M'.

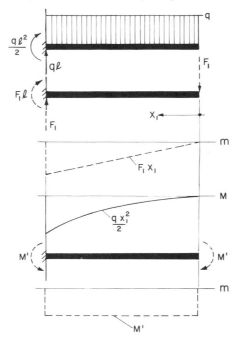

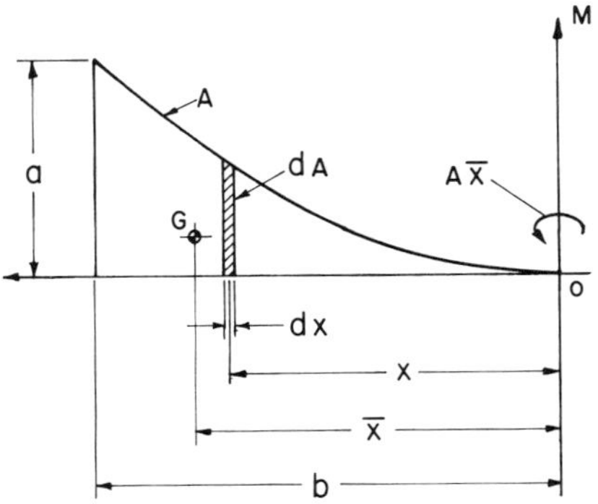

Figure 5.3 Area-moment integration factors involve the product of an area and a centroidal distance from a specified axis.

To obtain the end slope we apply an auxiliary couple M' (Fig. 5.2), and

$$M'\theta_1 = \int_0^\ell \frac{M'M}{EI} \, dx = \frac{M'}{EI} \int_0^\ell \left(\frac{qx_1^2}{2} \right) dx_1$$

$$\theta_1 = \frac{q}{2EI} \left(\frac{x_1^3}{3} \right)_0^\ell = \frac{q\ell^3}{6EI} \tag{5.3}$$

5.4 *Integration Factors*

In the previous solution [Eq. (5.2)] we have a prior relation

$$Fy = \frac{F}{EI} \int_0^\ell xM \, dx \tag{5.4}$$

where the integral represents the first moment of the actual bending moment diagram about an axis at 1 in the plane of the diagram and perpendicular to the beam and the base, or horizontal coordinate of the diagram. In Fig. 5.3 we see that the integral represents the summation of the products of the differential areas under the M diagram and their coordinate distances from the vertex of the m diagram. This term is also referred to as an *area-moment* characteristic, relating an area to an axis.

This term is equal to the product of the total area under the moment diagram and the distance from the centroid of the area to the axis, and these factors are tabulated in Table 5.1 for the fundamental types of loading. Area moments are given with respect to the centroidal distances from both ends of the moment diagrams, and these are different except for the constant moment (Table 5.1*a*). The smaller factors are at the left and the larger at the right. Centroids and areas are also indicated.

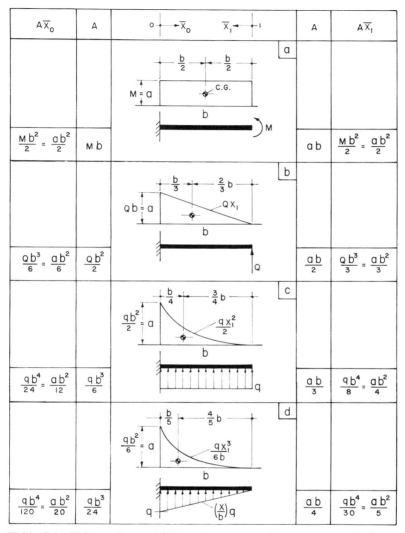

The table contents:

$A\overline{x}_0$	A	(diagram)	A	$A\overline{x}_1$
$\dfrac{Mb^2}{2} = \dfrac{ab^2}{2}$	Mb	a	ab	$\dfrac{Mb^2}{2} = \dfrac{ab^2}{2}$
$\dfrac{Qb^3}{6} = \dfrac{ab^2}{6}$	$\dfrac{Qb^2}{2}$	b	$\dfrac{ab}{2}$	$\dfrac{Qb^3}{3} = \dfrac{ab^2}{3}$
$\dfrac{qb^4}{24} = \dfrac{ab^2}{12}$	$\dfrac{qb^3}{6}$	c	$\dfrac{ab}{3}$	$\dfrac{qb^4}{8} = \dfrac{ab^2}{4}$
$\dfrac{qb^4}{120} = \dfrac{ab^2}{20}$	$\dfrac{qb^3}{24}$	d	$\dfrac{ab}{4}$	$\dfrac{qb^4}{30} = \dfrac{ab^2}{5}$

Table 5.1 Values of centroidal location, area, and area moment for beam span subjected to several basic types of loading. Left column relates to minimum area moment about the base end, and the right column about the apex, or maximum.

In the example of Figure 5.2, applying Table 5.1c,

$$y_1 = \frac{1}{EI} \int_0^\ell x_1 M \, dx_1 = \frac{1}{EI}(A\bar{x}_1) = \frac{1}{EI}\left(\frac{qb^4}{8}\right) = \frac{1}{EI}\left(\frac{q\ell^4}{8}\right) \quad (5.5)$$

since $b = \ell$.

For the end slope in Fig. 5.2, the integral represents simply the *area* under the actual bending moment diagram, and with the parabola (Table 5.1c),

$$\theta_1 = \frac{1}{EI} \int_0^\ell M \, dx = \frac{1}{EI}(A) = \frac{1}{EI}\left(\frac{ab}{3}\right) = \frac{1}{EI}\left(\frac{q\ell^3}{6}\right) \quad (5.6)$$

We will make extensive use of the information provided in Table 5.1 in practically all evaluations that follow. It greatly expedites solutions involving complementary energy and eliminates much repetitive time-consuming integration.

5.5 The Gravity Moment Analog

It is not completely correct to denote a point in a defined area as a *center of gravity*, since an area has no mass. The situation is similar to locating the center of gravity of a beam cross section to determine a principal neutral

Figure 5.4 Area-moment characteristics of areas are similar to weight-moment terms for horizontal plates in a gravity field.

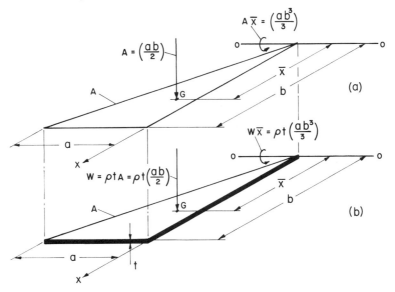

axis and to denoting the beam section as having a *moment of inertia*. For the section, $I = \int y^2 \, dA$ rather than $I = \int y^2 \, dm$, but this is obviously a similar expression.

In Fig. 5.4 we see the parallel of the massless area and the plate with uniformly distributed mass, illustrated by a triangle with respect to an axis at the vertex (Table 5.1*b*). The triangular area has identical geometric properties to the triangular plate (Fig. 5.4*b*), but the latter has a physical moment about the axis, required to support it in equilibrium at the apex when the plate is horizontal as shown in Fig. 5.4.

The similarity also applies if the rotational axis is taken coincident with the base *a*. In the solutions this property can be involved, and the direction of the $\bar{x}$ distance will be determined from the auxiliary bending moment from which the area–moment analogy is derived.

We are also concerned with only the centroidal distance, indicated in Fig. 5.4, parallel to the beam axis. Although there is also a centroidal location transversely, or perpendicular to *b*, this coordinate has no significance in the evaluation of strain energy.

5.6 *Continuous Deflection Functions*

Elastic energy deflection solutions as outlined here are most effectively and advantageously applied as a single solution for a particular system coordinate and for specific numerical parameters. Since a deflection distribution consists of multiple coordinates, this scheme can be used repetitively to compute a curve to any required accuracy.

For relatively simple beams, it is possible to develop expressions for complete deflection functions; however, the procedure tends to become unwieldy and is not generally recommended. To illustrate, the total elastic curve in Fig. 5.5 requires that F be applied at a general position on the beam (Fig. 5.5*b*). Using a single horizontal coordinate x ranging between 0 and x_0, the auxiliary moment equation is most simply expressed as a function of z, 0 at F:

$$m = F(x_0 - x) = Fz \qquad (5.7)$$

In terms of z, proceeding right to left, the actual-moment diagram consists of a truncated parabola and is so analyzed. This is done by resolving the geometry of the diagram into simple components (Fig. 5.5*d*). These components have physical counterparts (Fig. 5.5*c*), corresponding to the internal loading at F, or a shear and a couple induced by the section of the beam outboard of F. These produce the triangle and the rectangle in the M diagram, respectively.

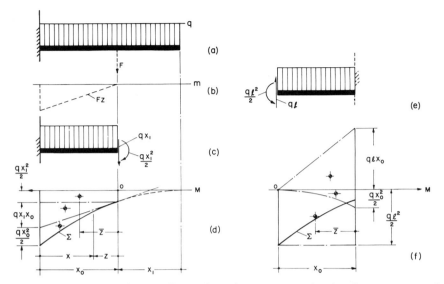

Figure 5.5 Solution for cantilever deflection at any point involves a truncated parabolic moment diagram analyzed by proceeding from F to the left (c) and (d). Alternately, the actual moment diagram can be considered from left to right.

Additionally, the continuing constant q loading results in a third moment component, parabolic and skewed. The M diagram is continuous at F, with a common tangent coincident with the hypotenuse of the triangle. Applying integration factors from Table 5.1,

$$Fy(x) = \int_0^{x_0} \frac{(Fz)M}{EI} \, dz$$

$$y(x) = \frac{1}{EI} \sum A\bar{z}$$

$$y(x) = \frac{1}{EI}\left[\left(\frac{qx_1^2}{2}\right)\left(\frac{x_0^2}{2}\right) + (qx_1)\left(\frac{x_0^3}{3}\right) + q\left(\frac{x_0^4}{8}\right)\right] \qquad (5.8)$$

where

$$x_0 = x + z \quad \text{and} \quad x_0 + x_1 = \ell$$

Equation (5.8) is valid but requires reduction to a single variable x and some algebraic effort.

By retaining $m = Fz$, requiring centroidal distances to the left, but by reversing the direction for developing the M diagram, we again have three moment components (Fig. 5.5e and f). Now, however, centroidal distances are from F, but the apexes of the M geometries lie at the wall, or at the origin of the x coordinate. Applying the Table 5.1 factors,

$$y(x) = \frac{1}{EI} \sum A\bar{z} = \frac{1}{EI}\left[\left(\frac{q\ell^2}{2}\right)\left(\frac{x^2}{2}\right) - \left(\frac{q\ell x^3}{6}\right) + \left(\frac{qx^4}{24}\right)\right] \quad (5.9)$$

These terms are at the left in the table, and the importance of selecting the appropriate directions when developing the diagrams has been demonstrated.

Equation (5.9) is reduced somewhat and tabulated (Table 5.2*a*). The elastic curve contains three parabolic components shown in Fig. 5.6. As just developed, they arise directly from the physically applied moments due to the couple at the wall, the shear reaction at the wall, and the distributed load. All three effects are superimposed as we progress across the beam from the left, generating the polynomial function.

Slope distribution is readily obtained from the deflection expression in the polynomial form:

$$\frac{dy}{d\left(\frac{x}{\ell}\right)} = \frac{q\ell^4}{24EI}\left[12\left(\frac{x}{\ell}\right) - 12\left(\frac{x}{\ell}\right)^2 + 4\left(\frac{x}{\ell}\right)^3\right] \quad (5.10)$$

Slope in terms of the absolute abscissa x is obtained by dividing both sides of the equation by ℓ (Table 5.2*a*). Having taken F downward and M' clockwise, these are the corresponding positive directions of deflection.

We note in this solution, and in Fig. 5.6, that identical results are obtained for deflection whether we develop the M diagram from the left or right. Obviously the $A\bar{z}$ terms are different in the two cases; however, complementary energy relates to the change in the total system internal energy, or to the $\sum A\bar{z}$ factor. Thus, although individual terms differ, by superposition the algebraic sum of all terms in a given length of beam *must be exactly equivalent*.

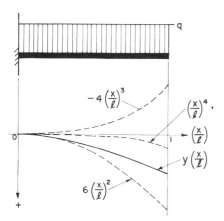

Figure 5.6 Deflection polynomial for cantilever deflection $y(x/\ell)$ contains three parabolic components.

5.7 Beams with Uniform Loading

Of all the types of distributed loadings, constant distribution is the most basic and the most frequent. Similarly, a beam of constant section is the most usual. Because of the importance of these common cases, a considerable number of relations are provided.

Table 5.2 indicates deflection and slope for various cantilever loading combinations. Elastic behavior can be visualized and numerical values estimated in Figs. 5.7 and 5.8.

Several loading combinations for the beam on two supports are given in Tables 5.3–5.5, with related functional characteristics shown in Figs. 5.9–5.12.

5.8 Loading Combinations by Superposition

The tabular and graphed results can be extended to a broad range of loading using superposition. For instance, in Fig. 5.13a we have a loaded interval or pulse function on a beam, a case not specifically tabulated; however, we can achieve this loading by using Table 5.3b.

As shown in Fig. 5.13b and c, we obtain the pulse by taking the total positive loading length a and then by adding a negative loading of length a'. The latter, having an equal load rate q, completely cancels the left section of the former loading as required. The deflection evaluation requires

Table 5.2 Polynomial functions for continuous deflections and slopes for the cantilever with various intervals of distributed loading.

	a		b			c
$y(x)$	$C[6(\frac{x}{\ell})^2 - 4(\frac{x}{\ell})^3 + (\frac{x}{\ell})^4]$	$y(x_0)$	$C[A_2(\frac{x}{\ell})^2 + A_3(\frac{x}{\ell})^3]$			$C[A_2(\frac{x}{\ell})^2 + A_3(\frac{x}{\ell})^3 + (\frac{x}{\ell})^4]$
y_m	$3C$	$y(x_1)$	$C[B_0 + B_1(\frac{x}{\ell}) + 6(\frac{x}{\ell})^2 - 4(\frac{x}{\ell})^3 + (\frac{x}{\ell})^4]$			$C[B_0 + B_1(\frac{x}{\ell})]$
$\theta(x)$	$D[12(\frac{x}{\ell}) - 12(\frac{x}{\ell})^2 + 4(\frac{x}{\ell})^3]$	$\theta(x_0)$	$D[2A_2(\frac{x}{\ell}) + 3A_3(\frac{x}{\ell})^2]$			$D[2A_2(\frac{x}{\ell}) + 3A_3(\frac{x}{\ell})^2 + 4(\frac{x}{\ell})^3]$
θ_m	$4D$	$\theta(x_1)$	$D[B_1 + 12(\frac{x}{\ell}) - 12(\frac{x}{\ell})^2 + 4(\frac{x}{\ell})^3]$			DB_1
		A_2	$6[1-(\frac{a}{\ell})^2]$	B_0	$(\frac{a}{\ell})^4$	A_2 $6(\frac{a}{\ell})^2$ B_0 $-(\frac{a}{\ell})^4$
		A_3	$-4[1-(\frac{a}{\ell})]$	B_1	$-4(\frac{a}{\ell})^3$	A_3 $-4(\frac{a}{\ell})$ B_1 $4(\frac{a}{\ell})^3$
	$C = \frac{q\ell^4}{24EI}$	$D = \frac{q\ell^3}{24EI}$	$0 \leq x_0 \leq a$			$a \leq x_1 \leq \ell$

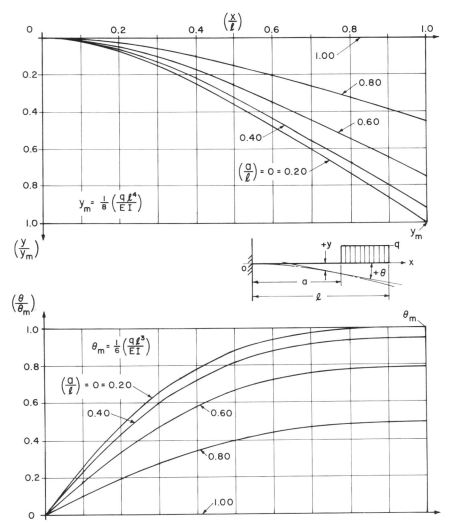

Figure 5.7 Distribution of deflection and slope in a cantilever as a function of the length of a partial uniform load contiguous with the free end.

determining the coefficients in the two cases (Fig. 5.13*b* and *c*) and then subtracting respective coefficients in the polynomials. For the plotted elastic curves, we similarly take Fig. 5.9 and subtract the curve for the shorter load from that of the longer. This is only one example, but shows the manner in which the organized results for uniform loading can be applied to many beam combinations.

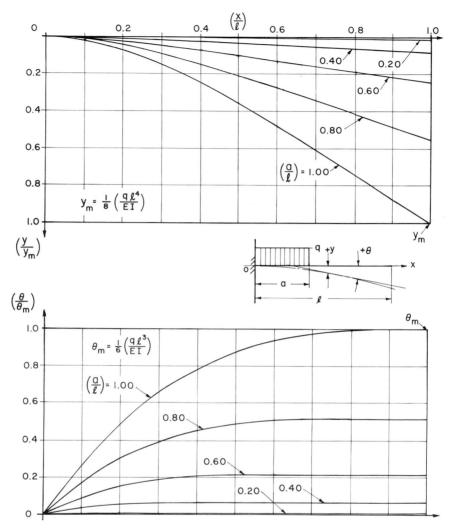

Figure 5.8 Similar to Fig. 5.7, but with the partial loading contiguous with the fixed end.

5.9 Axial Weight Loading

Although this chapter is mainly concerned with distributed transverse loads and their effects in developing flexural stresses and deflections, distributed axial loads are now briefly discussed. The most simple example is the suspended vertical bar of constant section acted upon by gravity (Fig. 5.14a).

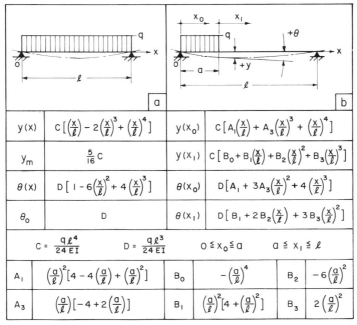

	(a)		(b)
$y(x)$	$C\left[\left(\frac{x}{\ell}\right) - 2\left(\frac{x}{\ell}\right)^3 + \left(\frac{x}{\ell}\right)^4\right]$	$y(x_0)$	$C\left[A_1\left(\frac{x}{\ell}\right) + A_3\left(\frac{x}{\ell}\right)^3 + \left(\frac{x}{\ell}\right)^4\right]$
y_m	$\frac{5}{16}C$	$y(x_1)$	$C\left[B_0 + B_1\left(\frac{x}{\ell}\right) + B_2\left(\frac{x}{\ell}\right)^2 + B_3\left(\frac{x}{\ell}\right)^3\right]$
$\theta(x)$	$D\left[1 - 6\left(\frac{x}{\ell}\right)^2 + 4\left(\frac{x}{\ell}\right)^3\right]$	$\theta(x_0)$	$D\left[A_1 + 3A_3\left(\frac{x}{\ell}\right)^2 + 4\left(\frac{x}{\ell}\right)^3\right]$
θ_0	D	$\theta(x_1)$	$D\left[B_1 + 2B_2\left(\frac{x}{\ell}\right) + 3B_3\left(\frac{x}{\ell}\right)^2\right]$

$$C = \frac{q\ell^4}{24\,EI} \qquad D = \frac{q\ell^3}{24\,EI} \qquad 0 \leq x_0 \leq a \qquad a \leq x_1 \leq \ell$$

A_1	$\left(\frac{a}{\ell}\right)^2\left[4 - 4\left(\frac{a}{\ell}\right) + \left(\frac{a}{\ell}\right)^2\right]$	B_0	$-\left(\frac{a}{\ell}\right)^4$	B_2	$-6\left(\frac{a}{\ell}\right)^2$
A_3	$\left(\frac{a}{\ell}\right)\left[-4 + 2\left(\frac{a}{\ell}\right)\right]$	B_1	$\left(\frac{a}{\ell}\right)^2\left[4 + \left(\frac{a}{\ell}\right)^2\right]$	B_3	$2\left(\frac{a}{\ell}\right)^2$

Table 5.3 Similar to Table 5.2, but for the constant beam on simple supports.

Table 5.4 Similar to Table 5.3, but with the beam extending beyond one support.

	(a)			(b)	
$y(x_0)$	$c\,a^4\left[A_1\left(\frac{x_0}{a}\right) + A_3\left(\frac{x_0}{a}\right)^3 + \left(\frac{x_0}{a}\right)^4\right]$			$c\,a^2 b^2\left[-2\left(\frac{x_0}{a}\right) + 2\left(\frac{x_0}{a}\right)^3\right]$	
$y(x_1)$	$c\,b^4\left[B_1\left(\frac{x_1}{b}\right) + 6\left(\frac{x_1}{b}\right)^2 - 4\left(\frac{x_1}{b}\right)^3 + \left(\frac{x_1}{b}\right)^4\right]$			$c\,b^4\left[B_2\left(\frac{x_1}{b}\right) + 6\left(\frac{x_1}{b}\right)^2 - 4\left(\frac{x_1}{b}\right)^3 + \left(\frac{x_1}{b}\right)^4\right]$	
$\theta(x_0)$	$c\,a^3\left[A_1 + 3A_3\left(\frac{x_0}{a}\right)^2 + 4\left(\frac{x_0}{a}\right)^3\right]$			$c\,a b^2\left[-2 + 6\left(\frac{x_0}{a}\right)^2\right]$	
$\theta(x_1)$	$c\,b^3\left[B_1 + 12\left(\frac{x_1}{b}\right) - 12\left(\frac{x_1}{b}\right)^2 + 4\left(\frac{x_1}{b}\right)^3\right]$			$c\,b^3\left[B_2 + 12\left(\frac{x_1}{b}\right) - 12\left(\frac{x_1}{b}\right)^2 + 4\left(\frac{x_1}{b}\right)^3\right]$	
A_1	$\left[1 - 2\left(\frac{b}{a}\right)^2\right]$	B_1	$\left(\frac{a}{b}\right)\left[4 - \left(\frac{a}{b}\right)^2\right]$	B_2	$4\left(\frac{a}{b}\right)$
A_2	$2\left[-1 + \left(\frac{b}{a}\right)^2\right]$		$C = \frac{q}{24\,EI}$		

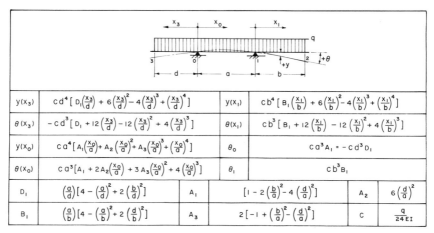

$y(x_3)$	$c\,d^4\left[D_1\left(\frac{x_3}{d}\right)+6\left(\frac{x_3}{d}\right)^2-4\left(\frac{x_3}{d}\right)^3+\left(\frac{x_3}{d}\right)^4\right]$	$y(x_1)$	$c\,b^4\left[B_1\left(\frac{x_1}{b}\right)+6\left(\frac{x_1}{b}\right)^2-4\left(\frac{x_1}{b}\right)^3+\left(\frac{x_1}{b}\right)^4\right]$		
$\theta(x_3)$	$-c\,d^3\left[D_1+12\left(\frac{x_3}{d}\right)-12\left(\frac{x_3}{d}\right)^2+4\left(\frac{x_3}{d}\right)^3\right]$	$\theta(x_1)$	$c\,b^3\left[B_1+12\left(\frac{x_1}{b}\right)-12\left(\frac{x_1}{b}\right)^2+4\left(\frac{x_1}{b}\right)^3\right]$		
$y(x_0)$	$c\,a^4\left[A_1\left(\frac{x_0}{a}\right)+A_2\left(\frac{x_0}{a}\right)^2+A_3\left(\frac{x_0}{a}\right)^3+\left(\frac{x_0}{a}\right)^4\right]$	θ_0	$c\,a^3A_1=-c\,d^3D_1$		
$\theta(x_0)$	$c\,a^3\left[A_1+2A_2\left(\frac{x_0}{a}\right)+3A_3\left(\frac{x_0}{a}\right)^2+4\left(\frac{x_0}{a}\right)^3\right]$	θ_1	$c\,b^3B_1$		
D_1	$\left(\frac{a}{d}\right)\left[4-\left(\frac{a}{d}\right)^2+2\left(\frac{b}{d}\right)^2\right]$	A_1	$\left[1-2\left(\frac{b}{a}\right)^2-4\left(\frac{d}{a}\right)^2\right]$	A_2	$6\left(\frac{d}{a}\right)^2$
B_1	$\left(\frac{a}{b}\right)\left[4-\left(\frac{a}{b}\right)^2+2\left(\frac{d}{b}\right)^2\right]$	A_3	$2\left[-1+\left(\frac{b}{a}\right)^2-\left(\frac{d}{a}\right)^2\right]$	c	$\dfrac{q}{24EI}$

Table 5.5 The simple beam extends beyond both supports with continuous uniform loading. Reversed reactions may be required for stability.

Axial loading is uniform, with q causing a maximum stress at the point of suspension, decreasing linearly to the free end. The actual tensile-load distribution is

$$Q(y)=\rho A\ell\left(1-\frac{y}{\ell}\right)=W\left(1-\frac{y}{\ell}\right) \qquad (5.11)$$

where ρ = density of the material
A = cross-sectional area

Stress is obtained by dividing by the area, also decreasing linearly:

$$\sigma(y)=\rho\ell\left(1-\frac{y}{\ell}\right) \qquad (5.12)$$

For axial deflection δ, we apply the auxiliary force F at any distance y, and have, using Eq. (1.3)

$$dU_s=FQ\frac{dy}{AE}$$

$$F\delta=\frac{1}{AE}\int_0^y F(\rho A\ell)\left(1-\frac{y}{\ell}\right)dy$$

$$\delta=\frac{\rho\ell^2}{2E}\left[2\left(\frac{y}{\ell}\right)-\left(\frac{y}{\ell}\right)^2\right] \qquad (5.13)$$

where $\rho\ell^2/2E$ is the end deflection, and the term in brackets indicates the distribution within the link (Fig. 5.14a).

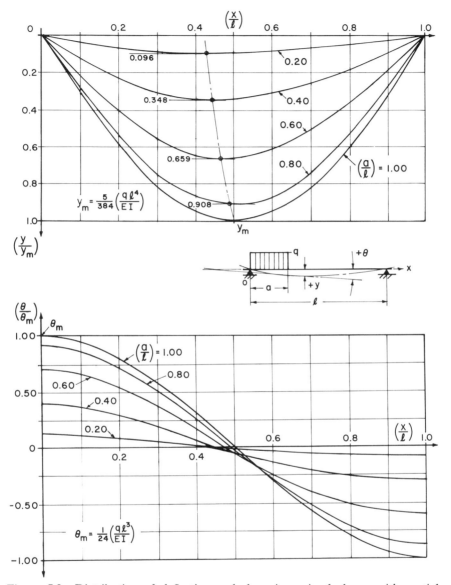

Figure 5.9 Distribution of deflection and slope in a simple beam with partial loading contiguous with one simple support.

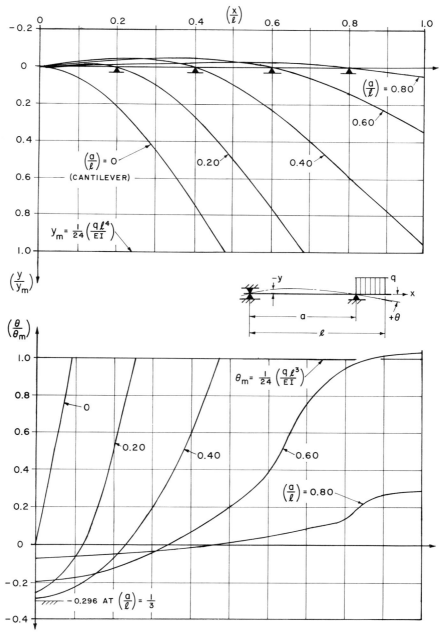

Figure 5.10 Similar to Fig. 5.9, but with uniform load beyond one support.

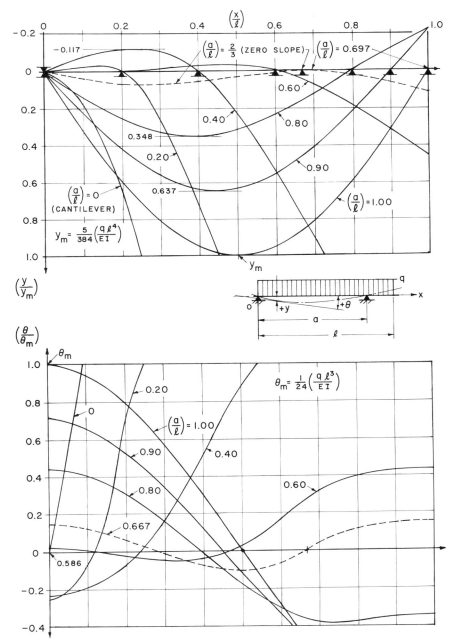

Figure 5.11 Deflection behavior with continuous distributed loading between supports and extending beyond one support.

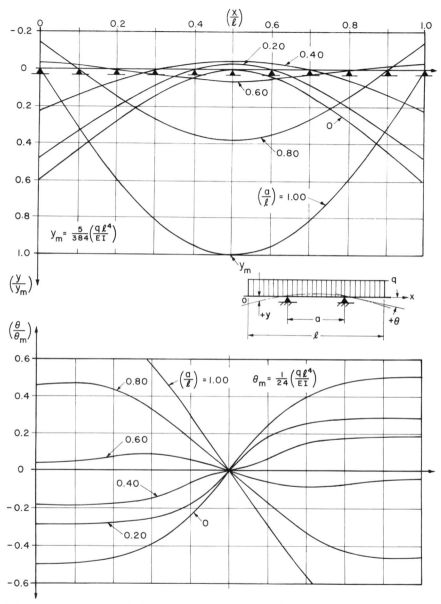

Figure 5.12 Symmetrical beam with uniform loading extending equal distances beyond each support.

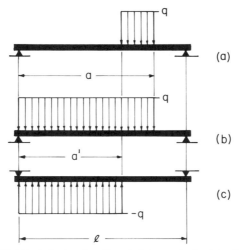

Figure 5.13 The distributed step loading (a) is equivalent to the sum of (b) and (c) by superposition.

5.10 *Axial Centrifugal Loading*

Given the rotation of a constant bar about one end (Fig. 5.14b), we have a variable distribution of internal inertia loading:

$$Q(y) = q\,dy = \frac{\rho A \Omega^2}{g_c} \int_y^\ell y\,dy = \frac{M\Omega^2 \ell}{2}\left[1 - \left(\frac{y}{\ell}\right)^2\right] \qquad (5.14)$$

where M is the total mass of the link.

Equation (5.14) also describes the stress distribution if divided by the constant area A.

Axial deflection is obtained by using complementary energy and the application of F, as in Sec. 5.9:

$$\delta = \frac{\rho \Omega^2 \ell^3}{6Eg_c}\left[3\left(\frac{y}{\ell}\right) - \left(\frac{y}{\ell}\right)^3\right] \qquad (5.15)$$

These dimensionless distributions are shown in Fig. 5.14b. For a variable section link having either weight or centrifugal loading, the analysis requires that the functional variation be introduced under the integral. As an alternative we can use discrete ordinates and numerical integration.

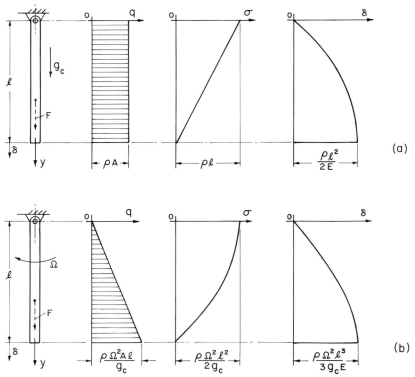

Figure 5.14 Distribution of load, stress, and deflection in a simple bar with vertical weight loading (a), and with centrifugal effects due to rotation about one end (b).

5.11 The Beam in Rotation

In Fig. 5.15 a simple beam rotates at Ω about the center 0. Radial accelerations are all proportional to the rotational radius, and the acceleration field is determined by constructing a line parallel to 1–3. Components of acceleration transverse to the beam are shown dashed. As seen, the *transverse acceleration components are all equal*, corresponding to the perpendicular distance between the lines A^T. If the beam is of constant section, the resulting beam loading is also constantly distributed, regardless of the angular attitude of the beam with respect to a radius. The significant acceleration is that of the point on the beam for which the radius of rotation is perpendicular to the beam 2. The loading is

$$q = \frac{\rho A \Omega^2}{g_c} R_2 \tag{5.16}$$

where A is the cross section of the beam.

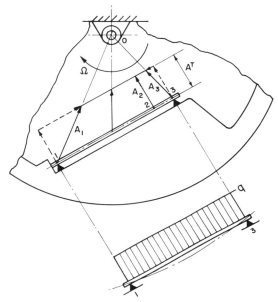

Figure 5.15 The simple beam subjected to rotation in an angular position is subjected only to a constant acceleration field transversely, and therefore to constant transverse inertia loading.

Stresses and deflections are readily calculated for the constant beam as indicated previously. If the mass and stiffness vary, these effects must be considered in the loading and deflection.

5.12   *Beam Loading in Linkages*

A straight beam in a planar mechanism is subjected to accelerations when in motion. Ends of the links are usually pinned and therefore simply supported. D'Alembert loads are proportional to the mass and the acceleration but opposite in direction.

The slider crank, or cylinder mechanism (Fig. 5.16), normally has a crank rotating with constant velocity. Radial inertia loads induced in the crank were discussed in Sec. 5.10, but the connecting rod (1–2) is in bending. Both points 1 and 2 have acceleration. A_1 is directed along the reciprocating axis (1–0), and A_2 is directed towards the center 0.

To determine the transverse loading producing bending we calculate the transverse components at each end (A_1^T and A_2^T). These are related to the translational acceleration normal to the link that causes A_1^T, and the angular acceleration accounting for the difference between A_2^T and A_1^T. The accelera-

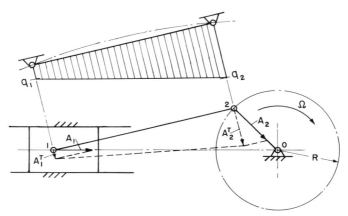

Figure 5.16 The connecting rod of a slider-crank mechanism has trapezoidally distributed transverse inertia loading related to the components of acceleration of the two end bearings.

tion field is *trapezoidal,* and the *transverse accelerations will always vary linearly from one end of a link to the other.* Although illustrated by the slider crank, this conclusion is applicable to any planar link, with results deriving from only the known magnitude and direction of the acceleration vectors for two pin constraints.

For beam analysis the trapezoidal loading is best treated as a combination of uniform and constantly increasing loading (Fig. 5.1 and Table 5.1).

Examples

5.1. A constant cantilever carries reversed uniform loading over half of its length. Determine the expression for the end deflection.

Solution:

$$F_2 y_2 = \frac{1}{EI} \int_0^\ell mM \, dx = \frac{1}{EI} \int_0^\ell (F_2 x_2) M \, dx_2 = \frac{F_2}{EI} \sum_2^0 A \bar{x}_2$$

$$y_2 = \frac{1}{EI}\left[\frac{qb^4}{8} + \left(\frac{qb^2}{2}\right) b \left(\frac{3}{2}b\right) + (qb^2)\left(\frac{b}{2}\right)\left(b + \frac{2}{3}b\right) \right.$$

$$\left. - \left(\frac{qb^2}{2}\right)\left(\frac{b}{3}\right)\left(b + \frac{3}{4}b\right)\right]$$

$$= \frac{qb^4}{EI}\left[\frac{1}{8} + \frac{3}{4} + \frac{5}{6} - \frac{7}{24}\right]$$

$$= \frac{17}{12}\left(\frac{qb^4}{EI}\right)$$

(In this and all following examples the integration factors for the various geometries are taken from Table 5.1.)

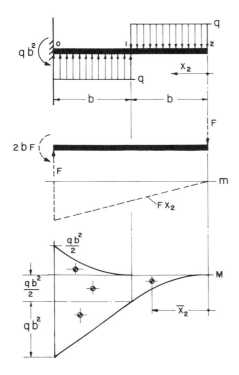

5.2. Repeat the solution of 5.1 using the equivalent distributed loading shown.

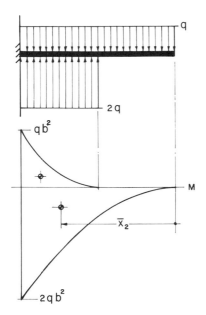

Solution:

$$y_2 = \frac{1}{EI} \sum A\bar{x}_2 = \frac{1}{EI}\left[\frac{q(2b)^4}{8} - (2q)\left(\frac{b^3}{6}\right)\left(b + \frac{3}{4}b\right)\right]$$

$$= \frac{qb^4}{EI}\left(2 - \frac{7}{12}\right) = \frac{qb^4}{EI}\left(\frac{17}{12}\right)$$

Note that in this problem the second method of loading superposition greatly simplifies the solution.

5.3. The cantilever beam is 40 in. long with a maximum triangularly distributed load of 60 lb/in. Find:
(a) The deflection at 2
(b) The deflection at 1
(c) The slope at 2
(d) The slope at 1

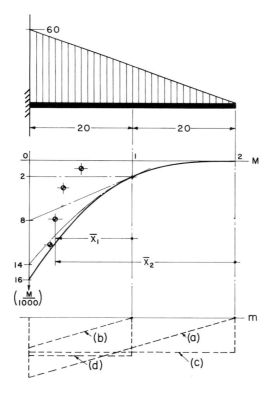

Solution: The development of auxiliary bending-moment diagrams depends upon a complete free-body equilibrium analysis of the beam under auxiliary loading by a force F or a couple M' (Fig. 5.2). Abbreviated geometric results for the four m diagrams are shown. Similarly, the deflection relations are now somewhat abbreviated since the source of these has been repeatedly demonstrated, deriving from energy equivalence. Results are obtained from numerical values of the M diagram shown. In the left interval the moment components are, progressing downward:

1. Effect of the couple at 1 caused by the outboard load.
2. Effect of the shear at 1 caused by the outboard load.
3. Distributed constant loading of 30 lb/in. on the inboard section.
4. Distributed triangular loading, inboard, from 0 at 1 to 30 lb/in. at 0:

(a) $y_2 = \dfrac{1}{EI} \sum_{0}^{40} A\bar{x}_2 = \dfrac{1}{EI}\left(\dfrac{ab^2}{5}\right) = \dfrac{1}{EI}\left(\dfrac{16,000(40)^2}{5}\right) = \dfrac{5.12(10)^6}{EI}$ in.

(b) $y_1 = \dfrac{1}{EI} \sum_{0}^{20} A\bar{x}_1$

$\qquad = \dfrac{1}{EI}\left[\dfrac{2,000(20)^2}{2} + \dfrac{6,000(20)^2}{3} + \dfrac{6,000(20)^2}{4} + \dfrac{2,000(20)^2}{5}\right]$

$\qquad = \dfrac{(10)^6}{EI}[0.40 + 0.80 + 0.60 + 0.16] = \dfrac{1.96(10)^6}{EI}$ in.

(c) $\theta_2 = \dfrac{1}{EI}\sum A = \dfrac{1}{EI}\left[\dfrac{16,000(40)}{4}\right] = \dfrac{0.16(10)^6}{EI}$ rad

(d) $\theta_1 = \dfrac{1}{EI}\sum A = \dfrac{1}{EI}\left[0.16(10)^6 - \dfrac{2,000(20)}{4}\right] = \dfrac{0.15(10)^6}{EI}$ rad

Note the slope at 1 is only slightly less than the slope at 2. This is related to the relatively small area under the toe of the parabola.

5.4. A stepped beam 140 cm long has a triangularly distributed load increasing to a maximum of 54 N/cm at 1. Beam inertia from 0–1 (I_0) is 4.6, and from 1–2 (I_2) is 6.7 cm⁴. Calculate, for $E = 20(10)^6$ N/cm²,

(a) The deflection at 1
(b) The slope at 0

Solution:

(a) Total, or resultant, load, using the area under the loading diagram is 2.16 kN, acting at the centroid of the triangular area at $x_0 = 53.33$ cm. For equilibrium of the beam on its supports,

$$R_0 = 1.337 \text{ kN} \qquad R_2 = 0.823 \text{ kN}$$

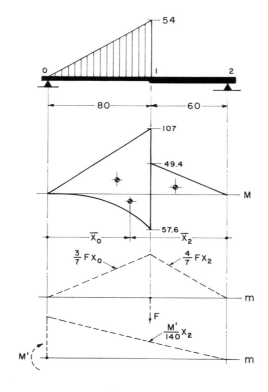

Ordinates shown in the M diagram are in units of kN · cm:

$$y_1 = \frac{3/7}{EI_0} \sum_0^{80} A\bar{x}_0 + \frac{4/7}{EI_2} \sum_0^{60} A\bar{x}_2$$

$$= \frac{0.093}{E} \left[\frac{(107)(80)^2}{3} - \frac{(57.6)(80)^2}{5} \right] + \frac{0.085}{E} \left[\frac{(49.4)(60)^2}{3} \right]$$

$$= 0.72 + 0.25 = 0.97 \text{ cm}$$

(b) $\theta_0 = \dfrac{1}{140\,EI_0} \sum_0^{80} A\bar{x}_2 + \dfrac{1}{140\,EI_2} \sum_0^{60} A\bar{x}_2$

$$= \frac{0.00155}{E} \left[\frac{(107)(80)}{2}\left(60 + \frac{80}{3} \right) - \frac{(57.6)(80)}{4}\left(60 + \frac{80}{5} \right) \right]$$

$$+ \frac{0.00107}{E} \left[\frac{(49.4)(60)}{2}(40) \right]$$

$$= 0.022 + 0.0032 = 0.025 \text{ rad}$$

6

Indeterminate Beams

As discussed in Sec. 2.1, an elastic structure with more constraints than required for basic stability cannot be solved using only the conditions of static equilibrium. Static conditions must be satisfied in all cases, but supplemental relations must be obtained from elasticity when redundancies are present. In fact, it is relative elastic behavior within a structure that determines how loads are distributed. The number of equations required from elasticity is equal to the number of defined redundancies, but this is often reduced if symmetry is present. At this point we limit the systems to planar loading of straight beams.

Conservation of system elastic energy has been shown to be a powerful tool for determining beam deflections. These same concepts are now extended to indeterminate solutions, and finally to the deflections of indeterminate flexural systems.

In addition to the conservation of complementary energy, redundancies can be solved by superposition of displacements, as in Sec. 2.4, with energy techniques available to evaluate the required deflections. As in previous systems, compliance effects related to axial stress can be combined with flexural if they are significant, with energies additive algebraically.

6.1 System Constraints

A typical beam is supported at specific points by constraints that prevent transverse deflection at these points, and these are termed *simple supports*. They are sometimes characterized as knife edges and assumed to be fixed points in space supplying no resistance to angular motion of the beam. Flexibility in the support can be negligible, but not 0. Small compliances exist related to Hertz contact deflection, bearing elasticity, and elastic deformation of the supporting structure. Although complete rigidity is impossible in engineering situations, the effects indicated are normally negligible in comparison with flexural contributions and are often different by an order of magnitude. This is true because the highly stressed localized regions involve small volumes and therefore, small total elastic energy quantities.

The pin joint considered in Chaps. 1 and 2 prevents displacement in any direction in a plane and provides both axial and transverse constraint if connected to a beam. Normally, the axial feature is not considered in straight beams with only transverse loading. Even a beam pinned at both ends will develop only small tensile effects unless the transverse deflections are extreme.

The completely clamped end, or cantilever type support, is one that provides total resistance to transverse shear deflection and also to rotation. It maintains zero slope of the tangent to the elastic curve. Again this idealized model is only approximated in practice.

6.2   *Multiple Spans*

In Fig. 6.1a there are three transverse supports for the beam and a single load. Given the nature of these supports, there are several options by which we can remove one, with the remaining two supporting the beam. The statically determinate cases that result from decoupling are indicated in Fig. 6.1b–d. They are solvable for reactions, bending moments, stresses and deflections by previously indicated methods. Figure 6.1c and d is more compliant and induces more bending stress than Fig. 6.1b.

With all three supports, the R_2 reaction is downward as superimposed on Fig. 6.1b, and thus R_2 can be identified as the redundant constraint. Similarly, R_1 can be added to Fig. 6.1c and R_0 to Fig. 6.1d. In this context, we view the redundancies as returning the displacements to 0, or as having prevented the potential displacements by their presence.

It is usually instructive to visualize or to sketch the nature of the continuous elastic curve. This helps to define the redundancies, as indicated, and to judge the probable directions in which they act; however, the sense of a reaction is sometimes difficult to determine before a complete solution. The directions must be assumed as in static analysis. Negative results then correspond to an incorrect assumption.

Supports for the beam in Fig. 6.1 can obviously be increased beyond 3 with no theoretical limit. In practice, there will only be rare cases in which

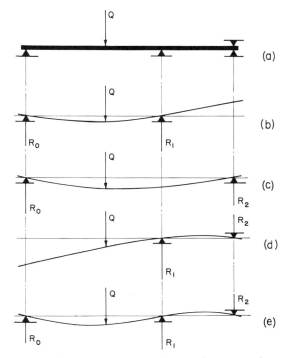

Figure 6.1 Any of the three vertical constraints can be removed to reduce the beam to static determinancy.

supported beams have a large number of simultaneous transverse constraints.

Unless otherwise specified, the transverse supports are assumed to be exactly aligned at assembly. If not, preload stresses are superimposed upon those due to external loads (Sec. 2.7).

6.3 *Elastic Energy Applied to Flexural Indeterminancy*

The beam in Fig. 6.1a can be preloaded by an auxiliary load before the application of Q. Then the equivalence of complementary energy becomes an important argument, as shown in Eq. (4.4c).

Removing the support and applying F at 2 (Fig. 6.2a), we develop an auxiliary-moment distribution and the end of the beam deflects downward under the preload, represented by the hanging weight. In addition to the m

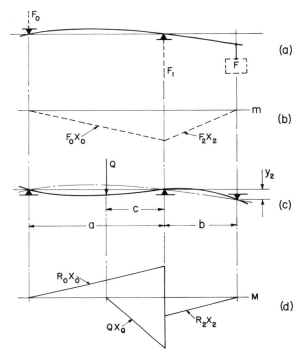

Figure 6.2 Auxiliary preloading for the Fig. 6.1 beam is applied at 2. Subsequently, the support is reinstated before actual loading.

distribution, there are now reaction forces (F_0 and F_1) at these constraints for equilibrium.

Next we reinsert the support, just contacting the beam in its deflected position and then add the actual load Q. This superimposes an M distribution and a second elastic deflection upon the beam. During the Q loading, the weight F at 2 is prevented from vertical displacement by its support. Similarly, neither F_0 nor F_1 can displace. *Thus the external work done by the auxiliary loads during the actual loading is* 0. In turn the associated change in complementary internal flexural energy *is also* 0 *by equivalence*. This important result is

$$F_0 y_0 + F_1 y_1 + F y_2 = \int_0^\ell \frac{mM}{EI}\, dx = 0 \qquad (6.1)$$

We have in effect utilized the methods of Chap. 4 in a situation in which there are zero displacements at the points of interest. In the process, we have generated a relation enabling us to quantify the indeterminate reaction. By summing system complementary energy (external and internal) to 0, we now have the analytical means for solving indeterminate structures.

6.4 *Solution for Transverse Reactions*

For the simple case of Fig. 6.1, we apply previous methods to determine the auxiliary- and actual moment distributions (Fig. 6.2*b* and *d*). The from Eq. (6.1)

$$0 = \int_0^a \frac{\left(\frac{b}{a}\right) Fx_0 M}{EI_a}\, dx_0 + \int_0^b \frac{Fx_2 M}{EI_b}\, dx_2 = \frac{(b/a)}{I_a}\sum A\bar{x}_0 + \frac{1}{I_b}\sum A\bar{x}_2 \tag{6.2}$$

where we routinely cancel F and E and also the I term if the beam has constant section.

The resulting equation is

$$0 = \frac{(b/a)}{I_a}\left[\frac{R_0 a^3}{3} + \frac{Qc^2}{2}\left(a - \frac{c}{3}\right)\right] + \frac{1}{I_b}\left[\frac{R_2 b^3}{3}\right] \tag{6.3}$$

or in reduced form

$$AR_0 + BR_2 = CQ \tag{6.4}$$

From statics

$$M_1 = aR_0 + bR_2 - cQ = 0 \tag{6.5}$$

Solving Eqs. (6.4) and (6.5) simultaneously, we determine R_0 and R_2, and the M diagram is now quantified for stress purposes. R_1 can be calculated finally from statics using $\sum F_y = 0$.

As indicated, negative results for the reactions mean incorrectly assumed directions; however, the assumed direction of F in Fig. 6.2 has no effect on the results. Reversing, the m diagram is equivalent to multiplying the energy relation [Eq. (6.2)] by -1, with no other effect.

6.5 *Preload by Support Displacement*

In the previous example, preload was induced (Fig. 6.2) by removing R_2 and applying an external weight F. The reaction was then replaced in the displaced position. This procedural concept is simplified by an equivalent displacement of the constraint. For instance in Fig. 6.3*b*, if a jack or wedge is applied at 1, a force F is developed as well as an m diagram. With the solution independent of the magnitude and direction of F, the displacement can be up or down and of any reasonable amount. In general, we will consider this as the preferred method of auxiliary preloading in redundant systems.

Auxiliary preload, as with deflection solutions, is employed as a purely analytical device, but the theoretical exercise is not without physical significance. Preloads by weights, turnbuckles, or jacks at supports can be applied to actual structures in a laboratory, with auxiliary moments sensed by means of strain gages for the related bending stresses. Preload corresponds to a *real force*. It is neither a dummy load nor a fictitious load, although it is introduced for the sole purpose of providing elasticity relationships.

On a philosophical level, we could argue that preload is negligibly small with respect to actual loading. Thus, although the auxiliary moments are superimposed on the actual in the final situation, their effect on final stresses and deflections is also negligible.

An alternate method of obtaining auxiliary preload in the cantilever beam (Fig. 6.3) is by angular displacement at the fixed end. In Fig. 6.3d this clockwise rotation induces F loads at 0 and 1. We then have the triangular m diagram (Fig. 6.3e), identical to Fig. 6.3c. The characteristic m diagram, yielding a ΣAx_1 expression, is the only relation possible for this redundant beam, regardless of the loading procedure. This is also demonstrated in Figs. 6.1 and 6.2, as the m diagram in Fig. 6.2b is obtained with preload applied at any one of the three supports.

Figure 6.3 Auxiliary preload can be developed by translational or rotational displacement of a constraint.

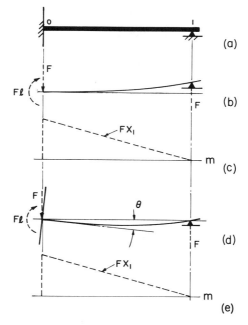

6.6 *Multiple Redundancies*

As the number of excess constraints increases, the number of required auxiliary loadings also increases. In fact, we need as many equations from elasticity as there are redundancies, and the source of each such equation is a related auxiliary loading.

For example, in Fig. 6.4 we can visualize the beam as a cantilever with the clamped end at 0. Redundant constraints then include shear reactions at 2 and 1, and a moment reaction at 2. The associated auxiliary loadings are shown in this order.

Note the beam is decoupled, or reduced to static determinancy during preloading; that is, when F_2 is applied we remove the vertical constraint at 1 and the moment constraint at 2. If these are not removed, the auxiliary loading also becomes indeterminate, thwarting a solution.

Figure 6.4 A multi-redundant beam requires as many decoupled preloading conditions as the number of redundancies.

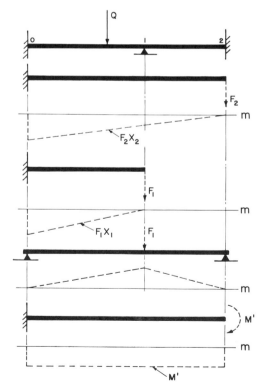

Although it might appear that such decoupling takes undue liberties with the system, *this is a legitimate procedure*. The complementary energy equation does not require system preloads at *all* constraints. Rather we require that an auxiliary moment distribution exist in the beam in static equilibrium with external loading at *any combination of system constraints*. The rules in this regard can be stated:

1. For purposes of system preload as many constraints can be removed as necessary to reduce the system to simple static stability.
2. No constraints can be added.

Four valid types of auxiliary preload are shown in Fig. 6.4, with associated decoupling and resulting *m* diagrams.

6.7 *Behavior of Basic Indeterminate Beams*

It is difficult to present complete data for the large number of indeterminate beam combinations that can exist; however, it is instructive to study several classical cases.

In Fig. 6.5 we can observe the effect of the position of a third support on the simple beam with uniform loading. Minimum bending moments and stresses are obtained at the center ($a/\ell = 0.50$) with maximum stress at the central support. At other positions the maximum moment also tends to occur at the support, but is nearly equaled in the larger span. The pair of reactions tend toward infinity as the intermediate support approaches either end, with minimum R_1 when this support is centered.

Figure 6.6 indicates the bending moment and thereby the stress distribution in a cantilever with uniform load and an auxiliary support at various axial positions. Maximum stress points tend to occur at the clamped end, at the support, or between the support and the ends. In all cases, bending stress must be 0 at the outboard end. The R_1 support carries the entire $q\ell$ load if located at $a/\ell = 0.48$, and, at this condition, $R_0 = 0$. At smaller values of a/ℓ, R_0 will be negative, or a downward shear.

In both figures approximate quantitative design data can be obtained by interpolation for these basic beam situations.

6.8 *The Characteristic Auxiliary Moment Diagram*

A brief review of the implications of the several types of moment diagrams is now in order. For the simple redundant case (Fig. 6.1), we have seen that an auxiliary load at the right end produces a negative triangular *m* diagram (Fig. 6.2*a* and *b*) when reacted at 0 and 1. This triangular geometry results *regardless of whether F is applied at* 0, 1, *or* 2. And since the redundant

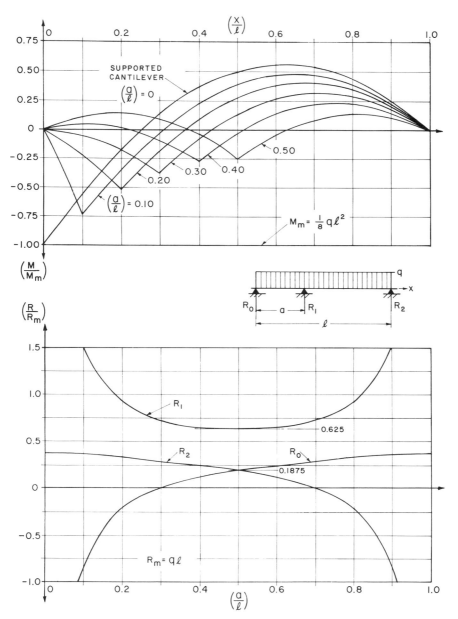

Figure 6.5 Characteristic distributions of bending moment and reactions for the constant beam with uniform loading as the position of the intermediate support is varied.

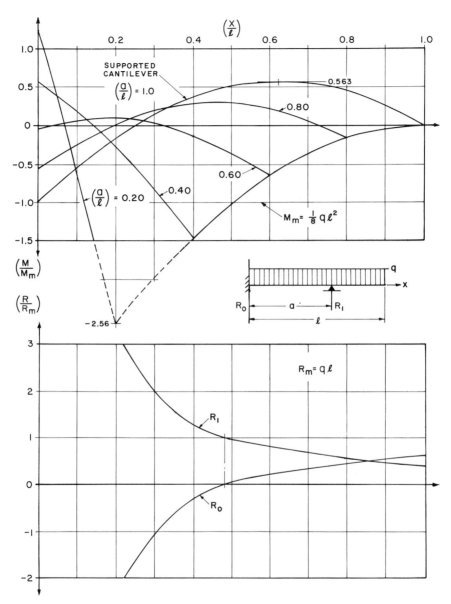

Figure 6.6 Similar to Fig. 6.5, but for a supported cantilever.

solution is independent of both the magnitude and sense of the auxiliary loading, we conclude that for this beam a triangular auxiliary moment diagram (Fig. 6.2*b*) is the *only possible geometric configuration relating the three simple supports*. This is true regardless of any subsequent actual loading, and it can be termed the *characteristic* auxiliary diagram.

6.9 *Moment Diagram by Load Superposition*

Although there is only one total moment diagram for a given beam, there are various representations using components. In Fig. 6.2*d* we have triangles generated by proceeding from 0 to 1 and 2 to 1, respectively. This approach is helpful in the use of geometric integration factors.

Another method is addition of the several component moment diagrams related to component loadings. The moment due to Q, with the R_2 constraint removed, yields the triangular M_1 diagram (Fig. 6.7*c*). The effect of the R_2 load, again with R_0 and R_1 reacting, develops the M_2 distribution. The combination ($M_1 + M_2$) represents the total moment diagram, equivalent to the summation in Fig. 6.2*d*. In fact, the M_2 diagram is analogous to the *m* diagram in Fig. 6.2*c* and *g*, respectively.

6.10 *Deflections of the Indeterminate Beam*

To analyze for deflections, we must first complete the redundancy solution, obtaining all external reactions. In Fig. 6.1, for example, although we are

Figure 6.7 Total moment distribution is the superposition of Q and R_1 loads reacted at 0 and 2. Displacement solution decoupling is shown in (*d*) and (*e*).

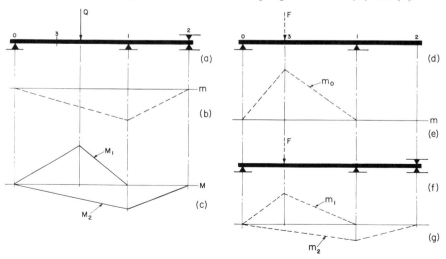

only interested in a particular displacement, our initial problem relates to the evaluation of R_0, R_1, and R_2. We then reapply complementary energy equivalence with auxiliary loading at the desired point or points for quantifying displacement.

It is important at this point to emphasize that the deflection solution is greatly simplified by decoupling the indeterminate beam before the auxiliary loading is applied. In Fig. 6.7*d* we remove the constraint at 2, obtaining the simple m_0 diagram (Fig. 6.7*e*). Unless we do this, and attempt instead to apply F_3 with all three supports present, we have to contend with another indeterminancy solution. This procedure violates no energy considerations and will yield correct results, but it is both unnecessary and impractical.

6.11 *Decoupling Rationale*

Although it appears that using a decoupled system somehow might alter the deflection behavior of the original complete beam, we now prove the legitimacy of decoupling. Taking the preload diagram for the original beam (Fig. 6.7*g*), there are two parts. The first is m_1 due to F_3 on supports at 0 and 1. The second is m_2 (the moment induced by F_3 as it creates a force at the restraint at 2). Then

$$F_3 y_3 = \int_0^\ell \frac{(m_1 + m_2) M}{EI} \, dx$$
$$= \frac{1}{EI} \int_0^\ell m_1 M \, dx + \frac{1}{EI} \int_0^\ell m_2 M \, dx \qquad (6.6)$$

But the second integral with m_2 (Fig. 6.7*g*) is equivalent to the m diagram (Fig. 6.7*b*), and this *has already been equated to 0 in the redundancy solution*:

$$\int_0^\ell \frac{mM}{EI} \, dx = \int_0^\ell \frac{m_2 M}{EI} \, dx = 0 \qquad (6.7)$$

Therefore, Eq. (6.6) reverts to

$$F_3 y_3 = \frac{1}{EI} \int_0^\ell m_1 M \, dx \qquad (6.8)$$

where m_1 represents the *decoupled* auxiliary bending moment. The decoupling can involve *any of the three reactions*. There are thus three possible m_1 diagrams, all of which lead to the same value of y_3.

Actually the preceding proof, although instructive, is unnecessary if we consider only the physical significance of the basic integral

$$U_s = \int_0^\ell \frac{mM}{EI} \, dx \qquad (6.9)$$

In this relation it is only required that there be an m distribution prior to the application of M. With the loading (Fig. 6.7d) providing a preload bending stress, we have a prestress feature between 0 and 1, and Eq. (6.9) is valid. There is no stipulation in the derivation that m is restricted to any particular version of preloading. Thus

$$F_0 y_0 + F_1 y_1 + F_3 y_3 = 0 + 0 + F_3 y_3 = \int_0^\ell \frac{mM}{EI}\, dx \qquad (6.10)$$

6.12 *Flexible Constraints*

If one or more transverse supports have known compliance, this feature will alter the load distribution and have a pronounced effect on the indeterminate solution. In Fig. 6.8 the constant beam has uniform loading and the end supports are rigid, but the center support has a spring rate of K_1. Auxiliary loading must involve both bending in the beam and in the spring, the latter equivalent to a tension–compression element in trusses. We therefore displace the lower end of K_1, developing a force F in the spring and on the center of the beam. Taking the displacement down, there is

Figure 6.8 Indeterminate beam has auxiliary load F applied through the spring support.

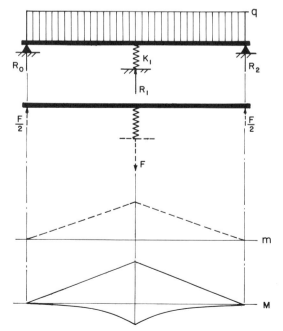

tensile or positive preload in K_1 and a positive auxiliary moment distribution m in the beam.

Both the auxiliary and actual moment diagrams are symmetrical, and with the base of the spring locked after preload, we add flexural and direct complementary energies:

$$0 = \int_0^\ell \frac{mM}{EI}\,dx + \sum CFQ$$

$$= 2\left[\frac{F\sum A\bar{x}_0}{EI} + \frac{1}{K_1}FR_1\right]$$

$$= \frac{1}{EI}\left[\frac{R_0(\ell/2)^3}{3} - \frac{q(\ell/2)^4}{8}\right] - \frac{R_1}{K_1}$$

$$2R_0 - \left(\frac{48EI}{K_1\ell^3}\right)R_1 = \frac{3}{8}q\ell \tag{6.11}$$

And from statics

$$2R_0 + R_1 - q\ell = 0 \tag{6.12}$$

Combining the simultaneous equations,

$$\begin{cases} 2R_0 - \alpha R_1 = \frac{3}{8}Q \\ 2R_0 + R_1 = Q \end{cases}$$

where

$$\alpha = \frac{\left(\dfrac{48EI}{\ell^3}\right)}{K_1} = \frac{K_B}{K_1} = \frac{\text{beam stiffness}}{\text{support stiffness}}$$

$$Q = q\ell = \text{total vertical load}$$

The generalized result is

$$\left(\frac{R_1}{Q}\right) = \frac{\left(\frac{5}{8}\right)}{1 + \alpha} \tag{6.13}$$

If the central support is rigid relative to the beam, $\alpha = 0$, and $\frac{5}{8}$ of the total load is carried by the spring (Fig. 6.5). If the spring stiffness is equal to the nominal beam stiffness, $\alpha = 1$, and $\frac{5}{16}$ of the total load is carried by K_1. For an extremely soft spring, the spring load obviously approaches 0 as α becomes large.

6.13 Constraint with Rotational Flexibility

A clamped end, as the base of a cantilever, can also involve angular compliance C_θ. Complementary torsional strain energy in the elastic element is

$$U_s = C_\theta M'M_0 = \frac{1}{K_\theta}M'M_0 \tag{6.14}$$

where C_θ = angular compliance of base (radians per unit couple)
K_θ = angular stiffness of base (moment per radian)
M' = auxiliary moment preload at base
M_0 = actual moment reaction at base

Then if we displace a fixed end rotationally to apply a preload to the structure as a couple through the base spring,

$$0 = \int_0^\ell \frac{mM}{EI}\, dx + C_\theta M' M_0 \qquad (6.15)$$

It is typically difficult to estimate the angular compliance factor at a fixed end, but Eq. (6.15) provides us with the means for taking this parameter into account.

6.14   *Nonlinearity at a Constraint*

As indicated in Sec. 2.7, a statically indeterminate structure can carry loads and stresses related to misalignment. A somewhat similar condition arises in beams if a support is not at its theoretical position. In Fig. 6.9 the transverse

Figure 6.9   Support below beam requires load Q_A before contact is made at 1. This results in two deflection phases (the latter involving redundancy).

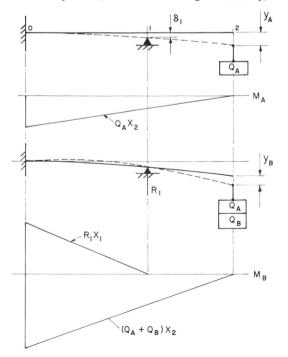

cantilever support at 1 is initially a distance δ_1 below the unloaded straight beam. The converse situation could involve a support R_1 aligned above the beam causing a preload moment between 0 and 1 prior to the application of Q.

In Fig. 6.9 as Q is increased at the end, the beam deflects as a simple cantilever until contact is made at 1. We can determine deflections at 1 and 2 using Table 4.1a corresponding to the contact condition Q_A, y_A, and δ_1.

If Q is increased beyond Q_A, beam compliance is reduced by the R_1 support with deflections and stresses conforming to the modified inde-terminate domain. The resulting load-deflection behavior at the end of the beam is *nonlinear* (Fig. 6.10), as deflection is not directly proportional to load as assumed in all previous developments. The two phases are from 0 to A, and from A to B, with the latter having a higher stiffness rate.

In Fig. 6.10 the external work done on the beam is represented by the area under the curve, and there are three components:

1. The initial work, or the triangular area $0AD = \frac{1}{2}Q_A y_A$, as Q_A is applied to cause contact at the reaction.
2. A second linear phase corresponding to the increase Q_B beyond Q_A, or the triangular area $ABC = \frac{1}{2}Q_B y_B$.
3. The complementary work, or the rectangular area $ACED = Q_A y_B$ representing the work done by the original load Q_A as point 2, further deflects with increasing load.

We note that (3) is similar to previous arguments in connection with complementary energy. In Sec. 1.7 and Fig. 1.4 a rectangular work area was generated by the preload F as it moved through a second deflection, fully acting, caused by Q. This now also occurs with two successive loads (Fig. 6.8), and the final strain energy in the beam includes a complementary quantity deriving from the successive loading.

Thus we have accommodated a nonlinear situation using our basic theory based upon conventional linear assumptions for which algebraic superposition applies.

Figure 6.10 The loading of the system in Fig. 6.9 results in an energy component analogous to complementary energy.

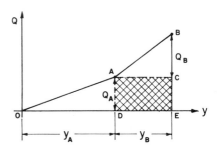

6.15 *Solution by Deflection Restoration*

As in Sec. 2.4, redundant constraints can be determined first by decoupling to find the deflection of the reduced loaded system. Secondly, the decoupled point is returned to the required end conditions, with deflections calculated with respect to the redundant constraints as simple loads and independent of any actual applied loads. This procedure is effectively that of superposition of deflections.

In Fig. 6.11 there are two redundancies corresponding to the shear and moment constraints at 1, since the cantilever supported at 0 is adequate for static stability. Decoupling by cutting the beam at the right end, the uniformly distributed load causes deflections of

$$y_1 = \frac{1}{8} \frac{q\ell^4}{EI} \qquad \theta_1 = \frac{1}{6} \frac{q\ell^3}{EI} \tag{6.16}$$

Considering the restoration effects of a shear at 1, we remove the q loading and apply a positive Q vertically, with deflections of

$$y_{1Q} = \frac{1}{3} \frac{Q\ell^3}{EI} \qquad \theta_{1Q} = \frac{1}{2} \frac{Q\ell^2}{EI} \tag{6.17}$$

Figure 6.11 The fixed end at 1 is equivalent to the restoration of cantilever deflection with simultaneous shear and couple loading at the end.

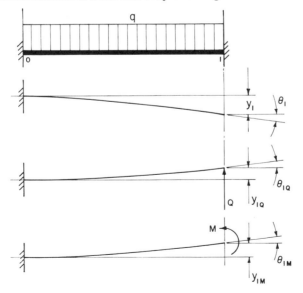

Moment constraint is related to the deflections produced by an end couple:

$$y_{1M} = \frac{1}{2}\frac{M\ell^2}{EI} \qquad \theta_{1M} = \frac{M\ell}{EI} \tag{6.18}$$

where these simple cantilever results are from Table 4.1a.

Returning the decoupled vertical deflection to 0 by means of the reactions,

$$\sum y = y_1 + y_{1Q} + y_{1M} = 0$$

$$-\frac{1}{8}\frac{q\ell^4}{EI} + \frac{1}{3}\frac{R_1\ell^3}{EI} + \frac{1}{2}\frac{M_1\ell^2}{EI} = 0$$

$$8R_1\ell + 12M_1 = 3q\ell^2 \tag{6.19}$$

Similarly for the slope,

$$\sum \theta = \theta_1 + \theta_{1Q} + \theta_{1M} = 0$$

$$-\frac{1}{6}\frac{q\ell^3}{EI} + \frac{1}{2}\frac{R_1\ell^2}{EI} + \frac{M_1\ell}{EI} = 0$$

$$3R_1\ell + 6M_1 = q\ell^2 \tag{6.20}$$

Solving Eqs. (6.19) and (6.20) simultaneously,

$$R_1 = \frac{1}{2}q\ell \qquad M_1 = \frac{-1}{12}q\ell^2 \tag{6.21}$$

Obviously, the shear is correct since for symmetry $R_0 = R_1$. The value of M_1 is also correct for this basic beam condition, with the negative sign indicating a reversed or clockwise sense for the couple, as it must to conform to the physical nature of this constraint.

If more constraints are present than in Fig. 6.11, additional equations are necessary, and higher order simultaneous equations must be developed and solved. The simultaneous behavior follows from the interplay of the coupled deflections at the several constraint points.

Although this example is solved using available deflection equations, elastic energy solutions are available if the loading distribution or the beam section varies in a more complex manner than in Fig. 6.11.

6.16   *Internal Auxiliary Loading*

As we have seen, there are two fundamental methods of inducing auxiliary bending moment to solve for redundancies in a beam. One is by removing a constraint, applying a preload at that point, and replacing the constraint

(Fig. 6.2). The other involves a simple displacement of a constraint (Fig. 6.3). In both situations, we argue that there can be no displacement at any constraint; therefore, the external preloads do no work as the actual loading is applied, and the auxiliary bending energy integral is also 0.

A third option is available in which the auxiliary loading is produced by a pair of equal and opposite moments or shears *at any point in a beam*. External work done during the applied actual loading by this type of *internal* loading must be 0. To illustrate, Fig. 6.12 shows the symmetrical, twice-redundant beam of Fig. 6.11 requiring two equations from elastic energy. We introduce internal M' and F loadings at any axial position by cutting the beam, applying the opposed loading pair, and reconnecting the preloaded beam. The equation for M' becomes

$$0 = \int_0^\ell \frac{mM}{EI}\, dx = \int_0^\ell \frac{M'M}{EI}\, dx$$

$$\sum A = \frac{q\ell}{2}\left(\frac{a^2}{2}\right) - \frac{qa^3}{6} - \frac{1}{12}q\ell^2 a + \frac{q\ell}{2}\left(\frac{b^2}{2}\right) - \frac{qb^3}{6} - \frac{1}{12}q\ell^2 b = 0$$

Figure 6.12 The twice-redundant beam of Fig. 6.11 can be solved by internal moment and shear preloads at any point in the beam.

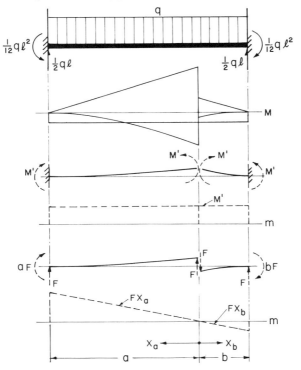

reducing to

$$3\left(\frac{a}{\ell}\right)^2 - 2\left(\frac{a}{\ell}\right)^3 - \left(\frac{a}{\ell}\right) + 3\left(\frac{b}{\ell}\right)^2 - 2\left(\frac{b}{\ell}\right)^3 - \left(\frac{b}{\ell}\right) = 0 \qquad (6.22)$$

Similarly for the F shear loads,

$$0 = \int_0^\ell \frac{mM}{EI}\, dx = \int_0^a \frac{Fx_a M}{EI}\, dx + \int_0^b \frac{Fx_b M}{EI}\, dx$$

$$\sum A\bar{x}_a + \sum A\bar{x}_b = \frac{q\ell}{2}\left(\frac{a^3}{6}\right) - \frac{qa^4}{24} - \frac{1}{12}q\ell^2\left(\frac{a^2}{2}\right)$$

$$- \frac{q\ell}{2}\left(\frac{b^3}{6}\right) + \frac{qb^4}{24} + \frac{1}{12}q\ell^2\left(\frac{b^2}{2}\right) = 0$$

reducing to

$$2\left(\frac{a}{\ell}\right)^3 - \left(\frac{a}{\ell}\right)^4 - \left(\frac{a}{\ell}\right)^2 - 2\left(\frac{b}{\ell}\right)^3 + \left(\frac{b}{\ell}\right)^4 + \left(\frac{b}{\ell}\right)^2 = 0 \qquad (6.23)$$

If the previous derivation is correct, and our solution is valid for any location of the auxiliary moment or shear force, Eqs. (6.22) and (6.23) must reduce to 0 for any combination of (a/ℓ) and (b/ℓ), *and this is the case.*

We conclude that the total moment diagram for the fixed–fixed beam satisfies geometric properties deriving from $\sum A = 0$ and $\sum A\bar{x} = 0$, the latter relative to any axial reference position, regardless of the nature of the transverse loading. Specifically, in this example, the total actual-moment distribution is parabolic (Fig. 6.13). As shown, taking an arbitrary x

Figure 6.13 From Fig. 6.12 there is a geometric equivalence of $\sum A\bar{x}$ relative to any axial position along the beam; that is, there is a balance of these terms about a transverse axis at a point.

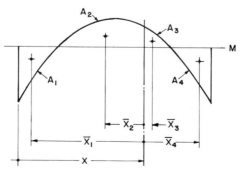

distance we divide the area under the diagram into four segments for which

$$\sum A\bar{x}_a - \sum A\bar{x}_b = 0$$
$$-A_1\bar{x}_1 + A_2\bar{x}_2 - A_3\bar{x}_3 + A_4\bar{x}_4 = 0$$
$$(A_2\bar{x}_2 + A_4\bar{x}_4) = (A_1\bar{x}_1 + A_3\bar{x}_3) \qquad (6.24)$$

That is, the gravity moments of the component parabolic areas viewed as gravity effects on horizontal plates about a horizontal axis (Fig. 5.4) have a balancing relationship, regardless of x. If we transpose A_4 to the left of x and A_1 to the right of x at the respective distances $\bar{x}_4$ and $\bar{x}_2$, the transposed system is *in static balance* about the horizontal x axis.

Examples

6.1. The cantilever carries an end load of 80 lb, 70 in. from the clamped end and has an additional fixed support at 1. Determine:
(a) The reaction R_1
(b) The maximum bending moment in the beam

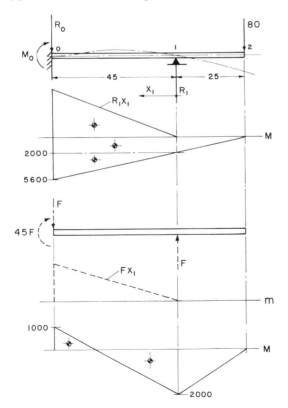

Solution:

(a) The auxiliary force F is applied at 1 in the direction of R_1 causing the dashed auxiliary bending moment m. The support is reestablished before applying Q. The M diagram is drawn from 2 to 0. No mM product exists between 2 and 1. With no external work done by any of the three locked auxiliary reactions during the Q loading, and with Table 5.1

$$0 = \int_0^\ell \frac{mM}{EI}\, dx = \frac{1}{EI}\int_0^{45}(Fx_1)\, M\, dx_1$$

$$0 = \sum_1^0 A\bar{x}_1 = \frac{R_1(45)^3}{3} - (2000)(45)\left(\frac{45}{2}\right) - (3600)\left(\frac{45}{2}\right)(30)$$

$$R_1 = 146.7\ \text{lb}$$

(b) From statics

$$\sum M_0 = M_0 + 80(70) - 146.7(45) = 0 \qquad M_0 = 1000\ \text{in. lb}$$

$$M_1 = 80(25) = 2000\ \text{in. lb} \quad (\text{maximum})$$

Since the final M diagram is equivalent to the original or component M diagram, we can make a final check of the results:

$$0 = \sum_1^0 A\bar{x}_1 = \frac{1000(15)}{2}(30 + 10) - \frac{2000(30)}{2}\frac{30}{3}$$

$$= 30{,}000 - 30{,}000 = 0$$

6.2. Verify the results of Example 6.1a by deflection superposition.

Solution: From Table 4.1a, deflection down at 1 due to Q_2 with no R_1 support is

$$y_{12} = Q_2\alpha_{12} = \frac{80(70)^3}{3EI}\left(\frac{45}{70}\right)^2\left[1 + \frac{1}{2}\left(\frac{25}{70}\right)\right]$$

$$= \frac{4.46(10)^6}{EI}$$

As a simply applied vertical load, we have a deflection y_{11} that is equated to

y_{12} as 1 is returned to zero displacement:

$$y_{11} = \frac{R_1(45)^3}{3EI} = \frac{0.030375R_1}{EI} = \frac{4.46(10)^6}{EI}$$

$$R_1 = 146.7 \text{ lb}$$

6.3. For the cantilever of Example 6.1, find:
(a) The deflection at the end
(b) The slope 1

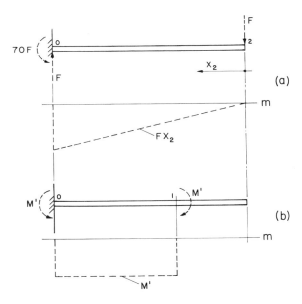

Solution:
(a) Removing the redundant reaction R_1, and applying F at 2

$$y_2 = \frac{1}{EI} \sum Ax_2 = \frac{1}{EI}\left[\frac{80(70)^3}{3} - 6600\left(\frac{45}{2}\right)(55)\right]$$

$$= \frac{0.98(10)^6}{EI}$$

(b) $$\theta_1 = \frac{1}{EI}\left[\frac{2000(30)}{2} - \frac{1000(15)}{2}\right] = \frac{22,500}{EI}$$

6.4. Verify the deflections in Example 6.3 using superposition and relations from Chap. 4.

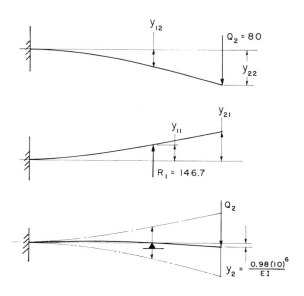

Solution:

(a) In the figure, we have the effects of both the Q_2 and R_1 loading, and from Table 4.1a

$$\text{Due to } Q_2, \quad y_{22} = 80\alpha_{22} = \frac{80(70)^3}{3EI} = \frac{9.15(10)^6}{EI} \quad (\text{down})$$

$$\text{Due to } R_1, \quad y_{21} = \left(\frac{146.7}{80}\right) y_{12} = 1.83\frac{4.46(10)^6}{EI} = \frac{8.17(10)^6}{EI} \quad (\text{up})$$

$$y_2 = \frac{(9.15 - 8.17)(10)^6}{EI} = \frac{0.98(10)^6}{EI} \quad (\text{down})$$

(b) From Table 4.1a

$$\theta_1 = M_1\gamma_{11} - V_1\beta_{11}$$

$$= \frac{2000(45)}{EI} - \frac{66.6(45)^2}{2EI} = \frac{22,500}{EI} \quad \text{rad CCW}$$

6.5. The constant diameter shaft is 24 in. long and supported by three self-aligning bearings. If there is a uniform transverse load of 10 lb/in., find all reactions.

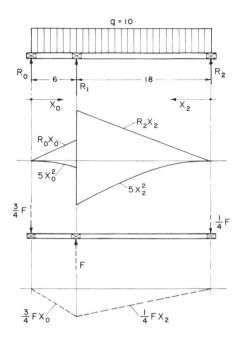

Solution: The auxiliary load F can be applied at any reaction. Taking F_1, a triangular negative m diagram results, providing the energy relationship,

$$0 = \frac{3}{4} F \sum_0^6 A\bar{x}_0 + \frac{1}{4} F \sum_0^{18} A\bar{x}_2$$

$$= 3\frac{R_0(6)^3}{3} - 3\frac{10(6)^4}{8} + \frac{R_2(18)^3}{3} - \frac{10(18)^4}{8}$$

$$216R_0 + 1944R_2 = 4860 + 131{,}220 = 136{,}080$$

$$R_0 + 9R_2 = 630 \tag{1}$$

A second equation from statics involving only R_0 and R_2 can be obtained using continuity of the M diagram at 1:

$$M_1 = 6R_0 - \frac{360}{2} = 18R_2 - \frac{3240}{2}$$

$$R_0 - 3R_2 = -240 \tag{2}$$

Solving (1) and (2) simultaneously,

$$R_0 = -22.5 \text{ lb} \qquad R_2 = 72.5 \text{ lb}$$

Summing forces in the vertical direction, $R_1 = 190$ lb. The negative reaction at 0 means that the end tends to lift because of the load in the longer span, 1–2.

6.6. A constant beam, 400 cm long, is fixed at both ends. It has a load of 5.0 kN at the midpoint and of 6.3 kN at the quarter point. Calculate the complete reactions.

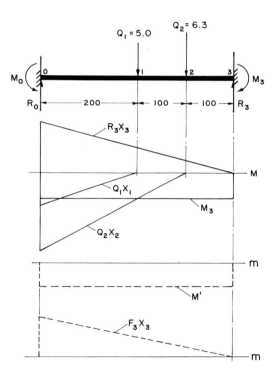

Solution: Applying the auxiliary moment M' at 3 induces a constant bending moment from 3 to 0, and the energy relation reduces to

$$0 = \sum A = R_3 \frac{(400)^2}{2} - 400 M_3 - \frac{(300)^2}{2}(6.3) - \frac{(200)^2}{2}(5.0)$$

$$R_3 - 0.005 M_3 = 4.794$$

With an auxiliary force F_3 in the sense of R_3, the triangular auxiliary moment diagram yields

$$0 = \sum A \bar{x}_3 = \frac{R_3(400)^3}{3} - \frac{M_3(400)^2}{2} - \frac{(300)^2(6.3)}{2}(200 + 100)$$

$$- \frac{(200)^2(5.0)}{2}\left(\frac{2}{3}200 + 200\right)$$

$$21.33 R_3 - 0.08 M_3 = 118.38$$

Solving the two elasticity equations simultaneously,

$$R_3 = 7.82 \text{ kN} \qquad M_3 = 604.4 \text{ kN} \cdot \text{cm}$$

The reactions at 0 are obtained from statics:

$$\sum M_0 = -M_0 - 7.82(400) + 5(200) + 6.3(300) + 604.4$$

$$M_0 = 366.4 \text{ kN} \cdot \text{cm}$$

$$\sum M_3 = R_0(400) + 604.4 - 366.4 - 6.3(100) - 5.0(200) = 0$$

$$R_0 = 3.48 \text{ kN}$$

6.7. In Example 6.6, find:
   (a) The deflection of the midpoint
   (b) The slope at the midpoint

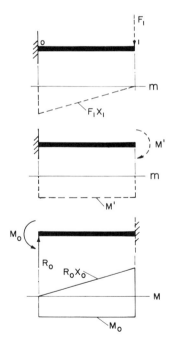

Solution:
(a) To apply F_1 for deflection purposes we decouple the twice-redundant beam by cutting it at 1 and consider only the complementary energy between 0 and 1. Although $m = F_1 x_1$ is taken to the left, it is most convenient to construct the M diagram from left to right, involving only

M_0 and R_0:

$$F_1 y_1 = \frac{F_1}{EI} \sum A x_1$$

$$y_1 = \frac{1}{EI}\left[366.4\frac{(200)^2}{2} - 3.48\frac{(200)^3}{6}\right] = \frac{(10)^6}{EI}(7.33 - 4.64)$$

$$= \frac{2.69(10)^6}{EI}$$

(For consistent units, E will be in kN/cm², I in cm⁴, and y in cm.)

(b) For θ_1 we similarly preload only the 0–1 section, using M'. Energy equivalence reduces to

$$\theta_1 = \frac{1}{EI}\sum A = \frac{1}{EI}\left[366.4(200) - \frac{3.48(200)^2}{2}\right] = \frac{3700}{EI} \text{ rad}$$

6.8. For the beam in Example 6.6, find the horizontal position of maximum deflection by means of zero slope.

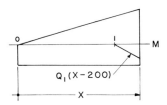

Solution: By moving the decoupling point to the right from 1, and also moving M' to the right, we involve Q_1 and have $\sum A = 0$:

$$3.48\frac{x^2}{2} = 366x + \frac{5(x - 200)^2}{2}$$

Solving the resulting quadratic, $x = 212$ cm.

7
_____

The Three-Moment
Theorem

INDETERMINATE BEAMS ARE READILY SOLVED by the methods of Chap. 6 using auxiliary preloads and by equating the complementary energy to 0. We now utilize this process to develop a more formalized technique that expedites the procedure. The relations produced focus on the internal bending moments occurring at the transverse supports. Although the support reactions are not present directly in the redundancy solution, they can be obtained after the moments are determined by applying static equilibrium to the individual spans.

The theorem, derived from elastic energy relations, involves considering successive pairs of adjacent spans and is based upon the continuity of internal bending moments at the supports. It is particularly suited to beams with multiple supports and can accommodate spans of different transverse sections as well as fixed ends. It can be adapted, however, to other continuous structures such as bent beams with straight elements. Various loading situations are treated by means of tabulated factors. Additionally, we can analyze for deflections after the three-moment solutions, obtaining the elastic curves in each span by superposition.

7.1 Moment Coupling in a Continuous Beam

In Fig. 7.1a a continuous horizontal beam carries arbitrary transverse loading in several spans and has multiple simple supports. The supports are assumed operable in either direction and will define an indeterminate system if there are more than two. Beam cross sections can vary but are restricted to a constant value in any span. Transverse loading can be applied either in the downward or upward direction. The beam is assumed to have conventional elastic behavior, with superposition applicable.

We note the continuity of the elastic curve (Fig. 7.1b) with loads in any span developing deflections and reactions throughout the system. The individual spans have zero deflection at each support with adjacent spans sharing a common slope at the supports. Internal bending moments react oppositely with respect to any neutral axis along the beam. At the beam supports, however, these moments become the variable parameters in the solution.

As seen in Fig. 7.1c and d, the beam can be decoupled by cutting at the supports. Each span is then an elementary beam having zero bending moment and deflection at the ends, for which formulas are usually available. Total moment distribution in the system is then obtainable by superposi-

Figure 7.1 The continuous beam can be decoupled at three successive supports. Constraint moment diagrams due to the diagrams as decoupled and due to the moment constraints at the three supports are combined in (g).

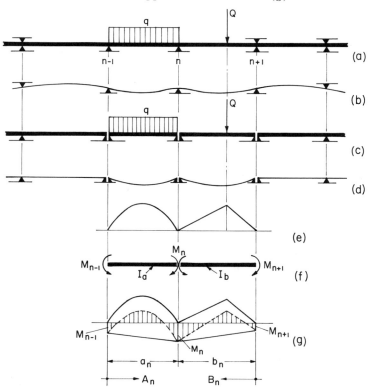

tion, or restoration of all the internal coupling moments at the supports (Fig. 7.1*f*).

Finally, the total moment distributions in each span result from addition of the uncoupled simply supported moments and the internal moments at each support (Fig. 7.1*g*). The latter produce a linear variation between ordinates at the supports. This trapezoidal geometry results from the superimposed end moments and the related superimposed shears as the beam is recoupled.

We have assumed these internal recoupling moments as normally negative; that is, the usual end moment supplied by a constraint is technically negative, developing tension in the top fibers of the beam. In the derivation to follow, we assume this normal constraint sense, and if a negative moment is found in a solution, it then signifies a reverse situation, or compression in the top fibers at the corresponding support.

7.2 Derivation of the Equation

In Fig. 7.1*a* we now proceed to analyze two adjacent spans $[(n - 1)$ to $(n + 1)]$, where n represents the center support. For purposes of preloading by auxiliary forces, we decouple the continuous beam, cutting at $(n - 1)$ and $(n + 1)$ but maintain continuous elastic coupling at n.

Figure 7.2 Auxiliary moment diagrams result from an auxiliary force applied at the n support (*b*). Actual moment constraint diagrams are shown in (*d*).

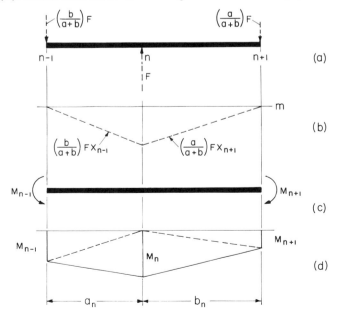

We next apply F at the n support (Fig. 7.2a) by raising this support a small distance, as in Fig. 6.3. Figure 7.2 shows the resulting negative auxiliary moment diagram and the equations for the auxiliary reactions at $(n - 1)$ and $(n + 1)$. With the beam recoupled and external loads applied, actual bending moments corresponding to M_{n-1}, M_n, and M_{n+1} (Fig. 7.1f) constitute the three negative ordinates in Fig. 7.1g.

For geometric analysis, we draw the dashed diagonals, transforming the trapezoidal elements into triangles for area–moment purposes (Fig. 7.2d).

We now equate the total complementary strain energy for the complete beam (Fig. 7.1a) after actual loading to 0:

$$0 = \int_{n-1}^{n} \frac{mM}{EI_a} \, dx + \int_{n+1}^{n} \frac{mM}{EI_b} \, dx \tag{7.1}$$

where there is only complementary energy in the two selected adjacent spans, as $m = 0$ in the rest of the beam. Then

$$0 = \frac{Fb}{I_a(a + b)} \int_0^a x_{n-1} M \, dx + \frac{Fa}{I_b(a + b)} \int_0^b x_{n+1} M \, dx$$

$$= \frac{b}{I_a} \sum_0^a A\bar{x}_{n-1} + \frac{a}{I_b} \sum_0^b A\bar{x}_{n+1}$$

$$= \frac{b}{I_a}\left[\frac{M_{n-1}a^2}{6} + \frac{M_n a^2}{3} - \sum_0^a A\bar{x}_a \right] + \frac{a}{I_b}\left[\frac{M_{n+1}b^2}{6} + \frac{M_n b^2}{3} - \sum_0^b A\bar{x}_b \right] \tag{7.2}$$

where the negative terms are the area–moment quantities for the applied *external* loads, downward assumed positive; that is, the areas of the moment diagrams in Fig. 7.1e are multiplied by the centroidal distances from the respective terminal ends of the beam pair.

Reducing Eq. (7.2), we have

$$M_{n-1} + 2(1 + \alpha)M_n + \alpha M_{n+1} = A_n + \alpha B_n \tag{7.3}$$

where $\alpha = [(b/a)(I_a/I_b)]_n$ = a dimensionless stiffness parameter relating the span pair
$A_n = (6/a^2)\sum_0^a A\bar{x}_a$ = the load factor for the left span
$B_n = (6/b^2)\sum_0^b A\bar{x}_b$ = the load factor for the right span

and A_n and B_n have the dimensions of a couple. If there are no external loads in either span, the right side of Eq. (7.3) is 0. If there is no external load in the left span, A_n is 0, and similarly B_n can be 0.

Equation (7.3) is a generalized form of the *Three-Moment Theorem* and is the formula that we will use extensively in solving for redundancies. It is an extremely versatile tool for this purpose.

7.3 Load Factors

As shown in Fig. 7.1g, the factor A_n is referred to a coordinate origin at the left of the left span and B_n to an origin at the right of the right span. The $\Sigma A\bar{x}$ quantities are geometric and related to the factors in Table 5.1. For the basic case of a center load (Fig. 7.3), we have, considering the two moment triangles each with an area of $Q\ell^2/16$,

$$A_n = \frac{6}{a^2}\sum A\bar{x}_a = \frac{6}{\ell^2}\left[\frac{Q\ell^2}{16}\left(\frac{\ell}{3}\right) + \frac{Q\ell^2}{16}\left(\frac{2}{3}\ell\right)\right] = Q\ell\left[\frac{1}{8} + \frac{1}{4}\right] = \frac{3}{8}Q \tag{7.4}$$

Similarly taking the entire area of the moment triangle,

$$A_n = \frac{6}{a^2}\sum A\bar{x}_a = \frac{6}{\ell^2}\left[\left(\frac{\ell}{2}\right)\left(\frac{Q\ell^2}{8}\right)\right] = \frac{3}{8}Q \tag{7.5}$$

And from symmetry

$$B_n = \frac{6}{b^2}\sum A\bar{x}_b = A_n \tag{7.6}$$

For a uniform load the total moment diagram is parabolic and symmetrical, with a maximum center ordinate of $q\ell^2/8$, and

$$A_n = B_n = \frac{6}{\ell^2}\sum A\bar{x} = \frac{6}{\ell^2}\left[\frac{2}{3}\ell\frac{q\ell^2}{8}\right]\frac{\ell}{2} = \frac{q\ell^2}{4} \tag{7.7}$$

Factors are provided in Table 7.1 for various loading combinations for the simple decoupled single-span beam, and these will suffice for most

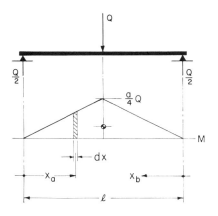

Figure 7.3 Load factors for the centrally loaded simple span follow from the geometry of the triangular moment diagram.

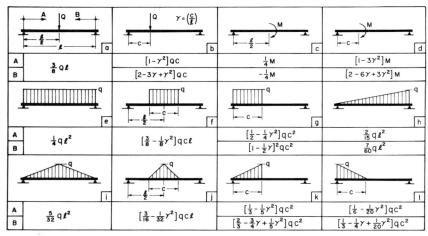

	a	b ($\gamma=(\frac{c}{\ell})$)	c	d
A	$\frac{3}{8}Q\ell$	$[1-\gamma^2]Qc$	$\frac{1}{4}M$	$[1-3\gamma^2]M$
B		$[2-3\gamma+\gamma^2]Qc$	$-\frac{1}{4}M$	$[2-6\gamma+3\gamma^2]M$

	e	f	g	h
A	$\frac{1}{4}q\ell^2$	$[\frac{3}{8}-\frac{1}{8}\gamma^2]qc\ell$	$[\frac{1}{2}-\frac{1}{4}\gamma^2]qc^2$	$\frac{2}{15}q\ell^2$
B			$[1-\frac{1}{2}\gamma]^2qc^2$	$\frac{7}{60}q\ell^2$

	i	j	k	l
A	$\frac{5}{32}q\ell^2$	$[\frac{3}{16}-\frac{1}{32}\gamma^2]qc\ell$	$[\frac{1}{3}-\frac{1}{5}\gamma^2]qc^2$	$[\frac{1}{6}-\frac{1}{20}\gamma^2]qc^2$
B			$[\frac{2}{3}-\frac{3}{4}\gamma+\frac{1}{5}\gamma^2]qc^2$	$[\frac{1}{3}-\frac{1}{4}\gamma+\frac{1}{20}\gamma^2]qc^2$

Table 7.1 Load factors A and B [Fig. 7.1g and Eq. (7.3)] are tabulated for a number of typical loadings for the basic span.

actual beams; however, if necessary they can be superimposed by algebraic addition. If the loading is upward, the factors are taken as negative.

7.4 The Calculation Process

Equation (7.3) is now applied to a double-span case (Example 6.5). We have zero end moments, take n at R_1, and $\alpha = \frac{18}{6} = 3$:

$$0 + 2(1 + 3)M_1 + 0 = \frac{1}{4}(10)(6)^2 + (3)\frac{1}{4}(10)(18)^2$$

$$M_1 = \frac{10}{32}(36 + 972) = +315$$

Since both spans are in equilibrium independently, we can obtain the reactions. From Fig. 7.4 and recognizing that the central reaction is the sum

Figure 7.4 Three-moment solution for the indeterminate beam of Example 6.5 involves only the unknown intermediate bending moment M_1.

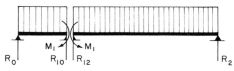

of the shears at the left and right beams,

$$\sum M_0 = M_1 + \frac{(10)(6)^2}{2} - 6R_{10} = 0$$

$$R_{10} = \frac{180 + 315}{6} = 82.5$$

$$\sum M_2 = -M_1 + 18R_{12} - \frac{(10)(18)^2}{2} = 0$$

$$R_{12} = \frac{1620 + 315}{18} = 107.5$$

$$R_1 = R_{10} + R_{12} = 82.5 + 107.5 = 190 \quad \text{(up)}$$

$$R_0 = 6q - R_{10} - R_{10} = 60 - 82.5 = -22.5 \quad \text{(down)}$$

$$R_2 = 18q - R_{12} = 180 - 107.5 = 72.5 \quad \text{(up)}$$

Checking,

$$\sum F_y = R_0 + R_1 + R_2 + q\ell = -22.5 + 190 + 72.5 - 240 = 0$$

7.5 Fixed Ends

Applying the theorem to the fixed–fixed beam of Example 6.6, we have two unknown bending moments at the ends, and Eq. (7.3) must be used twice. In Fig. 7.5 we revise the notation so these moments become M_0 and M_1 at the left and right wall, respectively.

Figure 7.5 Moment relations are established for spans -1, 0, 1, and 0, 1, 2, respectively in Example 6.6.

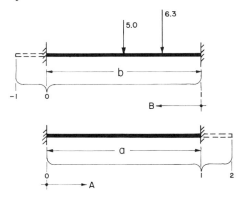

To proceed with the Three-Moment Theorem concept, we must have span pairs, so we assume the beam extends into the wall at the left, having infinite stiffness. Then

$$\alpha = \left(\frac{400}{0}\right)\left(\frac{\infty}{I_b}\right) = \infty$$

This presents a problem in Eq. (7.3). We resolve this difficulty by dividing the equation by α:

$$\left(\frac{1}{\alpha}\right)M_{n-1} + 2\left(\frac{1}{\alpha} + 1\right)M_n + M_{n+1} = \frac{1}{\alpha}A_n + B_n \qquad (7.8)$$

With $\alpha = \infty$, this reduces to the usable form

$$2M_n + M_{n+1} = B_n \qquad (7.9)$$

Then from Table 7.1, with $\gamma = 300/400 = 0.75$,

$$2M_0 + M_1 = \tfrac{3}{8}(5)(400) + \left[2 - 3(0.75) + (0.75)^2\right](6.3)(300)$$
$$= 1341 \qquad (7.10)$$

Shifting to the right pair with $n = 1$, the rigid beam extends into the right wall (Fig. 7.3) and we can use Eq. (7.3) directly since

$$\alpha = \left(\frac{0}{400}\right)\left(\frac{I_a}{\infty}\right) = 0$$

$$M_0 + 2M_1 = \tfrac{3}{8}(5)(400) + \left[1 - (0.75)^2\right](6.3)(300) = 1577 \quad (7.11)$$

Solving Eqs. (7.10) and (7.11) simultaneously, $M_0 = 367$ and $M_1 = 604$, which agree with the previous solution.

Although some calculations are required for the load factors, this example illustrates how the three-moment solution expedites the analysis, particularly if we are only interested in the bending moments for stress purposes. Total moments can be obtained without determining the vertical reactions. It has not been necessary to sketch actual or auxiliary moment diagrams, and we have used the load factors in Table 7.1 rather than having to sum a number of $\Sigma A\bar{x}$ terms algebraically. Another advantage resides in having an alternate solution available as a check on numerical results if desired.

7.6 Intermediate Deflections

Using the Three-Moment Theorem we resolve a complex beam into single beam elements. To obtain a transverse deflection at a specified axial location within a span, an auxiliary force can be applied in the span, decoupled at

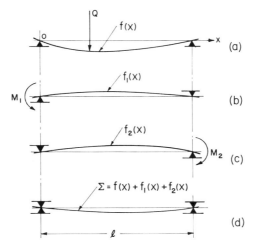

Figure 7.6 Deflection distributions are obtained by superposition of the effects of the two end couples on the deflection of the loaded simply supported beam.

the end supports (Fig. 7.1*d*). Then the auxiliary moment diagram is triangular and the actual moment diagram results from superposition (Fig. 7.1*g*).

With deflection information available for the simply supported beam in Chaps. 4 and 5, however, the deflection can be visualized as having component elastic curves resulting from the decoupled case and the end-moment loadings (Fig. 7.6). The latter can be taken from Tables 4.1*b* or 4.7, with the applied couples at the supports. Taking these as normally negative, for M_1 (Fig. 7.6*b*)

$$f_1(x) = \frac{M_1 \ell^2}{6EI} \left[-2\left(\frac{x}{\ell}\right) + 3\left(\frac{x}{\ell}\right)^2 - \left(\frac{x}{\ell}\right)^3 \right] \qquad (7.12)$$

and for M_2 (Fig. 7.6*c*)

$$f_2(x) = \frac{M_2 \ell^2}{6EI} \left[-\left(\frac{x}{\ell}\right) + \left(\frac{x}{\ell}\right)^3 \right] \qquad (7.13)$$

A similar approach can be used for the slope distribution.

7.7 *Overhung Deflections*

If deflections are of interest beyond an end support, these involve superposition of the cantilever deflection due to loading and the slope at the support (Fig. 7.7). As shown, with a clockwise slope of θ, the deflection curve is rotated by this amount at the base. All slopes in the cantilever are changed by this constant angle, determined as indicated in Sec. 7.6.

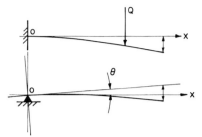

Figure 7.7 Deflection of an outboard section involves the basic cantilever deflection and the rotation related to the slope at the support.

For transverse deflections the alteration due to θ is related to the radial distance from the support:

$$f(x) = f(x) + x(\theta) \tag{7.14}$$

with proper algebraic sign maintained in the superposition.

7.8 Summary

The Three-Moment Theorem provides a convenient and methodical approach for analyzing continuous beams on multiple supports. In the calculations we proceed from left to right, considering paired spans incrementally, developing as many simultaneous equations as there are redundancies in the system. Solution of these equations quantifies the bending moments existing at the several transverse supports.

As a guide to the use of this method, it is necessary to consider:

1. *Fixed ends* as spans of infinite stiffness.
2. *Overhung ends* as *not* constituting a span. Rather these elements provide known bending moments at the support beyond which they extend.
3. *Sense* of the moment terms is the *reverse* of conventional, with $+M$ terms corresponding to compression in the bottom fibers of the beam.

Examples

7.1. Find the load factors for a span subjected to two concentrated forces.

Solution: From Table 7.1*b*

$$A = \left[1 - \left(\tfrac{9}{28}\right)^2\right](20)(9) + \left[1 - \left(\tfrac{21}{28}\right)^2\right](-25)(21)$$
$$= 161.4 - 229.7 = -68.3$$
$$B = \left[2 - 3\left(\tfrac{9}{28}\right) + \left(\tfrac{9}{28}\right)^2\right](20)(9) + \left[2 - 3\left(\tfrac{21}{28}\right) + \left(\tfrac{21}{28}\right)^2\right](-25)(21)$$
$$= 205.0 - 164.1 = +40.9$$

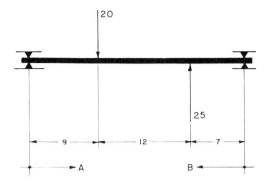

7.2. Determine the two load factors for the span with varying distributed loading.

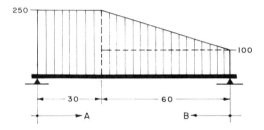

Solution: The loading diagram can be resolved into three geometric components for the application of Table 7.1 terms. For A we use, left to right successively, A from Table 7.1g, B from Table 7.1k, and B from Table 7.1g. It is necessary to use B terms for A because of the required reversal in sense, or origins:

$$A = \left[\tfrac{1}{2} - \tfrac{1}{4}\left(\tfrac{30}{90}\right)^2\right](250)(30)^2 + \left[\tfrac{2}{3} - \tfrac{3}{4}\left(\tfrac{60}{90}\right) + \tfrac{1}{5}\left(\tfrac{60}{90}\right)^2\right](150)(60)^2$$
$$+ \left[1 - \tfrac{1}{2}\left(\tfrac{60}{90}\right)\right]^2(100)(60)^2 = 404,250$$

For B we use, left to right successively, A from Table 7.1g, A from Table 7.1k, and B from Table 7.1g:

$$B = \left[\tfrac{1}{2} - \tfrac{1}{4}\left(\tfrac{60}{90}\right)^2\right](100)(60)^2 + \left[\tfrac{1}{3} - \tfrac{1}{5}\left(\tfrac{60}{90}\right)^2\right](150)(60)^2$$
$$+ \left[1 - \tfrac{1}{2}\left(\tfrac{30}{90}\right)^2\right](250)(30)^2 = 428,250$$

Alternatively, and more simply, we can use the entire rectangle ($q = 250$) and treat the triangle with $q = 150$ as a negative loading:

$$A = \tfrac{1}{4}(250)(90)^2 - \left[\tfrac{1}{3} - \tfrac{1}{4}\left(\tfrac{60}{90}\right) + \tfrac{1}{20}\left(\tfrac{60}{90}\right)^2\right](150)(60)^2 = 404,250$$

$$B = \tfrac{1}{4}(250)(90)^2 - \left[\tfrac{1}{6} - \tfrac{1}{20}\left(\tfrac{60}{90}\right)^2\right](150)(60)^2 = 428,250$$

7.3. For the general case of a supported cantilever with a distributed load, find:
(a) The end moment
(b) The maximum bending moment
(c) The deflection distribution

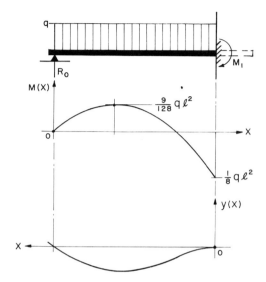

Solution:
(a) Considering the infinitely stiff span in the wall,

$$\alpha = \left(\frac{0}{\ell}\right)\left(\frac{I}{\infty}\right) = 0$$

From Eq. (7.3) and Table 7.1e,

$$0 + 2M_1 + 0 = A_n + 0 = \tfrac{1}{4}q\ell^2$$

$$M_1 = \tfrac{1}{8}q\ell^2$$

(b) We obtain R_0 from

$$\sum M_1 = \ell R_0 + M_1 - \frac{q\ell^2}{2} = 0 \qquad R_0 = \tfrac{3}{8}q\ell$$

The moment distribution becomes

$$M(x) = \left(\tfrac{3}{8}q\ell\right)x - \tfrac{1}{2}qx^2$$

Differentiating for the maximum value,

$$\frac{dM(x)}{dx} = \tfrac{3}{8}q\ell - qx = 0 \qquad x = \tfrac{3}{8}\ell$$

And at this position the maximum moment is $\tfrac{9}{128}q\ell^2$. Thus the maximum bending moment is at the support, from (*a*), distributed as shown.

(c) Reversing the beam coordinate and superimposing from Table 5.3a and β_{12} from Table 4.1b,

$$y(x) = \frac{q\ell^4}{24EI}\left[\left(\frac{x}{\ell}\right) - 2\left(\frac{x}{\ell}\right)^3 + \left(\frac{x}{\ell}\right)^4\right] - \left(\frac{1}{8}q\ell^2\right)\left(\frac{\ell^2}{6EI}\right)$$
$$\times\left[1 - \frac{x}{\ell}\right]\left[2\left(\frac{x}{\ell}\right) - \left(\frac{x}{\ell}\right)^2\right]$$
$$= \frac{q\ell^4}{48EI}\left[3\left(\frac{x}{\ell}\right)^2 - 5\left(\frac{x}{\ell}\right)^3 + 2\left(\frac{x}{\ell}\right)^4\right]$$

7.4. A fixed–fixed beam has the loading distribution of Example 7.2; that is, the span that was simply supported for load factor purposes now has both ends clamped. Find the moment constraint at each end.

Solution: Letting the left moment be M_1 and the right M_2, we apply Eq. (7.9) to the 0-1-2 spans, and then Eq. (7.3) to the 1-2-3 spans:

$$\begin{cases} 2M_1 + M_2 = 428,250 \\ M_1 + 2M_2 = 404,250 \end{cases}$$

Resulting moments are $M_1 = 150,750$ and $M_2 = 126,750$.

7.5. A continuous beam with different span stiffness carries an end couple M_0. Determine the complete bending-moment diagram for the beam.

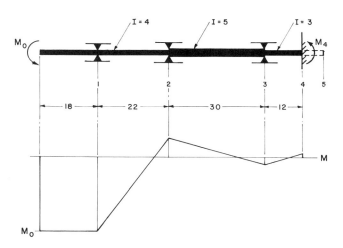

Solution: Applying the fundamental Three-Moment Equation (7.3),

1-2-3
$$\alpha = \left(\tfrac{30}{22}\right)\left(\tfrac{4}{5}\right) = 1.09 \qquad M_0 = M_1$$
$$M_0 + 2(1 + 1.09)M_2 + 1.09M_3 = 0$$
$$M_0 + 4.18M_2 + 1.09M_3 = 0 \tag{a}$$

2-3-4
$$\alpha = \left(\tfrac{12}{30}\right)\left(\tfrac{5}{3}\right) = 0.67$$

$$M_2 + 2(1 + 0.67)\,M_3 + 0.67M_4 = 0$$

$$M_2 + 3.33M_3 + 0.67M_4 = 0 \qquad\qquad\text{(b)}$$

3-4-5
$$\alpha = \left(\tfrac{0}{12}\right)\left(\tfrac{3}{\infty}\right) = 0$$

$$M_3 + 2M_4 = 0 \qquad\qquad\text{(c)}$$

Solving Eqs. (a), (b), and (c) simultaneously,

$$M_2 = -0.26M_0$$
$$M_3 = +0.09M_0$$
$$M_4 = -0.04M_0$$

In the moment diagram these senses are reversed, as the constraint moments were assumed negative in the derivation. Thus positive results indicate negative bending moment. Note also the decrease in the moment constraints as they become more remote from M_0.

8

Deflections of the
Planar Bent

ALTHOUGH THE STRAIGHT BEAM is often the only type analyzed in treatises concerned with strength and elastic behavior, there are many structures that are more complex geometrically. In this chapter we investigate the cases in which the system consists of beam elements subjected primarily to bending, but with the beams combined in a continuous manner. Typically, these are joined in a mutually perpendicular arrangement; however, the relative geometric orientation can involve any angle. At this point, we limit the scope of the discussion to the deflection of planar determinate systems, with indeterminate and out-of-plane situations analyzed in later chapters.

8.1 *Angular Transition of Loading*

The simple bent (Fig. 8.1a) carries an end load at 2, requiring an opposing force and a couple at 0. Within the bent we must also have equilibrium of the individual beam elements. As indicated in Fig. 8.1b, the bending moment Qb reacts from b to a. Also the transverse shear load Q on b becomes a direct compressive load on a at the corner; that is, with the perpendicular relation of the beams, the force reaction between them is also rotated by 90°, or transformed from a transverse to an axial sense in the transmission of load from b to a. Bending moment, however, is transmitted in the conventional manner (Fig. 8.1b).

In Fig. 8.1c corresponding bending-moment diagrams are constructed, with each leg treated as an ordinary beam. Sign convention in plotting the

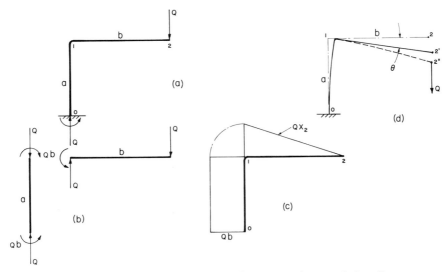

Figure 8.1 The basic cantilevered bent has a continuous b bending moment diagram developed from the beam elements. End deflection results from flexure of a and b (d).

diagrams is optional, as the relative sense of the bending between the actual and auxiliary moments is the only concern. Usually, it is most convenient to draw the diagrams externally rather than internally to prevent overlap, as shown.

Normally, the moment is transmitted unchanged at a corner, as indicated by the circular arc. If an external couple is applied at a corner, however, the continuity is altered.

8.2 Deflection of the Load

For the vertical displacement y_2, we again utilize the conservation of energy during the loading [Eq. (3.11)]:

$$\frac{Q_2 y_2}{2} = \int_0^\ell \frac{M^2}{2EI} \, ds = \int_0^b \frac{(Qx_2)^2}{2EI_b} \, dx_2 + \int_0^a \frac{(Qb)^2}{2EI_a} \, dx_1 \qquad (8.1)$$

where ds replaces dx in the general integral to accommodate the various

coordinate directions:
Solving,

$$y_2 = \frac{Q}{EI}\left[\frac{b^3}{3} + ab^2\right] = \frac{Qb^3}{3EI}\left[1 + 3\left(\frac{a}{b}\right)\right] \qquad (8.2)$$

Energywise the first integral in Eq. (8.1) is associated with the deflection caused by the flexure of b, with the resulting cantilever term. Energy stored in a produces the second term. In Fig. 8.1d we see the end moment on the vertical cantilever a resulting in the angle θ at the corner, or a tilt of b corresponding to a deflection to 2′ of ($b\theta$). If I_b were infinite, we would then only have the second term. Superimposed cantilever deflection in b adds the further deflection to 2″ because of a finite I_b. Thus from Eqs. (8.1) and (8.2), we can determine the contribution of each leg to the total displacement, and design alternatives can be assessed in terms of the length and inertia parameters as they bear on the deflection characteristic.

8.3 *Coupled Linear Deflections*

In addition to the vertical deflection y_2, Q produces a horizontal deflection of x_2 and a slope of the tangent to the elastic curve at 2. We find the former through complementary energy by applying a horizontal auxiliary load F

Figure 8.2 Auxiliary horizontal load F produces bending in a only.

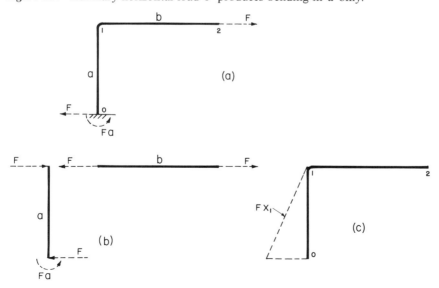

(Fig. 8.2). Since auxiliary moment only occurs in a, there is no contribution to the horizontal deflection by b. We plot m to the left of a to agree with the sense of M in Fig. 8.1c, as both produce compression on the inner or right fibers of a.

Applying Eq. (4.4c) and Table 5.1a,

$$Fx_2 = \int_0^\ell \frac{mM}{EI}\, ds = \int_0^\ell \frac{(Fx_1)M}{EI_a}\, dx_1$$

$$x_2 = \frac{1}{EI}\sum_0^1 A\bar{x}_1 = \frac{1}{EI}(Qb)\left(\frac{a^2}{2}\right) = \frac{Q}{2EI}(ba^2) \tag{8.3}$$

where the positive result confirms the horizontal displacement of 2 due to the vertical load.

8.4 Reciprocity

If we interchange the Q and F loadings, the effect is to interchange M and m, and if $Q = F$, or if these are unit loads, the mM integral in Eq. (8.3) is unchanged. This reconfirms the reciprocity relations (Sec. 4.7) in perpendicular directions:

$$y_{Qx} = x_{Qy} \tag{8.4}$$

where y_{Qx} = coupled vertical deflection produced by Q in the horizontal direction

x_{Qy} = coupled horizontal deflection produced by Q in a vertical direction

Reciprocal relations, incidentally, also exist relative to any two angular directions at 2, if they are not perpendicular. They also can be extended to forces and displacements at any two points on the system in any two selected directions. In addition, there is also reciprocity between load–slope and couple–displacement combinations (Sec. 4.7).

8.5 Coupled Slope Deflection

We determine the slope to the tangent to the elastic curve θ_2 by means of a couple M' that develops a constant auxiliary moment throughout the bent (Fig. 8.3):

$$M'\theta_2 = \int_0^\ell \frac{mM}{EI}\, ds = \frac{M'}{EI}\sum A$$

$$\theta_2 = \frac{1}{EI}\left[\frac{Qb^2}{2} + Qb(a)\right] = \frac{Qb^2}{2EI}\left[1 + 2\left(\frac{a}{b}\right)\right] \tag{8.5}$$

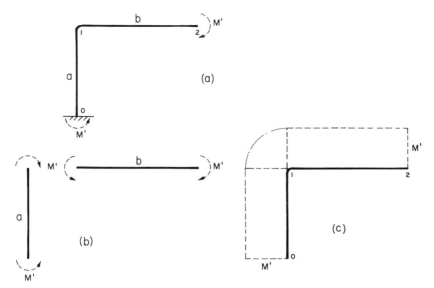

Figure 8.3 Auxiliary couple for end-slope calculation develops constant bending moment throughout the length.

Table 8.1 Compliance factors for the deflection of the constant cantilevered bent, U-bar to ground, and simple U-bar vary with the a/b ratio. In (c) displacements are relative between the ends.

		(a)	(b), (c)
α_x	$\dfrac{x}{Q_x}$	$\dfrac{a^3}{3EI}$	$\dfrac{a^2 b}{EI}\left[1 + \tfrac{2}{3}\left(\tfrac{a}{b}\right)\right]$
$\alpha_{xy}\ \dfrac{x}{Q_y}$ $\quad\alpha_{yx}\ \dfrac{y}{Q_x}$		$\dfrac{a^2 b}{2EI}$	$-\dfrac{a b^2}{2EI}\left[1 + \tfrac{a}{b}\right]$
α_y	$\dfrac{y}{Q_y}$	$\dfrac{b^3}{EI}\left[\tfrac{1}{3} + \tfrac{a}{b}\right]$	
β_x	$\dfrac{\theta}{Q_x}\quad\dfrac{x}{M}$	$\dfrac{a^2}{2EI}$	$\dfrac{ab}{EI}\left[1 + \tfrac{a}{b}\right]$
β_y	$\dfrac{\theta}{Q_y}\quad\dfrac{y}{M}$	$\dfrac{b^2}{EI}\left[\tfrac{1}{2} + \tfrac{a}{b}\right]$	$-\dfrac{b^2}{EI}\left[\tfrac{1}{2} + \tfrac{a}{b}\right]$
γ	$\dfrac{\theta}{M}$	$\dfrac{b}{EI}\left[1 + \tfrac{a}{b}\right]$	$\dfrac{b}{EI}\left[1 + 2\left(\tfrac{a}{b}\right)\right]$

The slope will be clockwise as assumed, with the contributions of each leg proportional to the respective terms. If the loading is in turn a couple M, a similar procedure is used.

Table 8.1 summarizes the characteristics of the simple grounded bent in terms of the influence ratios, or compliances, with respect to the end point. The grounded and ungrounded U-beams of constant section are also analyzed in Table 8.1.

8.6 *Extraneous Deflection Components*

The straight beam deflects primarily because of flexure, but can also have contributions from transverse elasticity at the supports and from shear effects (Sec. 4.9). Support flexibility is generally equivalent to the tension–compression elements of Chap. 1. With a bent, however, an additional component can be present as the individual beam elements are subjected to tensile or compressive stresses.

For instance, in Fig. 8.1b the vertical leg carries an axial load Q. Shortening of a tends to increase the total deflection of Q. Thus Eq. (8.1) becomes

$$\frac{Q_2 y_2}{2} = \int_0^\ell \frac{M^2}{2EI}\, ds + \sum \frac{C_i Q_i^2}{2} \tag{8.6}$$

Similarly, if both auxiliary and actual loadings produce axial forces in the same element, the deflection is given by

$$F\delta_F = \int_0^\ell \frac{mM}{EI}\, ds + \sum C_i F_i Q_i \tag{8.7}$$

Typically all other such effects are negligible with respect to the flexural. Both energies and the related deflection contributions are usually neglected without introducing errors of any practical significance.

8.7 *Principal Axes*

The subject of two mutually perpendicular uncoupled directions that exist for any point in an elastic structure was introduced in Sec. 1.13. Considering now the most fundamental example of the bent (Table 8.1a), we can illustrate this feature more fully. Taking the directional [Eq. (1.25)] and substituting from Table 8.1*a*,

$$\tan 2\theta = \frac{2C_{xy}}{C_x - C_y} = \frac{2\left(\dfrac{a^2 b}{2EI}\right)}{\dfrac{a^3}{3EI} - \dfrac{b^3}{EI}\left(\dfrac{1}{3} + \dfrac{a}{b}\right)} = \frac{3\left(\dfrac{a}{b}\right)^2}{\left(\dfrac{a}{b}\right)^3 - 3\left(\dfrac{a}{b}\right) - 1} \quad (8.8)$$

In Fig. 8.4 the bent has principal axes *p–p* and *q–q*, representing directions of maximum and minimum uncoupled compliances. Thus Eq. (8.8) defines four values of θ in the plane of the bent, with two complementary angles plotted against (a/b). The 45° condition occurs at $(a/b) = 1.88$,

Figure 8.4 Principal axes occur at θ_q and θ_p, respectively. Corresponding compliances are C_q and C_p.

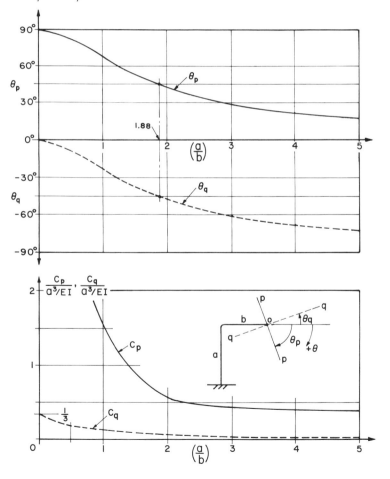

where this is the root of the cubic denominator in Eq. (8.8). A negative angle is counterclockwise as indicated.

Substituting values from Table 8.1a in Eqs. (1.24), we have the component compliances:

$$C_x = \frac{x}{Q} = \frac{a^3}{EI}\left[\frac{1}{3}\cos\theta + \frac{1}{2}\left(\frac{b}{a}\right)\sin\theta\right] \tag{8.9a}$$

$$C_y = \frac{y}{Q} = \frac{1}{EI}\left[b^3\left(\frac{1}{3} + \frac{a}{b}\right)\sin\theta + \frac{1}{2}\left(\frac{b}{a}\right)\cos\theta\right] \tag{8.9b}$$

Using the principal angular coordinates θ_p and θ_q, and calculating the vector resultants,

$$C_p = \frac{\delta p}{Q_p} = \sqrt{(C_x)_p^2 + (C_y)_p^2} \tag{8.10a}$$

$$C_q = \frac{\delta q}{Q_q} = \sqrt{(C_x)_q^2 + (C_y)_q^2} \tag{8.10b}$$

we have characteristics for the two-leg cantilevered bent in Fig. 8.4, referenced to a.

Maximum compliance C_p increases rapidly as (a/b) approaches 0 due to the increase in the length of b, corresponding to increased flexibility. Minimum compliance C_q is relatively small, indicating the much greater stiffness along the q–q axis. The $\frac{1}{3}$ factor at $(a/b) = 0$ signifies a limiting cantilever condition.

As (a/b) increases from 0, we observe the p and q axes rotate counterclockwise from $\theta_q = 0$, directed along b. In the limit, with (a/b) approaching infinity, the axes become interchanged, with the p–p axis eventually aligned with b.

8.8 *Zero-Slope Combinations*

The simple bent with an end force directed along a principal axis will deflect in a coincident direction, as just indicated; however, the terminus of the elastic curve (Fig. 8.5) will *rotate* during this type of displacement, as it will due to any other direction of end load. From Table 8.1a the resulting slope is

$$\begin{aligned}
\theta_2 &= \frac{a^2}{2EI}Q_x + \frac{b^2}{EI}\left(\frac{1}{2} + \frac{a}{b}\right)Q_y \\
&= \frac{Qa^2}{EI}\left[\frac{1}{2}\cos\theta + \left(\frac{b}{a}\right)^2\left(\frac{1}{2} + \frac{a}{b}\right)\sin\theta\right]
\end{aligned} \tag{8.11}$$

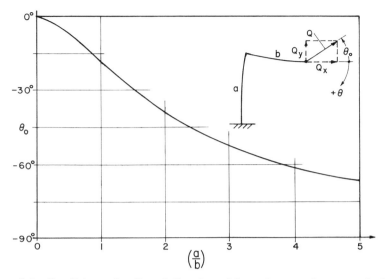

Figure 8.5 Conditions of a/b and direction of force for zero slope at end of bent.

Since the slope is a function of the (a/b) ratio and the direction of the applied force, we determine the condition of zero slope by equating θ_2 to 0:

$$\tan \theta_0 = \frac{-\frac{1}{2}\left(\frac{a}{b}\right)^2}{\left(\frac{1}{2} + \frac{a}{b}\right)} \tag{8.12}$$

8.9 Ungrounded Bents

The simple U-bar (Table 8.1c) or the bent bar (Fig. 8.6a) subjected to a force or couple, can only be reacted for equilibrium by an equal and opposite force or couple. With forces, the second external load must be colinear with the first. Then the deflections at the load points become *relative* and aligned with the forces. In Fig. 8.6 we obtain δ_{14} by equating $Q\delta_{14}/2$ to the M^2 energy integral. The M diagram is shown in Fig. 8.6b.

Only the total displacement occurring between 1 and 4 is obtained, or the net compliance of the structure. If 1 is in fact grounded to have zero displacement, δ_{14} would become δ_4 to the right and vice versa.

As we assume negligible tensile–compressive deflections, the lengths of a, b, and c are constant. This creates a slight ambiguity in a, which is shown shortened so that the Q loads are coaxial after deflection. More

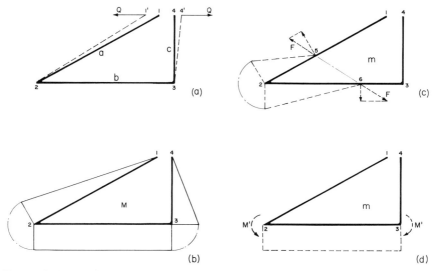

Figure 8.6 Deflection of angular bent derives from geometrically developed moment diagrams.

accurately, with the loading as the directional reference, there would be a slight counterclockwise rotation of the bent to accommodate this condition.

Similarly, the displacement between any two points (say δ_{56}) is obtained by applying twin F loading (Fig. 8.6c) and using the $F\delta_{56}$ energy relationship. Again the F loads must be equal and opposite for equilibrium, and the two chosen points move toward or away from each other as analyzed.

Relative slopes are obtained from opposite couples M'. In Fig. 8.6d the solution for this auxiliary loading yields θ_{23}, or the relative slope between the ends of b. This beam, incidentally, is subjected to constant couple loading Qc, with a flexural curve as given in Table 4.5 for M.

8.10 Internal Deflection Distribution

As indicated, the elements that compose a bent are individual straight beams. They carry moments and axial or shear loads at each end, and can also carry concentrated or distributed external transverse loads within a span. The complete elastic curve can thus be developed by isolating a beam element, treating the ends as simply supported to a reference ground axis connecting the two ends, and solving for the displacement coordinates by energy techniques. Also, basic beam deflection characteristics given in Chaps. 4 and 5 can be used to obtain this type of information by superposition. Construction of the complete system deflection curve then involves joining the elements with common slope and displacement coordinates at connected ends.

8.11 *Symmetry*

Figure 8.7 shows an ungrounded rectangular bent with the flexural deflections produced by separating forces Q at the top center edges. Although these forces could be applied *externally*, they can also be applied *internally*, maintaining equilibrium without any ground reactions. For internal horizontal loading (Fig. 8.7a), a wedge can be used, and for shear loading (Fig. 8.7b), a wedge or block can be used as shown. Loads and bending energies are thus locked into the system, regardless of any subsequent motion of the bent.

Neglecting extensional effects in the beams, all lengths are constant. In Fig. 8.7a, 2 and 5 therefore rotate about 3 and 4, respectively, with 3 and 4 a fixed distance 2a apart. In Fig. 8.7b with the corners 2, 3, 4, and 5 maintaining the basic rectangular geometry, it will be shown that x_{16} is 0.

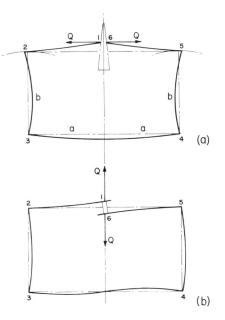

Figure 8.7 Deflection characteristics of rectangular bent due to loading at the juncture are symmetrical (*a*) and unsymmetrical (*b*).

Comparing the general characteristics, Fig. 8.7*a* is *symmetrical* about the vertical centerline; however, Fig. 8.7*b* represents *antisymmetrical* behavior. With the latter the two halves behave similarly, but one is reversed signwise from the other.

We also observe that corners 2, 3, 4, and 5 remain square after flexure; that is, tangents to the elastic curve remain mutually perpendicular at the corners during deformation. This is another manifestation of the several continuities that characterize elastic curves.

8.12 *Separating Loads*

Evaluating the horizontal separation compliance (Fig. 8.8*a*), free-body diagrams of the elements (Fig. 8.8*b*) lead to a symmetrical *M* diagram (Fig. 8.8*c*). From the M^2 integral we have the same result for the relative displacement x_{16} as in Table 8.1*c* for the U-bar.

Points 1 and 6 tend to rise relative to the structure due to flexure in the three lower beams. To provide a reference axis, we take a horizontal base

Figure 8.8 Horizontal separation causes actual and auxiliary moment diagrams that are symmetrical.

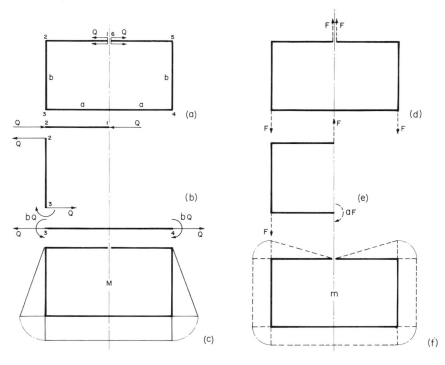

line through the corners at 3 and 4. The corresponding auxiliary loading (Fig. 8.8*d*) requires two pairs of *F* forces to maintain symmetry:

$$Fy_1 + Fy_3 + Fy_4 + Fy_6 = \int_0^\ell \frac{mM}{EI} \, ds \qquad (8.13)$$

From symmetry, and with $y_3 = y_4 = 0$ due to the chosen reference,

$$Fy_1 = Fy_6 = \int_0^{\ell/2} \frac{mM}{EI} \, ds \qquad (8.14)$$

where both the *m* and *M* diagrams (Fig. 8.8*c* and *f*) are symmetrical.

A horizontal reference axis could also be taken as a horizontal line 2–5, remaining coincident with these corners, or one remaining coincident with 1–6. In these cases, the *F* loads at 3 and 4 would be gravity weights acting vertically. All displacements with respect to the reference axes are *relative*; however, we may also visualize these as *absolute*, if the axes are grounded.

8.13 Shear Loads

Shear loading is indicated in Fig. 8.9*a*, with the moment diagram in Fig. 8.9*b*. Both the *M* diagram and the deflection distribution are *antisymmetrical*. The equation at the gap is

$$\frac{Qy_{16}}{2} = 2 \int_0^\ell \frac{M^2}{2EI} \, ds \qquad (8.15)$$

Figure 8.9 Actual moment diagram due to shear loading is antisymmetrical.

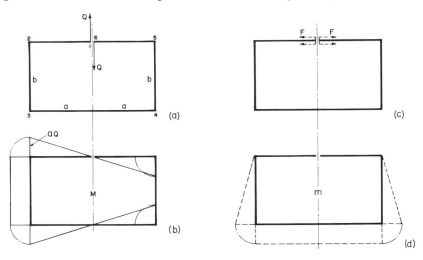

Strain energy determining the total compliance is distributed in all beam elements:

$$\frac{Qy_{16}}{2} = 4 \int_0^a \frac{M^2}{2EI} \, ds + 2 \int_0^b \frac{M^2}{2EI} \, ds \tag{8.16}$$

Alternatively, we can revert to the mM integral form by replacing the Q loads by F loads (Fig. 8.9a). This permits use of the integration factors (Table 5.1). Then

$$Fy_{16} = 4 \int_0^a \frac{(Fx_1) M}{EI} \, ds + 2 \int_0^b \frac{(Fa) M}{EI} \, ds$$

$$y_{16} = \frac{4}{EI} \sum_0^a Ax_1 + \frac{2a}{EI} \sum_0^b A$$

$$= \frac{Qa^3}{EI} \left[\frac{4}{3} + 2 \left(\frac{b}{a} \right) \right] \tag{8.17}$$

To determine the coupled horizontal effect, auxiliary loads are applied (Fig. 8.9c), producing a symmetrical m diagram (Fig. 8.9d). The product

$$\int_0^{\ell/2} \frac{mM}{EI} \, ds \neq 0$$

whether we consider the right or left half of the bent. Because m is symmetrical and M is antisymmetrical, however, the mM integrals are equal but of opposite sign. For the entire system,

$$Fx_{16} = \int_0^{\ell/2} \frac{mM}{EI} \, ds - \int_0^{\ell/2} \frac{mM}{EI} \, ds = 0 \tag{8.18}$$

We conclude from these complementary energy considerations that there is no horizontal change in the gap due to shear loading.

Examples

8.1. The rectangular bent has constant section and is loaded to separate the legs at 0 and 5. Determine the relative deflections due to bending:
   (a) Between 0 and 5
   (b) Between 1 and 4

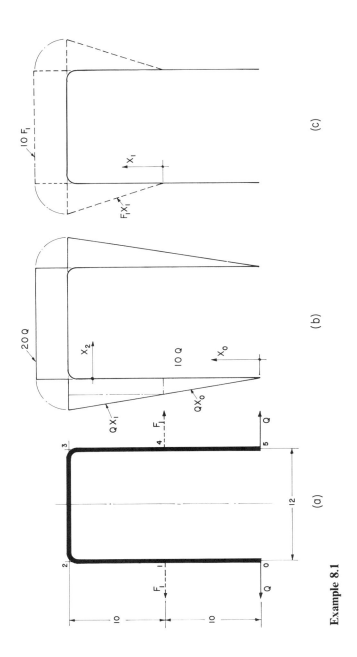

Example 8.1

Solution:

(a) The actual moment distribution is shown in (b) and is constant at the top:

$$\frac{Qx_{05}}{2} = \int_0^\ell \frac{M^2}{2EI}\,dx = \frac{1}{EI}\sum_0^{20}(Qx_0)^2\,dx_0 + \frac{1}{2EI}\sum_0^{12}(20Q)^2\,dx_2$$

$$x_{05} = \frac{Q}{EI}\left[\frac{16,000}{3} + 400(12)\right] = \frac{Q}{EI}(5333 + 4800)$$

$$= \frac{Q}{EI}10,130$$

Slightly more than one half the compliance is related to the bending of the legs.

(b) Applying auxiliary loading at the midpoints (*c*), we have

$$F_1 x_{14} = \int_0^\ell \frac{mM}{EI}\,dx = \frac{1}{EI}\left[2\int_0^{10}F_1 x_1 M\,dx_1 = \int_0^{12}(10F_1)M\,dx_2\right]$$

$$x_{14} = \frac{2}{EI}\left[\sum_0^{10}A\bar{x}_1 + \frac{10}{2}\sum_0^6 A\right]$$

$$= \frac{2}{EI}\left[10Q(10)(5) + Q(10)^3/3 + 20Q(10)(6)\right]$$

$$= \frac{Q}{EI}[1000 + 667 + 2400] = 4067\frac{Q}{EI}$$

8.2. Repeat Example 8.1 if the legs are rigid relative to the center span.

Solution:

(a) The moment diagrams remain unchanged, but there is no strain energy in the legs and we reduce to the second term only:

$$x_{05} = 4800\frac{Q}{EI}$$

(b) Similarly, we also now have only the last term:

$$x_{14} = 2400\frac{Q}{EI}$$

8.3. Repeat Example 8.1, but with the center span rigid.

Solution:

(a) Eliminating the 2–3 term,

$$x_{05} = 5333\frac{Q}{EI}$$

(b)
$$x_1 = 1667\frac{Q}{EI}$$

By adding one flexibility condition to the other we obtain the total actual deflections. This is another form of *superposition*, as we successively relax the elastic structure from its rigid state.

8.4. For the Example 8.1 bent, find the relative slope between the two ends.

Solution: Applying opposed auxiliary couples M' in a positive sense at 0 and 5 produces a constant M' moment distribution for the auxiliary loading. This carries throughout the length:

$$M'\theta_{05} = \frac{M'}{EI} \int_0^\ell M\, ds$$

$$\theta_{05} = \frac{1}{EI} \sum A = \frac{2}{EI}\left[\frac{Q(20)^2}{2} + (20Q)6\right]$$

$$= \frac{Q}{EI}(400 + 240) = 640\frac{Q}{EI}$$

8.5. The bent planar cantilever is loaded by a clockwise couple at the end. Find the induced vertical deflection at 2.

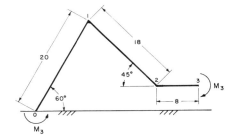

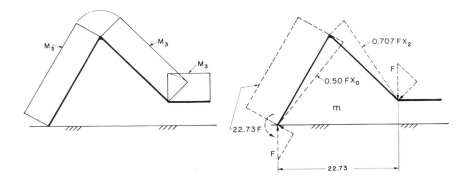

Solution: We apply F vertically at 2 and satisfy equilibrium with an opposed reaction and moment at the base. Constructing the auxiliary moment diagram requires resolving the F loads into transverse components. We proceed from 2 to 1 and from 0 to 1:

$$Fy_2 = \int_0^\ell \frac{M_3 m}{EI}\, ds = \frac{M_3}{EI} \int_0^\ell m\, ds = \frac{M_3}{EI} \sum A_m$$

$$= \frac{M_3}{EI} \left[\frac{0.707F(18)^2}{2} + 22.73F(20) - \frac{0.50F(20)^2}{2} \right]$$

$$y_2 = (114.5 + 454.6 - 100) = 469\frac{M_3}{EI}$$

Since the actual moment diagram is more simple than the auxiliary diagram, we reverse the usual procedure and factor out the M_3 term. Then we evaluate the total area under the m diagram.

8.6. A constant section bent is pivotally supported at 0, simply supported at 3, and loaded triangularly. Calculate the vertical deflection at the horizontal mid-point, 1.

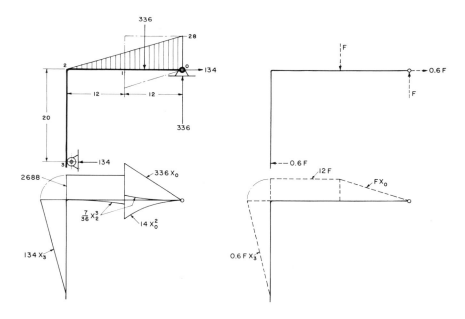

Solution: Considering the nature of the auxiliary diagram caused by F at 1, we note the origins at 0 and 3. It is therefore desirable to involve the same origins in the M diagram in order to apply the Table 5.1 factors without complication. This suggests progression from 0 to 3 with a break point at 1. We also resolve the trapezoidal loading in the 0–1 interval into an equivalent constant distributed loading downward and a triangular loading upward as indicated:

$$Fy_2 = \frac{1}{EI}\left[\int_0^1 (Fx_0)\,M\,dx_0 + \int_3^2 (0.6Fx_3)\,M\,dx_3 + \int_2^1 (12F)\,M\,dx_2\right]$$

$$y_2 = \frac{1}{EI}\left[\sum_0^{12} A\bar{x}_0 + 0.6\sum_0^{20} A\bar{x}_3 + 12\sum_0^{12} A\right]$$

$$= \frac{1}{EI}\left[\frac{336(12)^3}{3} + \frac{14(12)^4}{30} - \frac{28(12)^4}{8} + \frac{0.6(134.4)(20)^3}{3}\right.$$

$$\left. + 12(2688)(12) - \frac{12(14)(12)^3}{24}\right]$$

$$= \frac{1}{EI}[193{,}540 + 9680 - 72{,}580 + 215{,}040 + 387{,}070 - 12{,}100]$$

$$= \frac{1}{EI}(721{,}000)$$

9

Indeterminate Bents

CONCEPTS OF BENT ANALYSIS are now extended to indeterminate structures. The methods developed in Chaps. 6 and 7 are applicable, and using the principle of deflection restoration, we can also employ material from Chap. 8. Systems now analyzed include those with two or more grounded supports and indeterminate continuous or ungrounded cases composed of elementary straight-beam components. In two-dimensional geometry, however, the possible redundancies at a fixed end increase from two for the beam to three for the bent. These include the perpendicular force components and a couple by virtue of the constraint.

The basic approach to the solutions involves auxiliary loading, with complementary energies equated to 0 since the auxiliary loads at the reactions do not displace during the application of the actual loading. The Three-Moment Theorem can be applied advantageously to many redundant bent situations, but there are certain systems in which this method is not legitimate. Necessary tests will be developed.

9.1 Quantifying Bent Redundancies

The simple bent with two potential end support points, 0 and 3 (Fig. 9.1), has many alternative end constraints by means of which the load Q can be carried. In fact, there are 10 distinct possibilities, with statically determinate cases shown in Fig. 9.1a–d. In Fig. 9.1a and b there is a combination of pin and roller, and in Fig. 9.1c and d there is cantilever support.

One redundancy is introduced in Fig. 9.1e by converting the roller at 3 in Fig. 9.1a to a grounded pin, preventing horizontal displacement. Simi-

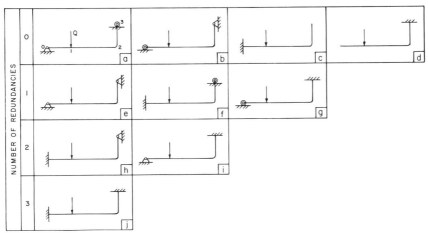

Figure 9.1 The basic bent with supports at the two ends has 10 potential combinations of support constraint.

larly, cantilevers in Fig. 9.1*c* and *d* now have vertical support assistance in Fig. 9.1*f* and *g*; that is, end deflection is prevented. Two redundancies exist in cantilevered cases (Fig. 9.1*h* and *i*), with the pinned ends providing both horizontal and vertical end-deflection constraints. Finally, the ultimate fixation occurs with both ends clamped (Fig. 9.1*j*). There are now three displacement constraints added at what was initially a free end of a cantilevered bent.

These basic examples are provided to indicate the numerous alternatives that exist for indeterminate combinations, even for a simple structure. They also illustrate how the degree of redundancy is determined by deflection constraints imposed successively upon the minimum, or statically determinate condition. A redundant force or couple is associated with each additional constraint, linear or rotational.

9.2 Once-Redundant Bent

Taking the pinned–pinned bent (Fig. 9.1*e*), we use the fundamental auxiliary approach as for beams. Reactions at the pins (Fig. 9.2*a*) include vertical and horizontal components. From statics

$$\sum F_x = H_0 - H_3 = 0 \tag{9.1a}$$

$$\sum F_y = V_0 + V_3 - Q = 0 \tag{9.1b}$$

$$\sum M_2 = 30V_0 + 10H_3 - 20Q = 0 \tag{9.1c}$$

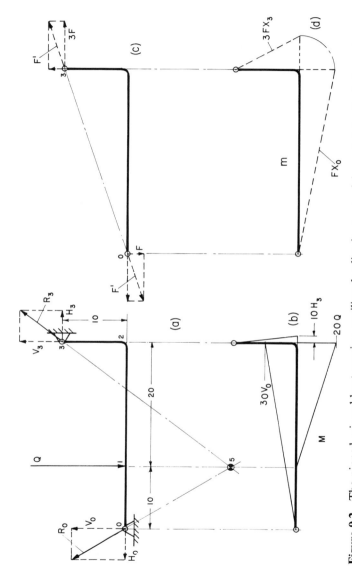

Figure 9.2 The pinned–pinned bent requires auxiliary loading by opposed forces (c) for the indeterminate solution.

In addition, we need one supplementary equation from elasticity to solve for the four unknown reactions. Incidentally, also from statics, the three external forces on the bent must be *concurrent* at a single point, about which $M = 0$. As shown (Fig. 9.2a), the line of action for R_0 and R_3 must intersect on the directional of Q. While there are limitless points that satisfy statics, related to the indeterminacy, elastic behavior determines the unique distance of the intersection 5 below the line 0–2. This, in turn, depends on the relative lengths of the legs and the location of Q.

Applying auxiliary loading at the pins (Fig. 9.2c), the forces F' must be equal and opposite and *colinear* for equilibrium. Resolving into components for purposes of the m diagram (Fig. 9.2d), we obtain F_{3x} and F_{0y}. Since the magnitude of F' is arbitrary, we simplify the calculation by taking $F_{0y} = F$, and from the geometry, $F_{3x} = 3F$. With the pins 0 and 3 locked after the required separation to induce F', we have

$$0 = \int_0^\ell \frac{mM}{EI}\, ds = \frac{1}{EI} \int_0^{30} (Fx_0) M\, ds + \int_0^{10} (3Fx_3) M\, ds$$

$$0 = \sum_0^{30} A\bar{x}_0 + 3\sum_0^{10} A\bar{x}_3 = -90V_0 + 10H_3 + 46.67Q \qquad (9.1d)$$

Solving Eqs. (9.1c) and (9.1d) simultaneously,

$$V_0 = 0.555Q$$

and from statics

$$V_3 = 0.444Q \qquad H_0 = H_3 = 0.333Q$$

These place the concurrent point 5 a distance 16.67 below the bent, and the solution is complete with $M_1 = 5.55Q$ and $M_2 = 3.33Q$.

9.3 *Rectifying the Moment Diagram*

The moment diagrams (Fig. 9.2b) are constructed using components and plotted on the bent as a base, requiring a 90° rotation at 2. Although there is no avoiding the two-dimensional geometry basic to the force and moment analysis, it is often helpful to refer the moment distribution to a common straight axis. We rectify the bent (Fig. 9.3) by rotating the leg 2–3 from vertical to horizontal, making the base length of the M diagram the total length of the bent. The M triangle on 2–3 is similarly rotated. The resulting diagram is then continuous and in conventional beam form. Net moment between 1 and 2 is shown dashed.

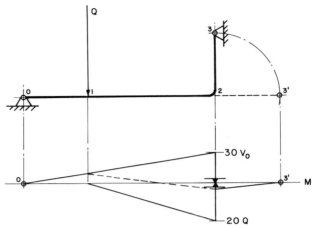

Figure 9.3 The pinned–pinned bent is rectified to an equivalent straight beam for constructing the moment diagram.

9.4 *Displacement Superposition*

The pinned–pinned bent (Fig. 9.2*a*) can be achieved physically by starting with the decoupled case (Fig. 9.1*a*) and allowing 3 to deflect freely horizontally. We then convert to a pin at 3 by applying a horizontal force H to return the pin to its original position (Fig. 9.4) as indicated in Sec. 6.11. Solution then reverts to compliance determinations.

In Fig. 9.4 we find x_Q for the Q loading (Fig. 9.4*a*) and the M diagram (Fig. 9.4*b*). Auxiliary loading corresponds to Fig. 9.2*c* and *d*. We anticipate deflection x_Q to the left, so we reverse the sense of F and m. Then

$$3Fx_Q = \int_0^\ell \frac{mM}{EI}\, ds = \frac{F}{EI} \sum_0^{30} Ax_0$$

$$x_Q = 444.4 \frac{Q}{EI}$$

Compliance at 3 with respect to the H loading relates to Fig. 9.4*c* and *d*:

$$\frac{Hx_H}{2} = \int_0^\ell \frac{M^2}{2EI}\, ds = \frac{1}{2EI}\left[\int_0^{30}\left(\frac{H}{3}x\right)^2 dx + \int_0^{10}(Hx)^2\, dx \right]$$

$$x_H = 1333.3 \frac{H}{EI}$$

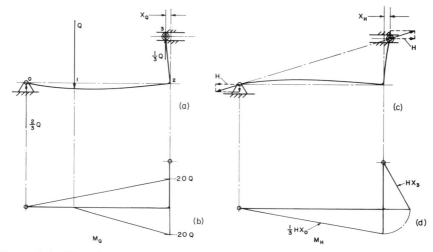

Figure 9.4 The pinned–pinned bent can be analyzed by horizontal displacement superposition at 3.

Equating the absolute displacements,

$$x_Q = x_H = 444.4 \frac{Q}{EI} = 1333.3 \frac{H}{EI}$$

$$H = \tfrac{1}{3} Q$$

9.5 *Three-Moment Applicability*

A fixed–fixed bent (Fig. 9.5) involves three redundancies and can be solved by energy methods. This requires three auxiliary loads at one end, including F_x, F_y, and M', to establish the three auxiliary moment diagrams. Although not difficult technically, the solution becomes somewhat cumbersome and susceptible to numerical and sign errors. We now illustrate the methods of Chap. 7 applied to the rectified bent.

In Fig. 7.1a the Three-Moment Theorem is developed for the continuous beam on fixed simple supports. As indicated in the proof (Fig. 7.2a), if there is no transverse deflection at the supports, work done by the F loading is 0 during the actual loading.

Returning to the bent (Fig. 9.5), the corner at 2 has zero horizontal and zero vertical deflection, neglecting direct axial strain. Thus the rectified bent is effectively a straight beam fixed at 0 and 3′ and simply supported at 2. The Three-Moment Theorem can then be applied, as *all junctures of the beam elements have zero linear displacement in any direction.* With this condition satisfied, any bent can be rectified as each junction becomes a

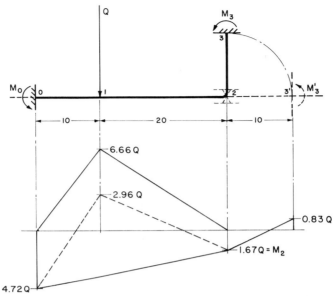

Figure 9.5 The fixed–fixed bent, if rectified to a straight beam with zero displacement at 2, can be solved by the Three-Moment Theorem with 0–2 and 2–3 as the actual spans.

simple support. In the case of a continuous bent, the theorem must include the continuity of bending moment at the first and last points; that is, even though the rectification results in a straight beam, the two extremities must again be considered joined, as in the bent. It is not necessary that successive beam elements be perpendicular to each other.

9.6 Three-Moment Solution

Analysis of the bent (Fig. 9.5) by the Three-Moment Theorem follows the techniques of Sec. 7.5. We write three equations, proceeding from left to right:

$$
\begin{cases}
2M_0 + M_2 = \frac{100}{9}Q & (9.2a) \\
M_0 + \frac{8}{3}M_2 + \frac{1}{3}M_3 = \frac{80}{9}Q & (9.2b) \\
M_2 + 2M_3 = 0 & (9.2c)
\end{cases}
$$

Solving for M_0 using determinants,

$$\frac{M_0}{Q} = \frac{\begin{vmatrix} 11.11 & 1 & 0 \\ 8.88 & 2.66 & 0.33 \\ 0 & 1 & 2 \end{vmatrix}}{\begin{vmatrix} 2 & 1 & 0 \\ 1 & 2.66 & 0.33 \\ 0 & 1 & 2 \end{vmatrix}} = \frac{59.26 - 21.48}{10.66 - 2.66} = 4.72 \qquad (9.3)$$

Similarly, $M_2 = 1.66Q$ and $M_3 = -0.83Q$, as seen in Fig. 9.5. With further application of statics equations to the beam components and to the bent,

$$H_3 = 0.25Q \qquad V_3 = 0.23Q$$
$$H_0 = 0.25Q \qquad V_0 = 0.77Q$$

9.7 Deflection Analysis

If we are interested in the deflection behavior of the bent (Fig. 9.5), we proceed beyond the necessary first phase of reaction solutions. Continuing to ignore direct effects, 0–2 has no horizontal deflection and 2–3 has no vertical deflection, and points 0, 2, and 3 are fixed in space. All displacements are caused by flexure, and slopes at 0 and 3 are equal to 0.

To find y_1 we could apply the $M^2/2EI$ integral, but this becomes unwieldy because of squaring multiple terms in the M diagram. As explained in Secs. 6.8–6.11, we greatly simplify the problem by introducing an F load at 1, and further *decouple the structure to simple static stability*. This can occur in several ways. Specifically, we can reduce to any of the cases in Fig. 9.1a–d, with F replacing Q. Figure 9.1a and b will have the same m diagram, but Fig. c and d is different. All three m diagrams are legitimate, although each produces quite different numerical combinations. The three possibilities are shown in Fig. 9.6a–c.

At this point we select the option resulting in the minimum complexity in the integration, particularly when also considering the related M diagram (Fig. 9.5). Least terms are involved if we choose Fig. 9.6c. Then we have

$$Fy_1 = \int_0^\ell \frac{mM}{EI}\,dx_1 = \frac{1}{EI}\int_0^\ell (Fx_1)M\,dx_1$$

$$y_1 = \frac{1}{EI}\sum A\bar{x}_1 \qquad (9.4)$$

There is a complication in the M diagram, but from the geometry we find the point at which it crosses the axis at $x_0 = 6.15$. Then numerically,

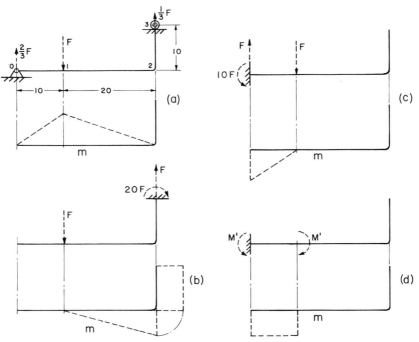

Figure 9.6 After the indeterminate solution, three decoupling arrangements are possible for the displacement solution at 1. The simplest preload for the slope at 1 is shown in (d).

using Table 5.1

$$y_1 = \frac{1}{EI}\left[4.72Q\left(\frac{6.15}{2}\right)\left(3.85 + \left(\tfrac{2}{3}\right)6.15\right) - 2.96Q\frac{(3.85)^2}{6}\right]$$

$$= \frac{1}{EI}\left[4.72Q\left(\frac{6.15}{2}\right)3.85 + 4.72Q\frac{(6.15)^2}{3} - 7.31\right] = \frac{108.1}{EI} \quad (9.5)$$

where we see the $\Sigma A\bar{x}_1$ of the larger triangle is the sum of the product of the area of the triangle times the offset of the apex plus the $A\bar{x}$ of the large triangle about its apex 6.

Similarly, the simplest diagram for m to determine the slope of the elastic curve at 1 is that resulting from decoupling to the cantilever support at 0 (Fig. 9.6d). Then we have only the area summation from 0 to 1:

$$M'\theta_1 = \int_0^\ell \frac{M'M}{EI}\,dx_1 = \frac{M'}{EI}\Sigma A$$

$$\theta_1 = \frac{8.82}{EI}Q \quad (9.6)$$

With the positive area exceeding the negative, our assumption of a clockwise sense for θ_1 is validated.

9.8   *The Continuous Bent*

A square structure of constant section (Fig. 9.7a) with a concentrated diagonal loading is only once redundant. This is determined by visualizing pin joints at 2 and 4, with $Q/2$ carried at each pin, and with static stability. The actual constraint or moment in the continuous corners, obviously equal at 2 and 4, becomes the single indeterminancy. This structure is *bisymmetrical*; that is, it is symmetrical about both centerlines 1–3 and 2–4.

To determine the actual bending-moment diagram, we take free-body diagrams for the two vertical halves (Fig. 9.7b). From symmetry, a shear of $Q/2$ is carried by each half at 1 and 3. There are no horizontal forces at these points because if there were a force H_1 on one half, there would be an opposite H_3 on the same half for equilibrium. But then the fibers at one corner would be in tension and at the other corner in compression horizontally, violating symmetry about the 2–4 axis; therefore, $H_1 = H_3 = 0$.

Using a slightly different argument, there are also no horizontal forces at the corners 2 and 4 (Fig. 9.7c). The upper half can carry opposite shears and

Figure 9.7   A continuous square bent can be separated vertically or horizontally for constructing the M' diagram. In (d) the internal preload travels the entire length as a constant moment.

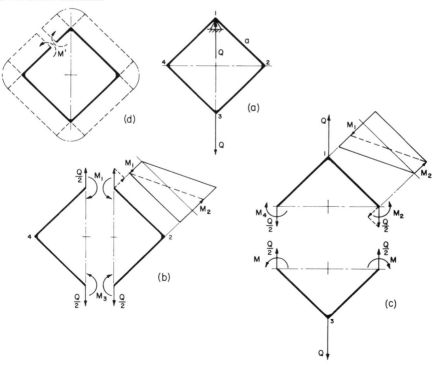

also satisfy symmetry about the vertical axis; however, they would have to be reacted oppositely by the lower half. With one half tending to open and the other tending to close due to these shears, behavior of one half would be different from the other and unsymmetrical. The vertical load is divided equally as shown.

Actual M diagrams follow the force considerations in Fig. 9.7b. Working from 1 with M_1 at the end, we have the transverse component of $Q/2$ causing a positive triangle and M_1 a negative rectangle, leading to M_2 at 2. In Fig. 9.7c, progressing from 2 to 1, M_2 is the initial moment. Total moment in 1–2 is shown by the dashed lines, with the same result by either approach.

We determine the equal moments M_1 and M_3 by introducing an auxiliary couple M' at any point on the bent. Actually, the constant M' preload that travels the complete loop is caused by two opposite angular displacements applied after cutting the beam, and with the cut reunited by welding to lock in the preload (Fig. 9.7d). Taking the M diagram from Fig. 9.7b, we have

$$0 = \int_0^\ell \frac{M'M}{EI}\,ds = \frac{4}{EI}\sum_0^a A = \sum_0^a A$$

$$0.707\left(\frac{Q}{2}\right)\frac{a^2}{2} - M_1 a = 0$$

$$M_1 = 0.177Qa \tag{9.7}$$

Applying statics to 1–2 with $\sum M_2 = 0$, $M_2 = 0.177Qa$. Thus bending moments and stresses are equal and maximum at the four corners, but with alternately reversed sense.

Considering axial effects in the four sides, all carry a tensile load of $(Q/2)\cos 45° = 0.354Q$ and this results in deflection; however, this has no bearing on the results obtained based upon bending energy. The auxiliary moment M' (Fig. 9.7d) produces no axial preload, with the CFQ products all 0.

This problem cannot be solved by the Three-Moment Theorem, as 2 and 4 deflect in space, as discussed in Sec. 9.5. Its attempted use leads to erroneous results.

9.9   Deflections of the Continuous Bent

Again with Fig. 9.7, we now determine the change in the vertical diagonal due to load. Symmetrical decoupling can be achieved by assuming pins at 1 and 3 or 2 and 4 (Fig. 9.8a and b). In fact, the m distributions are bisymmetrical, resulting in identical energy terms in all four legs, and with

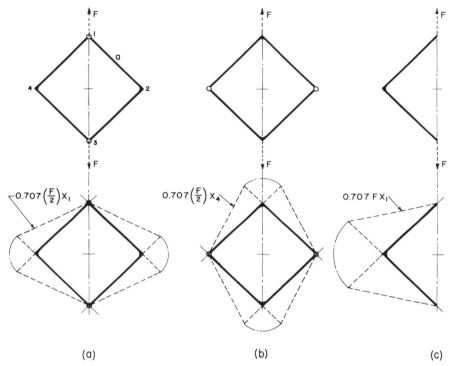

Figure 9.8 Deflection solution for the square bent can be obtained by any of three decoupling schemes.

F divided equally relative to the vertical centerline. We have, with Fig. 9.8a or b,

$$Fy_3 = Fy_{13} = \int_0^\ell \frac{mM}{EI}\, ds = \frac{4(0.707)}{2EI} \int_0^a xM\, dx$$

$$y_3 = \frac{1.414}{EI} \sum_0^a A\bar{x} = 0.042 \frac{Qa^3}{EI} \tag{9.8}$$

where x can be x_1 or x_4.

In Fig. 9.8c we simply remove the right half for preload purposes, with m equal to 0 in this half during actual loading. Then

$$Fy_3 = Fy_{13} = \frac{2(0.707)F}{EI} \sum A\bar{x}$$

$$y_3 = 0.042 \frac{Qa^3}{EI} \tag{9.9}$$

Results in Eqs. (9.8) and (9.9) are identical, as they must be. Note the m diagrams are half as large in Fig. 9.8a and b as in Fig. 9.8c; however, the integration encompasses four sides in Fig. 9.8a and b, but only two sides in Fig. 9.8c. One effect compensates for the other with either procedure correct.

9.10 Direct Stress Effects

Although the redundant solution is independent of axial stress contributions, the deflection is not. The equation is

$$Fy_3 = \int_0^\ell \frac{mM}{EI} ds + \sum CFQ$$

$$y_3 = \frac{1.414}{EI} \sum_0^a Ax_1 + \frac{4a}{AE}\left(\frac{0.707}{2}\right)\left(\frac{Q}{2}\right)(0.707)$$

$$= 0.042 \frac{Qa^3}{EI} + 0.500 \frac{Qa}{EA} \tag{9.10}$$

Taking the ratio of the direct contribution to the bending term,

$$\frac{y_{D3}}{y_{M3}} = 12.05\left(\frac{I}{Aa^2}\right) \tag{9.11}$$

If the beam cross section is square, this ratio reduces to $(h/a)^2$. Thus with a sectional dimension $\frac{1}{10}$ of the length of a side, the direct contribution is only 1 percent with respect to the bending. While the exact ratio will vary depending on the cross section, the negligible effect of the axial deformation has been demonstrated.

9.11 Secondary Deflections

Calculation of the relative displacement between 2 and 4 (Fig. 9.7a) requires preload consisting of equal and opposite F forces in the horizontal direction on the decoupled bent. This is equivalent in Fig. 9.8 to rotating the F direction 90° in Fig. 9.8a or b. With m from Fig. 9.9 and M from Fig. 9.7c,

$$F\delta_{24} = \int_0^\ell \frac{mM}{EI} ds = \frac{4}{EI} \int_0^a \left(0.707\left(\frac{F}{2}\right)x_2\right) M dx_2$$

$$\delta_{24} = -0.042 \frac{Qa^3}{EI} \tag{9.12}$$

Having naively taken F as corresponding to a separation of 2 from 4, the negative sign reminds us that they, in fact, deflect toward each other. In this rather special case of square geometry, the horizontal contraction is exactly equal to the vertical elongation [Eq. (9.8)].

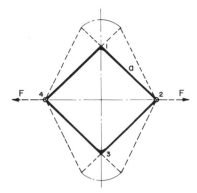

Figure 9.9 Horizontal displacement solution for δ_{24} can employ pins at 2 and 4 during auxiliary loading.

9.12 *Multiple Loading*

The square bent can also be loaded by pairs of equal and opposed forces Q_x and Q_y (Fig. 9.10), independent of ground. Having solved the case of single coaxial loading, we can now use superposition of these results [Eq. (9.7)]. Solving for the total bending moment at the corners,

$$M_1 = -0.177Q_x a + 0.177Q_y a \tag{9.13a}$$

$$M_2 = 0.177Q_x a - 0.177Q_y a \tag{9.13b}$$

If the forces are equal, all corner moments are 0, and there is no bending throughout. The bent is equivalent structurally to four bars with four pins and has only tensile loading.

Deflections are similarly superimposed:

$$\delta_{13} = -0.42\frac{Q_x a^3}{EI} + 0.042\frac{Q_y a^3}{EI} \tag{9.14a}$$

$$\delta_{24} = 0.042\frac{Q_x a^3}{EI} - 0.042\frac{Q_y a^3}{EI} \tag{9.14b}$$

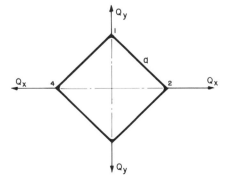

Figure 9.10 Superposition is applied for stresses and deflections caused by simultaneous horizontal and vertical loading.

For equal loads there is complete cancellation of flexural deflection, with relative displacements across the corners 0. Direct effects can be obtained as indicated in Chap. 1.

Examples

9.1. The bent beam is loaded by its own weight. Find the corresponding reactions if $a = 30$ and $b = 20$.

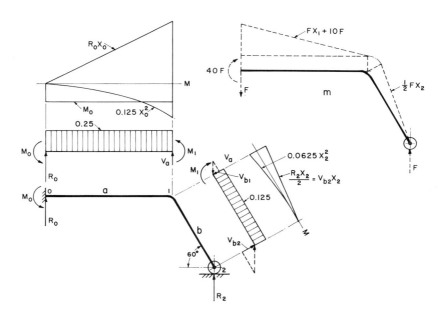

Solution: Taking R_2 as the redundancy we apply the auxiliary load F at 2 and with the same sense as a preload, resulting in a modified cantilever bending-moment diagram. Application of the distributed actual loading then yields

$$0 = \int_0^\ell \frac{mM}{EI}\, ds = \frac{1}{EI}\left[\frac{F}{2}\int_2^1 x_2 M\, dx_2 + F\int_1^0 x_1 M\, dx_1 + 10F\int_1^0 M\, dx_1 \right]$$

$$= \frac{1}{2}\left[\frac{R_2(20)^3}{(2)(3)} - \frac{0.125(20)^4}{8} \right] + \left[\frac{R_0(30)^3}{6} - \frac{M_0(30)^2}{2} - \frac{0.25(30)^4}{24} \right]$$

$$+ 10\left[\frac{R_0(30)^2}{2} - 30M_0 - \frac{0.25(30)^3}{6} \right]$$

Combining this reduced equation from elasticity with two relations from static equilibrium, we have three simultaneous linear equations:

$$\begin{cases} 12R_0 + 0.889R_2 - M_0 = 27.92 \\ R_0 + \quad R_2 \quad = 12.5 \\ \quad 40R_2 + M_0 = 287.5 \end{cases}$$

$$R_0 = 6.77 \qquad R_2 = 5.73 \qquad M_0 = 58.5$$

Note the distributed vertical gravity loading in *b* is equivalent to a transverse beam loading reduced by the factor sin 30° or 0.50.

9.2. In Example 9.1 determine the horizontal deflection of the roller 2.

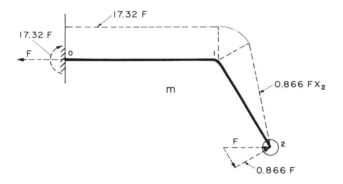

Solution: The redundant system, having already been solved for reactions, is now reduced to determinacy by removing the constraint R_2 and F is applied horizontally. Using the resulting auxiliary moment distribution in conjunction with the previous actual moment diagram,

$$Fx_2 = \frac{1}{EI}\left[(0.866F)\sum Ax_2 + (17.32F)\sum A\right]$$

$$x_2 = \frac{1}{EI}\left[0.866\left(\frac{5.73}{2}\frac{(20)^3}{3} - \frac{0.25}{2}\frac{(20)^4}{8}\right)\right.$$

$$\left. + 17.32\left(\frac{6.77(30)^2}{2} - (58.46(30)) - \frac{0.25(30)^3}{6}\right)\right]$$

$$= \frac{1541}{EI}$$

9.3. For the constant section structure calculate all reactions.

Solution: The Q loading has a horizontal component that is shared by H_0 and H_3, and this is a once-redundant case for which the conservation of

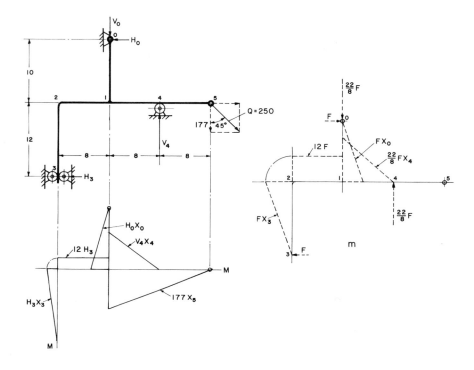

complementary energy requires that

$$0 = \frac{1}{EI}\int_0^\ell mM\,ds = \frac{1}{EI}\left[F\sum_3^2 A\bar{x}_3 + 12F\sum_2^1 A + F\sum_0^1 A\bar{x}_0 + \frac{22}{8}F\sum_4^1 A\bar{x}_4\right]$$

$$0 = \frac{H_3(12)^3}{3} + (12)(8)(12)\,H_3 - \frac{H_0(10)^3}{3}$$

$$+ \frac{22}{8}\left[\frac{V_4(8)^3}{3} - \frac{1414(8)^2}{2} - \frac{176.8(8)^3}{3}\right]$$

Reducing this equation and obtaining two more from statics,

$$\begin{cases} -H_0 + 5.18H_3 + 1.408V_4 = 622.2 \\ H_0 + H_3 = 176.8 \\ -2.75H_3 + V_4 = 132.6 \end{cases}$$

$$H_0 = 115.9 \qquad H_3 = 60.9 \qquad V_4 = 300$$

9.4. In Example 9.3 calculate the horizontal deflection of the horizontal section 2–5.

Solution: Applying the F_5 load to the right and removing V_4 we have horizontal reactions of $F_0 = \frac{12}{22}F$ and $F_3 = \frac{10}{22}F$. There are then triangular auxiliary diagrams from 0 to 1 and 3 to 2, and a constant $m = \frac{120}{22} F$ from 2 to 1:

$$
\begin{aligned}
x_5 &= \frac{1}{EI}\left[\frac{12}{22} \sum_0^1 A\bar{x}_0 + \frac{10}{22} \sum_3^2 A\bar{x}_3 + \frac{120}{22} \sum_2^1 A \right] \\
&= \frac{1}{EI}\left(\frac{10}{22} \right)\left[1.2\frac{(115.9)(10)^3}{3} + \frac{(60.9)(12)^3}{3} + (60.9)(8)(12)^2 \right] \\
&= \frac{68,900}{EI}
\end{aligned}
$$

9.5. The continuous triangular tube has internal gas pressure p. Neglecting any end constraints, it is a two-dimensional problem. Find the maximum bending moment and the maximum bending stress.

Solution: This is a twice-redundant case. We test by cutting at 1, and assuming no gas leakage, the circumferential triangular beam could support the pressure loading. Reapplication of the corner constraint requires a horizontal tension and a bending moment at 1 to maintain the square corner. The associated preloads could both be at 1, but locating F loads at 3 allows us to have no auxiliary moment in 0–2. Free-body diagrams are drawn for $\frac{1}{2}$ of the tube since there is symmetry about the vertical centerline 1–3 and thus no possible vertical shears at 1 or 3. For M' the energy relations become

$$
0 = \frac{2M'}{EI}\int_0^{\ell/2} M\,ds = \sum_1^3 A = \frac{-V_1 a^2}{2} + M_1 a_1 + \frac{qa^3}{6}
$$

$$
+ \frac{q(0.707a)^3}{6} - M_3(0.707a)
$$

$$
0.500aV_1 - M_1 + 0.707M_3 = 0.226qa^2 \tag{1}
$$

And with F

$$
0 = 0.707F\int_0^a x_0 M\,dx_0 \quad \text{or} \quad \sum_0^1 A\bar{x}_0 = 0
$$

$$
0 = \frac{-V_1 a^3}{6} + \frac{M_1 a^2}{2} + \frac{qa^4}{24}
$$

$$
0.333aV_1 - M_1 = 0.833qa^2 \tag{2}
$$

From statics

$$
\sum M_3 = M_3 + M_1 - (0.707a)H_1 + (0.707a)^2 q/2 = 0
$$

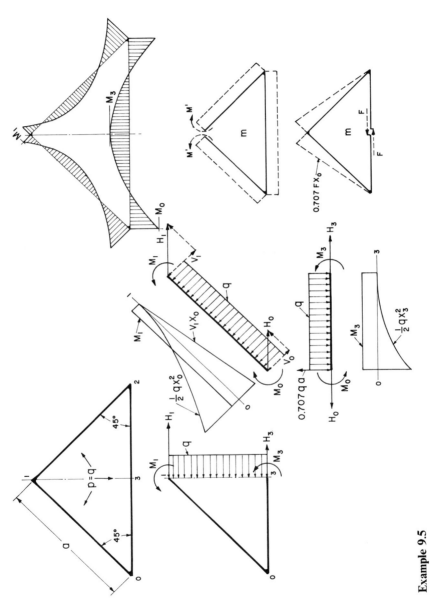

Example 9.5

Since $H_1 = V_1/0.707$, $M_3 = (V_1 a - M_1 - 0.25 qa^2)$. Substituting in (1), we obtain two simultaneous linear equations involving only V_1 and M_1:

$$\begin{cases} 1.207 a V_1 - 1.707 M_1 = 0.403 qa^2 \\ 0.333 a V_1 - M_1 = 0.0833 qa^2 \end{cases}$$

$$V_1 = 0.409 qa \qquad H_1 = 0.578 qa \qquad M_1 = 0.053 qa^2$$

$$M_3 = 0.105 qa^2 \qquad M_0 = 0.145 qa^2 \qquad H_0 = H_3 = 0.131 qa$$

Resulting moment distributions are shown for the complete tube, with possible maximum values at 0, 1, 3, and between 0 and 1. The actual maximum is $M_0 = 0.145 \ qa^2$ per unit length. The loaded beam section I/c is also per unit length, and for the rectangular geometry is $bt^2/6$ or $t^2/6$. The bending stress is

$$\sigma = \frac{0.145 qa^2}{t^2/6} = 0.87 q \left(\frac{a}{t} \right)^2$$

where t is the nominal wall thickness.

9.6. Repeat Example 9.5 using the Three-Moment Theorem.

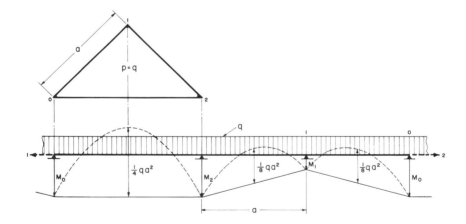

Solution: We obtain the bending moments M_0, M_1, and M_2 directly from the three-moment equations. They apply because the corners are effectively fixed in space, neglecting tensile extensions:

0-1-2 $\alpha = 1$

$$M_0 + 4M_1 + M_0 = \frac{qa^2}{4} + \frac{qa^2}{4} \quad (1)$$

1-2-0 $\alpha = \sqrt{2}$

$$M_1 + 2(2.414)\, M_0 + 1.414 M_0 = \frac{qa^2}{4} + \frac{(1.414)(1.414a)^2 q}{4}$$

$$\begin{cases} 2M_0 + 4M_1 = 0.500\, qa^2 \\ 6.243 M_0 + \quad M_1 = 0.957 qa^2 \end{cases}$$

$$M_0 = 0.145\, qa^2 \qquad M_1 = 0.053\, qa^2$$

These results confirm those of Example 9.5, and a considerable saving of time and effort is apparent using the Three-Moment Theorem, and we are now also able to calculate the maximum bending stress on the basis of M_0. To obtain complete moment distribution, however, the free-body analyses of the previous example would be required, but the total undertaking is still much less.

10

Fundamentals of Planar Curved Beams

THE CIRCLE IS ONE of the most common geometric constituents of engineering design, either as an arc or a complete circle. A cylindrical pressure vessel or pipe is probably the most simple example, with hoop stress distributed constantly as a tangential tensile condition. In many situations, however, flexure is the main component of both deflection and stress. Circular beams can be broadly classified as determinate or indeterminate, and in following chapters we use strain energy to analyze these cases. We now study the nature of bending, direct, and shear loading developed by various types of loading.

We assume the beam to be slender, relative to its length and curvature, thereby avoiding the nonlinear stress distribution described by *curved-beam theory*. Within this restriction, the elastic behavior of a beam with a curved neutral axis is similar to that of a straight beam, but analytically it is necessary to convert from rectilinear to polar coordinates in expressing moments and energies. Direct stress effects can be included by integration, as indicated in Chap. 1.

The circular arc is by far the most common geometry and the most tractable mathematically; however, the beam with an arbitrarily curved shape does arise. This type of solution requires numerical integration, and this procedure is outlined in subsequent chapters.

All arguments advanced in Chap. 3 relative to bending behavior apply to the slender curved beam. These include bending, bending stress, and bending strain energy relations. Similarly, direct loads correspond conceptually to the effects of axial loading indicated in Chap. 1. Shear effects are

205

negligible in slender beams, but equations for longitudinal shear stress and strain energy are presented in Chap. 13.

10.1 Concentrated End Loading

In several subsequent chapters we will have numerous occasions to develop expressions evolving from bending moment due to loading in the plane of the ring. These functions in turn usually assume a free end as the initial condition at $\theta = 0$. This corresponds to a coordinate origin at the free end of a cantilever (Fig. 3.4). Taking a circular quadrant cantilever with a radial load (Fig. 10.1), we revert to free-body diagrams to study the loading distribution. The mean radius from the center to the neutral axis is R, and the end reactions Q and Q_r are shown.

To determine the internal loading we break the beam at θ, the polar angle, placing each part in static equilibrium. An internal vertical force Q is present at the break and an internal bending moment, each reacting from one part to the other. Resolving the force into axial and transverse components,

$$M(\theta) = Q_y R \sin \theta \qquad (10.1a)$$

$$Q_A(\theta) = Q_y \sin \theta \qquad (10.1b)$$

$$Q_T(\theta) = Q_y \cos \theta \qquad (10.1c)$$

Figure 10.1 A radial force on a circular cantilever produces sinusoidally varying bending moment and axial loading (M and Q_A). Shear load Q_T is the transverse or shear force.

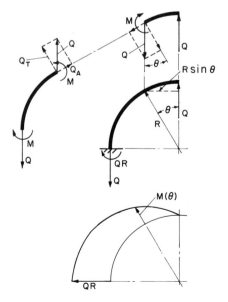

Table 10.1 Expression for distributed bending moment, axial and shear loading developed in a circular beam subjected to various loading conditions.

CONCENTRATED LOAD

a

M	$Q_y R \sin\theta$
Q_A	$Q_y \sin\theta$
Q_T	$Q_y \cos\theta$

b

M	$Q_x R(1-\cos\theta)$
Q_A	$-Q_x \cos\theta$
Q_T	$Q_x \sin\theta$

COUPLE

c

M	M
Q_A	0
Q_T	0

UNIFORM PRESSURE

d

M	$qR^2(1-\cos\theta)$
Q_A	$qR(1-\cos\theta)$
Q_T	$qR\sin\theta$

HYDROSTATIC PRESSURE

e

M	$\rho R^3\left(1 - \frac{\theta}{2}\sin\theta - \cos\theta\right)$
Q_A	$\rho R^2\left(1 - \frac{\theta}{2}\sin\theta - \cos\theta\right)$
Q_T	$\frac{1}{2}\rho R^2(\sin\theta - \theta\cos\theta)$

f

M	$\frac{1}{2}\rho R^3(\sin\theta - \theta\cos\theta)$
Q_A	$\rho R^2\left(\frac{1}{2}\sin^3\theta + \cos^2\theta - \cos\theta\right)$
Q_T	$\rho R^2\left(\frac{\theta}{2}\sin\theta\right)$

DISTRIBUTED WEIGHT

g

M	$-\rho AR^2(\theta\sin\theta + \cos\theta - 1)$
Q_A	$-\rho AR\sin\theta$
Q_T	$-\rho AR\cos\theta$

h

M	$-\rho AR^2(\sin\theta - \theta\cos\theta)$
Q_A	$\rho AR\theta\cos\theta$
Q_T	$-\rho AR\theta\sin\theta$

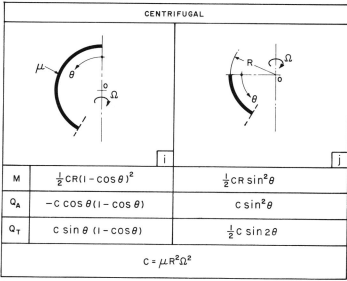

	CENTRIFUGAL	
M	$\frac{1}{2}CR(1-\cos\theta)^2$	$\frac{1}{2}CR\sin^2\theta$
Q_A	$-C\cos\theta(1-\cos\theta)$	$C\sin^2\theta$
Q_T	$C\sin\theta\,(1-\cos\theta)$	$\frac{1}{2}C\sin2\theta$
	$C=\mu R^2\Omega^2$	

Table 10.1 (*Cont.*)

Q_A represents an axial-tensile loading and Q_T a radial shear relative to the principal axes of the beam at this angular position. The loads produce bending, direct, and shear stresses, respectively.

With tangential loading at the end (Table 10.1*b*), the equations become

$$M(\theta) = Q_x R(1 - \cos\theta) \qquad (10.2a)$$

$$Q_A(\theta) = -Q_x\cos\theta \qquad (10.2b)$$

$$Q_T(\theta) = Q_x\sin\theta \qquad (10.2c)$$

where the negative sign at Q_A indicates that compressive stress exists in the first quadrant.

Although Fig. 10.1 shows a cantilever for illustrative purposes, the angular parameter can continue for a complete revolution, or beyond. If θ is greater than 360°; however, the ring must depart slightly from planar.

10.2 *Moment Due to Internal Pressure*

The most probable source of constant pressure is fluid loading. A long thin tank or tube can be subjected to gas or liquid effects, usually internally (possibly externally). Loading is technically caused by the differential between the two, with the latter commonly ambient atmospheric. In the

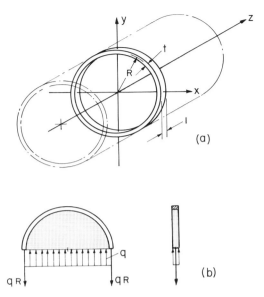

Figure 10.2 Analysis of a long cylindrical shell with internal pressure involves a typical strip of unit axial length. Simple hoop stresses are developed.

simplest case with internal pressure on a complete cylinder (Fig. 10.2), stresses are direct tensile, or *hoop stress* throughout the circumference, with no bending. However, even in this situation there is bending moment distribution (Table 10.1*d*) due to the pressure loading because of the arbitrary termination of the ring in order that this component effect can be isolated.

For the long thin cylinder, we take a ring of unit axial width bounded by two planes perpendicular to the central *z–z* axis (Fig. 10.2*a*). The ring cross section is then 1 × *t*. Neglecting constraints related to any end structures, this model now corresponds to the constant bar (Table 10.1*d*), and the internal unit pressure corresponds numerically and physically to the circumferentially distributed load *q*.

A circular bar can also be subjected to uniform radial loading if rotated at a significant speed about its central axis. This centrifugal effect creates a uniform outward loading per unit cylindrical area.

10.3 *Moment Derivation*

For the constant distributed radial loading, results in Table 10.1*d* follow from Fig. 10.3, in which the elemental moment about the neutral axis at θ is

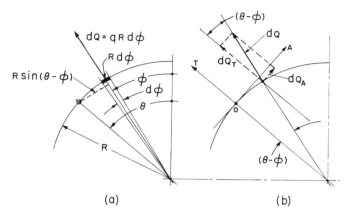

Figure 10.3 Distributed moment in the arcuate beam is derived using the elemental load, the elemental moment, and two angular variables.

caused by an elemental force at ϕ having the dashed moment arm:

$$dM = q(R\,d\phi)(R\sin(\theta - \phi))$$

$$M = qR^2 \int_0^\theta \sin(\theta - \phi)\,d\phi$$

$$= qR^2 \left[\sin\theta \int_0^\theta \cos\phi\,d\phi - \cos\theta \int_0^\theta \sin\phi\,d\phi \right]$$

$$= qR^2 [\sin^2\theta + \cos^2\theta - \cos\theta]$$

$$= qR^2 [1 - \cos\theta] \tag{10.3}$$

where θ = a constant for purposes of the integration

ϕ = the variable function ranging internally from 0 to θ

This procedure is typical of moment derivation for any distributed loading in polar coordinates. The differential moment is expressed as the product of the differential force and the geometric moment arm. One integration is required.

10.4 Force Distribution Due to Constant Pressure

As with moment, there is a cumulative effect of the differential components from 0 to θ, leading to Q_A and Q_T, the axial and transverse components, respectively. In Fig. 10.3b the vector force dQ is resolved into components parallel to the tangential and radial directions at θ. As shown, the important directional angle is $(\theta - \phi)$ and we obtain for the axial force

$$dQ_A = \sin(\theta - \phi) \, dQ$$

$$Q = qR \int_0^\theta \sin(\theta - \phi) \, d\phi$$

$$= qR(1 - \cos\theta) \tag{10.4a}$$

and for the transverse component

$$dQ_T = \cos(\theta - \phi) \, dQ$$

$$Q_T = qR \int_0^\theta \cos(\theta - \phi) \, d\phi$$

$$= qR \sin\theta \tag{10.4b}$$

10.5 *Equilibrium Including Internal Fluid*

In the preceding analysis the beam was assumed with a free end at the initiation of loading ($\theta = 0$), and to have a constantly distributed unit loading applied directly to the ring. Alternatively, if the loading is caused by

Figure 10.4 Bending moment is obtained from free-body diagrams including a portion of the fluid under internal pressure.

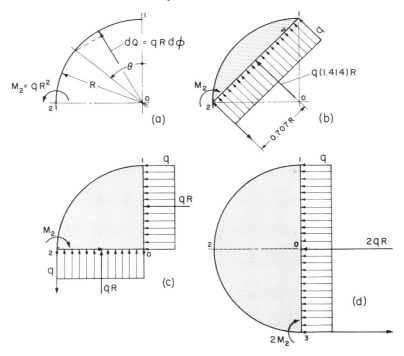

constant internal fluid pressure, the free-body diagram can include a portion of the fluid. Pressure on the fluid boundaries then replaces contact pressure on the shell. This procedure can serve as a check on results, and in some instances this technique has advantages in the development of moment distribution.

Figure 10.4 illustrates the inclusion of the fluid within a chordal plane at 90° (Fig. 10.4b), producing a moment about 2 due to the resultant of qR^2. Taking the entire 90° quadrant (Fig. 10.4c), we obtain the same moment from the two resultants. For the complete semicircle (Fig. 10.4d), the moment reaction is $2qR^2$. All moment and force results agree with values from the derived results in Table 10.1d.

10.6 *Hydrostatic Pressure*

The pressure in a contained liquid in the earth's gravitational field is directly proportional to the density of the fluid and the vertical distance from the surface of the fluid, or head. For the cylindrical shell, the triangular distribution is 0 at the top (Table 10.1e). If there is an additional external head, a further constant pressure must be superimposed; however, the triangular variation to zero constitutes a component loading, and the associated characteristics are analyzed separately.

Removing the fluid, but taking the normal pressure exerted on the inner surface by the fluid, the differential force (Fig. 10.4a) becomes

$$dQ = \rho R(1 - \cos\phi)(R\,d\phi) \qquad (10.5)$$

where ρ is the density of the fluid.

With hydrostatic conditions, a long, enclosed shell is implied, and a unit strip is selected for analysis (Fig. 10.2). Having the force dQ, we proceed to integrate the differential moment with ϕ ranging from 0 to θ as before.

10.7 *Diagrams Including Fluid*

As in Sec. 10.5, we can visualize the problem with the fluid in place. The free-body diagram for one half of a cylinder has external fluid pressure applied horizontally on the vertical central plane, with a resultant fluid force resultant at the centroid of the triangle on the unit strip (Fig. 10.5).

With fluid loading, and the fluid having significant weight, we must include the vertical weight of the enclosed fluid acting at its centroid W. This requires a vertical supporting reaction for static equilibrium and

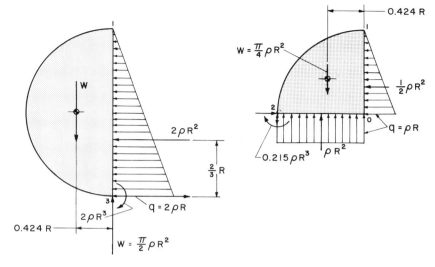

Figure 10.5 Similar to Fig. 10.4, but with the weight of the contained fluid causing hydrostatic pressure.

contributes a moment about the neutral axis at the bottom of the shell:

$$M_3 = \frac{\pi}{2}\rho R^2(0.424R) + 2\rho R^2\left(\frac{2}{3}R\right) - M_3 = 0$$

$$M_3 = 2\rho R^3 \tag{10.6}$$

Horizontally, the resultant of the distributed pressure is reacted by tension in the shell at the bottom.

Taking a 90° quadrant including the enclosed liquid (Fig. 10.5), on a similar basis we have

$$M_2 = \frac{1}{2}\rho R^2\left(\frac{1}{3}R\right) + \frac{\pi}{4}\rho R^2(1 - 0.424)R + \rho R^2\left(\frac{1}{2}R\right) + M_2 = 0$$

$$M_2 = \left(1 - \frac{\pi}{4}\right)\rho R^3 = 0.215\rho R^3 \tag{10.7}$$

Here we have a constant pressure acting on the bottom plane as shown contributing an additional moment about the neutral axis.

These illustrations indicate methods of obtaining specific moment and force values by means of free-body diagrams.

10.8   Distributed Weight

A uniform gravitational field acting on a straight horizontal beam creates a constant distributed transverse loading, but if the beam is rotated to vertical, the bending loads are 0. Thus the transverse loading on a vertical ring (Table 10.1g) varies from q to 0 as we move from the point of horizontal tangency to that of vertical tangency.

As shown in Fig. 10.6, the differential moment for weight loading is

$$dM = AR^2(\sin\theta - \sin\phi)\, d\phi \qquad (10.8)$$

Integration then results in the Table 10.1g equation.

Taking the free-body diagram of the complete semicircle, the weight acts at the centroid of the arc, producing one moment, and

$$M_3 = \pi(0.637)\rho AR^2 = 2\rho AR^2 \qquad (10.9)$$

For $\theta = 90°$, $M_2 = 0.571\rho AR^2$. Note these two moments have reverse sense. The latter is negative with compression induced on the inner fibers.

Weight loading is normally not a problem unless the ring is rather large in diameter.

Although gravity effects are the principal source of distributed loading, translational acceleration of a ring will also induce similar loads because of a parallel, uniform acceleration field. With acceleration replacing gravity, the effective loading is directly proportional to the acceleration and of reversed direction.

Figure 10.6 Moment developed by the weight loading of the beam by integration is confirmed by complete free-body diagrams involving the centroidal location.

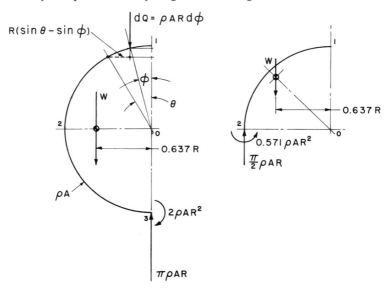

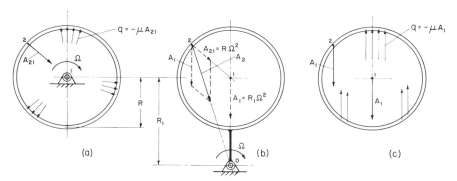

Figure 10.7 The ring rotating about its central axis generates a centrifugal radial pressure (*a*). With rotation about a displaced axis rotational loading is the superposition of rotation (*a*), translation due to the displacement (*c*) with the resultant effects in (*b*).

10.9 *Rotation about the Cylindrical Axis*

A full or partial ring revolving about the central axis develops uniformly distributed outward loading around the circumference (Fig. 10.7*a*). For a constant section

$$q = \mu R \Omega^2 \tag{10.10}$$

where μ = mass per unit circumferential length
R = mean radius of rotation
Ω = angular velocity

A common example of a ring rotating in its own plane is the rim of a pulley or wheel. Another is a circumferential retaining ring in a slot in a rotating shaft, in which the C-ring is not continuous. The shell, or longitudinal version, is a long, hollow rotating shaft. With the *q* loading determined, Table 10.1*d* is then applicable.

10.10 *Rotation about a Displaced Axis*

A circular ring can be rotated about an axis perpendicular to the plane of the ring, but eccentric relative to the central axis (Fig. 10.7*b*). The axis 0 results in a radius R_1 that can also be less then *R*.

As a result of rotation, 2 has a radial acceleration $A_2 = R_2\Omega^2$, and this multiplied by the element of mass at 2 becomes *dQ*. We must further multiply *dQ* by the geometric moment arm to any coordinate θ on the ring.

The resulting relations are quite involved. Viewed as superposition, however, the moment distribution is relatively simple.

In Fig. 10.7*b* the acceleration of a point on the ring is

$$A_2 = A_1 \nrightarrow A_{21} \tag{10.11}$$

where A_2 = resultant absolute acceleration of 2
A_1 = absolute radial acceleration of the center
$= R_1 \Omega^2$ towards 0
A_{21} = relative acceleration of 2 with respect to 1
$= R \Omega^2$ toward 1

and the addition is *vectorial*. Obtaining the magnitude and direction of the right side of Eq. (10.11), we have determined the actual acceleration A_2.

This exercise indicates that the rotation of the offset ring is composed of component accelerations:

1.   The center of the ring about 0.
2.   The ring about the center 1.

The first effect is a *translation* toward 0 in the direction of R_1 with a translatory acceleration of A_1 (Fig. 10.7*c*). D'Alembert, or reversed effective, loads are $-\mu A_1 = -\mu R_1 \Omega^2$.

The second effect, or the spin about 1, corresponds to uniform radial internal loading, for which we apply Table 10.1*d*. This accounts for the effect of *rotation*.

Thus we obtain the total bending moment for the offset ring by the superposition of the moment distributions in Table 10.1 for distributed weight and for uniform pressure, respectively. The vector addition in Eq. (10.11) is satisfied by the superposition technique. Shear and axial loads follow similarly.

10.11   *Rotation about a Diametral Axis*

Although less likely, a ring or ring segment can be rotated about a diameter, perpendicular to the cylindrical axis. If the ring has appreciable axial length, the problem is three dimensional and becomes involved. With a slender bar, the analysis is very close to two dimensional, resulting in the equations of Table 10.1*i* and 10.1*j*.

Centrifugal loading now involves a varying radius and radial forces perpendicular to the rotational axis (Fig. 10.8). Then

$$dQ = \rho A \Omega^2 (R \sin \phi) R \, d\phi \tag{10.12a}$$

$$dM = R(\cos \phi - \cos \theta) \, dQ \tag{10.12b}$$

and we proceed to integrate the equation for the differential moment with respect to ϕ for $M(\theta)$.

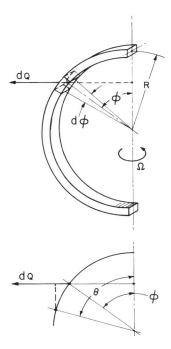

Figure 10.8 A narrow ring rotating about a diameter is loaded perpendicularly to the diametral axis.

Total free-body diagrams indicated for other cases are not shown for this combination. The complication lies in the centroid of the circumferential arc not being a meaningful point for locating a resultant. This is because the centrifugal field although parallel, is not uniform, and centers of gravity derive from a uniform gravitational field. There is a scheme for representing the rotating ring on an equivalent dynamic basis, but it is rather too lengthy to include here.

10.12   *Dimensional Considerations*

Various combinations of circular structures and loading lead to different dimensional situations. The principal distinction is between the slender ring with finite cross section and the thin cylindrical shell having considerable length. The latter is most apt to occur where fluid loading is involved, and we then revert to the unit strip analysis (Fig. 10.2a).

Table 10.2 is provided to facilitate the selection of proper units and to indicate other factors to be incorporated for proper numerical results in both the SI and British domains. Table 10.2a and c applies to the beam of finite section and Table 10.2b and d to the axial shell. The results are especially pertinent to Table 10.1, with the $[f(\theta)]$ nondimensional function indicating the nature of the angular distribution from Table 10.1.

The unit of mass in the SI system is kg, and the unit of force is the Newton (or N):

$$N = mg_c = kg\left(\frac{m}{\sec^2}\right) \qquad (10.13)$$

where g_c = acceleration of gravity
= 9.81 m/sec²

Thus for weight loading, g_c is introduced. Mass units are satisfied using kg for centrigual loading, but a factor of 0.01 is required with the radius in cm rather than m.

In British notation, $g_c = 386$ in./sec², and this term must be used to convert weight in lb to mass for centrifugal loading.

Table 10.2 Dimensional combinations required in circular beam calculations in SI units [(a) and (b)], and the British system [(c) and (d)], relating to Table 10.1.

		CONSTANT SECTION [a]		CONTINUOUS SHELL [b]	
UNIFORM PRESSURE	M	$qR^2[f(\theta)]$	N·cm	$q_1R^2[f(\theta)]$	N·cm/cm
	Q	$qR[f(\theta)]$	N	$q_1R\ f(\theta)]$	N/cm
	q	—	N/cm	p	N/cm/cm
HYDROSTATIC PRESSURE	M	—	—	$9.81\rho R^3[f(\theta)]$	N·cm/cm
	Q	—	—	$9.81\rho R^2[f(\theta)]$	N/cm
	W	—	—	$9.81\rho A$	N/cm
DISTRIBUTED WEIGHT	M	$9.81\rho AR^2[f(\theta)]$	N·cm	$9.81\rho tR^2[f(\theta)]$	N·cm/cm
	Q	$9.81\rho AR[f(\theta)]$	N	$9.81\rho tR[f(\theta)]$	N/cm
	W	$9.81\rho AR\theta$	N	$9.81\rho tR\theta$	N/cm
CENTRIFUGAL	M	$0.01\rho AR^3\Omega^2[f(\theta)]$	N·cm	$0.01\rho tR^3\Omega^2[f(\theta)]$	N·cm/cm
	Q	$0.01\rho AR^2\Omega^2[f(\theta)]$	N	$0.01\rho tR^2\Omega^2[f(\theta)]$	N/cm
	q	—	—	$0.01\rho tR\Omega^2$	N/cm/cm
	σ_M	$M/(I/c)$	N/cm²	$6M_1/t^2$	N/cm²
	σ_Q	Q/A	N/cm²	Q_1/t	N/cm²
	I	—	cm⁴	$t^3/12$	cm⁴/cm
		R,t = cm	A = cm²	ρ = kg/cm³	

		CONSTANT SECTION [c]		CONTINUOUS SHELL [d]	
UNIFORM PRESSURE	M	$qR^2[f(\theta)]$	lb in.	$q_1R^2[f(\theta)]$	lb in./in.
	Q	$qR[f(\theta)]$	lb	$q_1R[f(\theta)]$	lb/in.
	q	—	lb/in.	p	lb/in./in.
HYDROSTATIC PRESSURE	M	—	—	$\rho R^3[f(\theta)]$	lb in./in.
	Q	—	—	$\rho R^2[f(\theta)]$	lb/in.
	w	—	—	ρA	lb/in.
DISTRIBUTED WEIGHT	M	$\rho AR^2[f(\theta)]$	lb in.	$\rho tR^2[f(\theta)]$	lb in./in.
	Q	$\rho AR[f(\theta)]$	lb	$\rho tR[f(\theta)]$	lb/in.
	w	$\rho AR\theta$	lb	$\rho tR\theta$	lb/in.
CENTRIFUGAL	M	$\left(\frac{\rho A}{386}\right)R^3\Omega^2[f(\theta)]$	lb in.	$\left(\frac{\rho t}{386}\right)R^3\Omega^2[f(\theta)]$	lb in./in.
	Q	$\left(\frac{\rho A}{386}\right)R^2\Omega^2[f(\theta)]$	lb	$\left(\frac{\rho t}{386}\right)R^2\Omega^2[f(\theta)]$	lb/in.
	q	—	—	$\left(\frac{\rho t}{386}\right)R\Omega^2$	lb/in./in.
	σ_M	$M/(I/c)$	lb/in.²	$6M_1/t^2$	lb/in.²
	σ_A	Q/A	lb/in.²	Q_1/t	lb/in.²
	I	—	in.⁴	$t^3/12$	in.⁴/in.
R, t = in.		A = in.²		ρ = lb/in.³	

Table 10.2 (*Cont.*)

Centrifugal loading in conjunction with the continuous shell implies rotation about the cylindrical axis. The subscript 1 indicates loading per unit length.

10.13 *Integration*

Energy solutions in polar coordinates invariably lead to combinations of trigonometric functions with various limits. Some of those frequently encountered are summarized in Table 10.3, with typical angular values. Numerical values of definite integrals are obtained by subtracting the lower limit from the upper if necessary. For limits not given, substitution can be made in the basic integral forms provided.

Other integrals can usually be found in tables or developed algebraically. If intractable, the plot of a function provides an area equal to the integral.

ψ $\int f(\theta)d\theta$	$\int f(\theta)d\theta$	$\int_{0}^{\psi} f(\theta)d\theta$				
		$\frac{\pi}{4}$	$\frac{\pi}{2}$	π	$\frac{3\pi}{2}$	2π
$\int \sin\theta\, d\theta$	$-\cos\theta$	0.2929	1	2	1	0
$\int \cos\theta\, d\theta$	$\sin\theta$	0.7071	1	0	-1	0
$\int \sin^2\theta\, d\theta$	$\frac{\theta}{2} - \frac{1}{4}\sin 2\theta$	0.1427	0.7854	1.5708	2.3562	3.1416
$\int \cos^2\theta\, d\theta$	$\frac{\theta}{2} + \frac{1}{4}\sin 2\theta$	0.6427	0.7854	1.5708	2.3562	3.1416
$\int (1-\cos\theta)\, d\theta$	$\theta - \sin\theta$	0.0783	0.5708	3.1416	5.7124	6.2832
$\int (1-\cos\theta)^2\, d\theta$	$\frac{3}{2}\theta - 2\sin\theta + \frac{1}{4}\sin 2\theta$	0.0139	0.3562	4.7124	9.0686	9.4248
$\int \sin\theta(1-\cos\theta)\, d\theta$	$-\cos\theta - \frac{1}{2}\sin^2\theta$	0.0429	0.5000	2	0.5000	0
$\int \cos\theta(1-\cos\theta)\, d\theta$	$-\frac{\theta}{2} + \sin\theta - \frac{1}{4}\sin 2\theta$	0.0644	0.2146	-1.5708	-3.3562	-3.1416
$\int \sin\theta\cos\theta\, d\theta$	$\frac{1}{2}\sin^2\theta$	0.2500	0.5000	0	0.5000	0
$\int (\sin\theta+\cos\theta-1)^2\, d\theta$	$2\theta + \sin^2\theta + 2\cos\theta - 2\sin\theta$	0.0708	0.1416	2.2832	10.4248	12.5664
$\int \theta\sin\theta\, d\theta$	$\sin\theta - \theta\cos\theta$	0.1517	1	3.1416	-1	-6.2832
$\int \theta\cos\theta\, d\theta$	$\cos\theta + \theta\sin\theta$	0.2625	0.5708	-2	-5.7124	0
$\int \theta\sin^2\theta\, d\theta$	$\frac{1}{4}\theta^2 - \frac{\theta}{4}\sin 2\theta + \frac{1}{4}\sin^2\theta$	0.0829	0.8669	2.4674	6.5517	9.8696
$\int \theta\cos^2\theta\, d\theta$	$\frac{1}{4}\theta^2 + \frac{\theta}{4}\sin 2\theta + \frac{1}{4}\cos^2\theta$	0.2256	0.3669	2.4674	5.5517	9.8696
$\int \theta\sin\theta\cos\theta\, d\theta$	$-\frac{\theta}{4} + \frac{1}{8}\sin 2\theta + \frac{\theta}{2}\sin^2\theta$	0.1250	0.3927	-0.7854	1.1781	-1.5708
$\int \sin\theta\cos^2\theta\, d\theta$	$-\frac{1}{3}\cos^3\theta$	-0.1179	0	0.3333	0	-0.3333
$\int \cos^3\theta\, d\theta$	$\frac{2}{3}\sin\theta + \frac{1}{3}\sin\theta\cos^2\theta$	0.5893	0.6667	0	-0.6667	0
$\int (1-\cos\theta)^3\, d\theta$	—	0.9457	1.5937	7.8540	14.1143	15.7080

Table 10.3 Integrals and definite integral values for typical functions encountered in the analysis of circular beams.

Functions are dimensionless with θ in radians; therefore, units present no difficulty in this procedure.

10.14 *Limits of Integration*

A word of caution is in order regarding the evaluation of lower limits, usually 0. Although the majority of functions are 0 in this situation, a notable exception is $\cos\theta = 1$. Unless this factor is recognized, substantial numerical errors will result.

Also, integration can proceed continuously until interrupted by a load; that is, in a ring forming a complete circle, limits applying to the complete moment function can proceed from 0 to 2π if there are no intervening loads or supports. Obviously radians must be used rather than degrees when the angle parameter is present in the integrated result.

In Table 10.3 numerical values are provided for the four quadrant positions and 45°, with 0 as the lower limit. Thus we can often avoid the integration phase of a solution completely by direct substitution of numerical values of the definite integrals provided. This is usually the fastest and most accurate procedure.

Examples

10.1. The U-shaped bar has separating forces at the ends. With British dimensions determine:
  (a) The maximum bending stress
  (b) The bending stress at 3

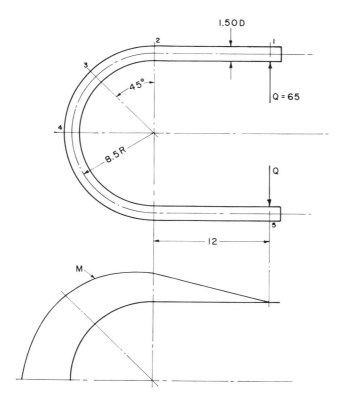

Solution:

(a) Maximum moment arm exists at 4, and is $(8.5 + 12) = 20.5$ in.

$$M = 65(20.5) = 1333 \text{ lb in.}$$

$$I/c = \frac{\pi d^3}{32} = 0.33 \text{ in.}^3$$

$$\sigma_{M4} = \frac{M}{I/c} = \frac{1333}{0.33} = 4020 \text{ lb/in.}^2$$

(b) At 2, or 45°, there is a moment of $65(12) = 780$ lb in. and a transverse force of 65 lb. Superimposing from Table 10.1a and c,

$$\sigma_{M3} = \frac{1}{0.33}[780 + 65(8.5)(0.707)] = 3550 \text{ lb/in.}^2$$

10.2. Repeat Example 10.1, but interpreting the units as SI.

Solution:

(a) $\sigma_{M4} = \dfrac{1333}{0.33} = 4020 \text{ N/cm}^2 = 4.02 \text{ kN/cm}^2$

(b) $\sigma_{M3} = 3550 \text{ N/cm}^2 = 3.55 \text{ kN/cm}^2$

10.3. A retaining ring has a constant rectangular section. If the shaft rotates at 6000 rpm, find the maximum total centrifugally induced stress.

Solution: From Tables 10.1d and 10.2a,

$$M = 0.01(0.008)(0.24)(5)^3\left[6000\left(\frac{2\pi}{60}\right)\right]^2 2 = 1895 \text{ N} \cdot \text{cm}$$

$$I/c = \frac{1}{6}bh^2 = \frac{(0.3)(0.8)^2}{6}$$

$$= 0.032 \text{ cm}^3$$

$$\sigma_M = \frac{1895}{0.032} = 59,200 \text{ N/cm}^2$$

$$Q_A = \frac{1895}{5} = 380 \text{ N}$$

$$Q_T = 0$$

$$\sigma_A = \frac{380}{(0.3)(0.8)} = 1580 \text{ N/cm}^2$$

$$\sum \sigma = 59,200 + 1580 = 60,800 \text{ N/cm}^2$$

at the inner fibers at the bottom of the ring.

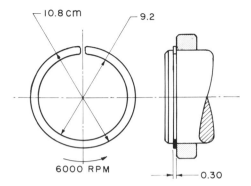

6000 RPM

10.4. The ring of Example 10.3 rotates about its transverse axis of symmetry at 6000 rpm. Calculate the maximum bending stress.

Solution: From Tables 10.1*i* and 10.2*a*, the maximum stress occurs at the bottom as the gap tends to open:

$$M = \left[0.01(0.008)(0.24)(5)^3(628.3)^2\right]\frac{1}{2}(2)^2$$

$$= [947.5]2 = 1895 \text{ N} \cdot \text{cm}$$

$$\sigma_M = \frac{1895}{0.032} = 59,200 \text{ N/cm}^2$$

These stresses are identical in both rotational conditions. Similarly the tensile separating force is 379 N for the same superimposed stress.

10.5. The continuous shell is partially filled with water, and has a wall thickness of $\frac{5}{16}$ in. Find the maximum stress
(a) Due to bending
(b) Due to direct tension

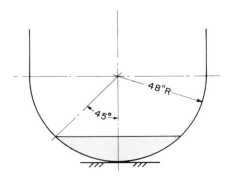

48"R

45°

Solution:

(a) Taking the angular variable ϕ from the horizontal at the left, the pressure is

$$ph = p(R \sin \phi - R \sin 45°)$$

$$dQ = phR\,d\phi$$

$$dM = (dQ)R\sin(\theta - \phi)$$

$$= pR^3[\sin \phi - 0.707][\sin \theta \cos \phi - \cos \theta \sin \phi]$$

$$M = pR^3 \left[\sin \theta \int_{\pi/4}^{\pi/2} \sin \phi \cos \phi \, d\phi - \cos \theta \int_{\pi/4}^{\pi/2} \sin^2 \phi \, d\phi \right.$$

$$\left. - 0.707 \sin \theta \int \cos \phi \, d\phi + 0.707 \cos \theta \int \sin \phi \, d\phi \right]$$

$$= pR^3[0.25 - 0.707(0.293)] = 0.043 pR^3$$

Using Table 10.2*d*

$$\sigma_M = \frac{(0.036)(48)^3(0.043)(6)}{(0.313)^2} = 10,500 \text{ lb/in.}^2$$

(b) The depth of the water is $0.293R$, with a triangular pressure distribution from 0 at the surface. The total separating force is

$$Q_A = p(0.293R)\left(\frac{0.293R}{2} \right) = 3.56 \text{ lb/in.}$$

due to the maximum pressure of 0.51 lb/in.^{2}

$$\sigma_A = \frac{3.56}{0.313} = 11.4 \text{ lb/in.}^2$$

Note the predominance of the bending stress.

10.6. The shell of Example 10.4 becomes half full of water. Recalculate the stresses.

Solution:

(a) From Tables 10.1*f* and 10.2*d*,

$$M_1 = \tfrac{1}{2}(0.036)(48)^3(1 - 0) = 1990 \text{ lb/in./in.}$$

$$\sigma_M = \frac{1990(6)}{(0.313)^2} = 117,600 \text{ lb/in.}^2$$

(b)

$$Q_A = 0.036(48)^2(\tfrac{1}{2}) = 41.5 \text{ lb/in.}$$

$$\sigma_A = \frac{41.5}{0.313} = 133 \text{ lb/in.}^2$$

10.7. If the shell of Example 10.5 extends vertically upward a distance of 26 in., find the maximum bending stress due to the weight of the shell.

Solution: The vertical sides provide a moment about the neutral axis at the bottom of

$$M_V = \rho t(26) R \text{ lb in./in.}$$

From the circular quadrant, Tables 10.1h and 10.2d,

$$M_C = \rho t R^2 = \rho t R (48)$$

By superposition

$$M_V + M_C = \rho t R (26 + 48) = (0.29)(0.313)(48)(74)$$
$$= 322 \text{ lb in./in.}$$
$$\sigma_M = \frac{(322)6}{(0.313)^2} = 19,700 \text{ lb/in.}^2$$

11

Planar Deflections of Circular Beams

WITH THE TECHNIQUES AND RELATIONSHIPS DEVELOPED in Chap. 10, particularly with respect to bending-moment behavior in polar coordinates and values of definite integrals, we are now prepared to solve for deflections in arcuate beams. Since slender curved beams behave essentially as a conventional beam in bending, our procedures involve only energy balance, and are similar to those indicated in Chaps. 4 and 5. In this chapter we attack only statically determinate cases. As there are literally an unlimited number of beam and loading combinations that can be encountered, tabular design data are only provided for several of the most basic and representative arrangements.

An excellent book dealing with circular beams has been written by Blake (1979) (see Bibliography, page 430), detailing analysis for slender beams, as well as methods for the thick beam with nonlinear stress distribution, delving into the actual *curved beam theory*. This latter subject is not attempted here as it tends to violate the basic energy relationships underlying the entire scope of analysis based upon linearity and superposition; however, departure from classical beam theory is moderate and gradual as beam depth increases with respect to the radius of curvature. Such approaches, in any case, usually yield fairly good first approximations.

11.1 *Primary Deflection of the Semicircular Beam*

This elementary deflection solution includes the semicircle (Fig. 11.1*a*) with one end pinned to ground as a reference point. The horizontal diameter,

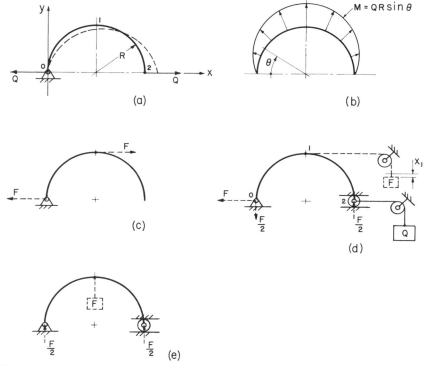

Figure 11.1 Solution for deflections of the semicircular beam requires auxiliary physical loading arrangements. A reference ground coordinate is located at 0.

assumed to remain in this direction, is a reference axis x. Moment distribution is sinusoidal (Table 10.1a) and symmetrical (Fig. 11.1b)

For energy balance

$$\frac{Qx_2}{2} = \int_0^\ell \frac{M^2}{2EI}\, ds = \frac{1}{2EI}\int_0^\pi (QR\sin\theta)^2 R\, d\theta$$

$$x_2 = \frac{QR^3}{EI}\left(\frac{\theta}{2} - \frac{1}{4}\sin 2\theta\right) = \frac{\pi}{2}\frac{QR^3}{EI} \tag{11.1}$$

where $R\, d\theta$ is the elementary arc length ds.

Although this is not a difficult integration, the complete procedure requires integration, substitution of limits, and numerical evaluation. Using Table 10.3 we obtain the result much more directly, reducing the time required and the possibility for error:

$$x_2 = \frac{QR^3}{EI}\int_0^\pi \sin^2\theta\, d\theta = 1.5708\frac{QR^3}{EI} \tag{11.2}$$

The same result is obtained by doubling the integral obtained from 0 to $\pi/2$, because of the symmetrical moment diagram.

From the moment equation, or from statics, maximum moment occurs at 1, or $\theta = 90°$, and is simply QR. It is of positive sense, usually defined here as corresponding to compressive bending stress in the outer fibers of a circular beam.

11.2   Secondary Linear Deflections

The beam in Fig. 11.1*a* assumes a noncircular elastic curve relative to the original semicircle. By energy equivalence, we can now obtain any desired deflection by applying a related auxiliary loading. For instance, the horizontal deflection component of 1 requires the F load (Fig. 11.1*c*). As shown, the beam is hardly in equilibrium as reacted at 0, with an unbalanced clockwise couple. Point 2 is ungrounded, but has been assumed to remain on the x axis during loading; therefore we achieve equilibrium by confining 2 to a horizontal slot. This permits a vertical reaction $F/2$ upward at the slot, with static equilibrium as indicated. The cable and weight loading analogy for applying F as a preload emphasizes the physical or laboratory significance of the prestressing scheme. The actual load Q, applied after F, can be similarly applied by a gravity weight, with the descent of F corresponding to the complementary external energy Fx_1.

In Fig. 11.1*d* there are four dashed auxiliary loads, each a potential source of work done with Q loading; however, only the F load at 1 is permitted to deflect. The other three can do no work if they do not displace:

$$Fx_1 = \int_0^\ell \frac{mM}{EI}\,ds = \frac{1}{EI}\int_0^1 \left[FR \sin\theta - \frac{F}{2}R(1 - \cos\theta) \right](QR \sin\theta)\,R\,d\theta$$

$$+ \int_2^1 \frac{F}{2}R(1 - \cos\theta)(QR \sin\theta)\,R\,d\theta$$

$$x_1 = \frac{QR^3}{EI}\int_0^{\pi/2}\left(\sin^2\theta - \frac{1}{2}\sin\theta + \frac{1}{2}\sin\theta\cos\theta \right)d\theta$$

$$+ \int_0^{\pi/2}\left(\frac{1}{2}\sin\theta - \frac{1}{2}\sin\theta\cos\theta \right)d\theta$$

$$= \frac{QR^3}{EI}\left(0.7854 - \frac{1}{2} + \frac{1}{4} + \frac{1}{2} - \frac{1}{4} \right) = 0.7854\frac{QR^3}{EI} \qquad (11.3)$$

As expected, this is exactly one half of x_2 from Eq. (11.1).

To obtain the coupled vertical displacement of 1 with respect to the horizontal diameter, we apply F vertically (Fig. 11.1*e*), reacting at 0 and 2, providing the relation

$$Fy_1 = \int_0^\ell \frac{mM}{EI}\,ds = \frac{1}{EI}2\int_0^{\pi/2}\frac{1}{2}FR(1-\cos\theta)(QR\sin\theta)R\,d\theta$$

$$y_1 = \frac{QR^3}{EI}\int_0^{\pi/2}(\sin\theta - \sin\theta\cos\theta)\,d\theta = \frac{1}{2}\frac{QR^3}{EI} \qquad (11.4)$$

In Eq. (11.4) we are *not* able to use a continuous integration from 0 to π, as there is an intervening load at 1. We have therefore integrated from the ends 0 and 2 and invoked symmetry.

The vector displacement in space of 1 is now the resultant of the two components obtained, to the right and down:

$$\delta_1 = x_1 \nrightarrow y_1 = \sqrt{x_1^2 + y_1^2} = 0.9310\frac{QR^3}{EI} \qquad (11.5)$$

11.3   *Relative Symmetrical Deflections*

Whereas we have proceeded with some logic to determine absolute displacements with 0 pinned to ground, we are dealing with a symmetrical beam, and this feature usually provides a number of analytical simplifications. To illustrate, we select central axes on the semicircle (Fig. 11.2a). On this basis the previous x_2 in Eq. (11.1) now is the relative displacement x_{23} of the ends, composed of two equal radial displacments.

Using symmetry, we obtain the relative radial deflections at 4 and 5 (Fig. 11.2b). In polar coordinates we apply twin F loads. Horizontal components balance, and equilibrium is provided by reacting the vertical components at

Figure 11.2   The free semicircular beam is analyzed with respect to several coordinate deflections utilizing symmetry.

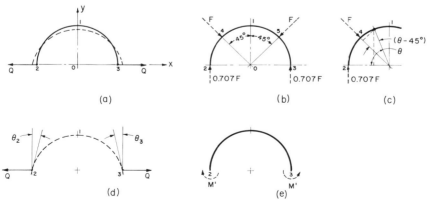

2 and 4. The latter do no external work as the Q loading is applied, and the deflection relation is

$$F\delta_{R4} + F\delta_{R5} = 2F\delta_R = \int_0^\ell \frac{mM}{EI} \, ds$$

Auxiliary moment from 2 to 4 is

$$m = 0.707FR(1 - \cos\theta)$$

From 4 to 1 (Fig. 11.2c)

$$m = 0.707FR(1 - \cos\theta) - FR\sin(\theta - 45°)$$
$$= 0.707FR(1 - \sin\theta)$$

Auxiliary moment, incidentally, is 0 at $\theta = 90°$, or at 1.

The complete displacement expression is then

$$\delta_{R3} = \frac{QR^3}{EI}(0.707)\left[\int_0^{45°}(\sin\theta - \sin\theta\cos\theta)\,d\theta + \int_{45°}^{90°}(1 - \sin\theta)\sin\theta\,d\theta\right]$$
$$= 0.0758\frac{QR^3}{EI} \tag{11.6}$$

With a positive result, deflection is directed inward as assumed. Apparently there is an angle θ that will result in no radial displacement, greater than 45°, as eventually the deformation is outward at the ends.

11.4 Symmetrical Slopes

As a further extension of the analysis we solve for the end slopes θ_2 and θ_3 (Fig. 11.2d), equal from symmetry. We proceed to induce a constant preload moment M' (Fig. 11.2e):

$$2M'\theta_2 = \int_0^\ell \frac{M'M}{EI}\,ds$$

$$\theta_2 = \frac{1}{EI}\int_0^{\pi/2}QR\sin\theta R\,d\theta = \frac{QR^2}{EI} \tag{11.7}$$

The slopes θ_2 and θ_3 occur relative to the initial perpendicular directions to the x axis, and in that sense are not absolute. If we consider one tangent fixed, or reference (say at 2), the slope equation is

$$M'\theta_3 = \int_0^\ell \frac{M'M}{EI}\,ds$$

$$\theta_3 = \frac{1}{EI}\int_0^\pi MR\,d\theta = 2\frac{QR^2}{EI} \tag{11.8}$$

This slope is twice that of Eq. (11.7), as it represents the total relative slope of the tangent at 3 with respect to the tangent at 2.

11.5 Direct Stress Components

Loaded curved beams are usually subjected to continuous moments, with direct load and shear load varying over the length of the beam (Fig. 10.1). Direct effects have been referred to as *axial*, and shear as *transverse*. The latter energy contributions are discussed briefly in Chap. 13; however, deformations associated with direct stress follow from developments in Chap. 1 and are now indicated.

Total direct strain energy for simple loading is given by Eq. (1.5). For the circular cantilever (Table 10.1a),

$$U_s = \int_0^\ell \frac{Q_A^2}{2EA}\, ds = \frac{Q_y^2}{2EA} \int_0^\psi \sin^2\theta R\, d\theta \tag{11.9}$$

and for Table 10.1b,

$$U_s = \frac{Q_x^2}{2EA} \int_0^\psi \cos^2\theta R\, d\theta \tag{11.10}$$

where ψ is the total central angle of the beam.

For coupled deflections involving complementary energy, auxiliary loading causes a distributed F_A, obtained similarly. From Eq. (11.2) the CFQ term for a curved beam becomes

$$(U_s)FQ = \int_0^\ell \frac{F_A Q_A}{EA}\, ds \tag{11.11}$$

and the complete deflection energy relation [Eq. (11.3)] becomes

$$Fx_1 = \int_0^\ell \frac{mM}{EI}\, ds + \int_0^\ell \frac{F_A Q_A}{EA}\, ds \tag{11.12}$$

In Fig. 11.1a the additive deflection resulting from bending [Eq. (11.2)] and direct effects is then given by

$$\frac{Qx_2}{2} = \frac{1}{2EI} \int_0^\pi (QR\sin\theta)^2 R\, d\theta + \frac{1}{2EA} \int_0^\pi (Q\sin\theta)^2 R\, d\theta$$

$$= \frac{1.5708}{EI} QR^3 \left[1 + \left(\frac{r}{R}\right)^2\right] \tag{11.13}$$

The direct term involves the square of the ratio of the radius of gyration of the cross section to the mean radius of the beam. Taking this as, say, one to ten, the inclusion of this effect only increases the calculated diametral deflection by 1 percent.

11.6 *The Noncircular Curved Beam*

Most of the preceding analysis has been confined to the circular beam of constant section. Real design problems, however, often entail distributed parameters defined by numerical coordinates, and any type of closed solution by classical integration is usually precluded. Further, if the geometry of the beam is noncircular, simple polar coordinates are not possible and some type of numerical integration is required.

To illustrate, consider the constant planar beam having a general curved shape with opposed colinear Q loading (Fig. 11.3a). Separation of A and C requires determination of the M^2 function with respect to the beam axis. For the incremental calculation, we divide the length into 15 equal intervals. These are measured chordally, but deviation due to the curvature for these intervals is negligible.

At each of these points, we measure the vertical distance from the load line, representing the moment arm. Then $M = Qy$ and

$$\frac{Q}{2}\delta_{AC} = \int_0^\ell \frac{M^2}{2EI}\,ds = \frac{Q^2}{2EI}\int_0^\ell y^2\,ds$$

$$\delta_{AC} = \frac{Q}{EI}\int_0^\ell y^2\,ds \qquad (11.14)$$

Figure 11.3 A beam of general curvature requires numerical integration of rectilinear plots involving functions of the coordinate distances that describe the curvature.

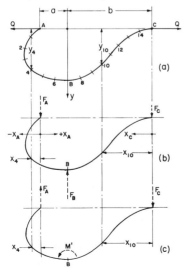

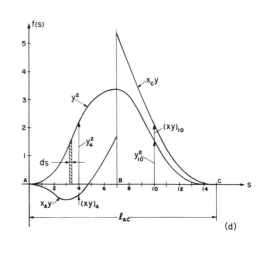

The y^2 function is plotted in Cartesian coordinates with the horizontal base corresponding to the rectified curved beam length (Fig. 11.3*d*), resulting in a continuous curve maximum at Band 0 at A and C. Greatest strain energy, and consequently, the greatest local contribution to δ_{AC}, occurs in the beam flexure in the vicinity of B. By the same token B is the point of maximum bending stress.

Evaluation of a definite integral [Eq. (11.14)] has been indicated in Example 4.13. It requires the determination of the total area under the curve as accurately as necessary. Dimensionally, the integral represents distance cubed, and the numerical value *must be to full scale*. This requires scale factors unless incorporated in the area determination.

11.7 *Transverse Deflection*

To obtain the deflection of B toward AC we apply F_B, with reactions F_A and F_C (Fig. 11.3*b*). For static equilibrium

$$F_A = \left(\frac{b}{a+b}\right)F_B \quad \text{and} \quad F_C = \left(\frac{a}{a+b}\right)F_B \tag{11.15}$$

The corresponding auxiliary moments are

$$m_{AB} = F_A x_A \quad \text{and} \quad m_{BC} = F_C x_C \tag{11.16}$$

with maximum at B. Defining positive moment as producing compression in the exterior surface as in Fig. 11.3*a*, there is negative moment for the first section of the beam progressing from A. This is indicated by the negative horizontal coordinate x_A. Directly below A on the extended line of action of F_A, $m = 0$. Beyond this point, m is always positive. The energy relation is

$$F_B y_B = \int_0^\ell \frac{mM}{EI}\, ds = \frac{1}{EI}\int_A^B (F_A x_A)(Qy)\, ds + \int_C^B (F_C x_C)(Qy)\, ds$$

$$y_B = \frac{Q}{EI}\left[\left(\frac{b}{a+b}\right)\int_A^B x_A y\, ds + \left(\frac{a}{a+b}\right)\int_C^B x_C y\, ds\right] \tag{11.17}$$

These product curves are plotted (Fig. 11.3*d*) with the integration applying to one negative and two positive segments. Typical coordinates are identified at points 4 and 10 on the beam and on the curves. A net positive total area confirms the assumed upward deflection of B.

11.8 *Slope Determination*

For rotation of the tangent to the elastic curve at B we apply a counterclockwise couple M' at B, reacting at A and C. The latter two equal and opposite forces will contribute no external work with the reference coordinate AC remaining horizontal (Fig. 11.3c). For complementary equivalence,

$$M'\theta_B = \frac{1}{EI}\left[\int_A^B\left(-\frac{M'}{a+b}\right)x_A(Qy)\,ds + \int_C^B\left(\frac{M'}{a+b}\right)x_C(Qy)\,ds\right]$$

$$\theta_B = \frac{Q}{EI(a+b)}\left[-\int_A^B x_A y\,ds + \int_C^B x_C y\,ds\right] \qquad (11.18)$$

Note the horizontal coordinates for m are identical in Fig. 11.3b and c.

Integral functions in Eqs. (11.14), (11.17), and (11.18) contain only coordinates derived from the geometry of the beam curvature. They are paired at simultaneous points on the curve. As such, they convey elastic behavioral information concerning the bending response of the beam under auxiliary and actual loading, and this stems from the relationship of the geometry to the loading.

The coordinate products are assumed to full scale, and thus reflect the absolute size of the structure. We could, however, factor out absolute dimensions and reduce the variables under the integral signs to dimensionless ratios. The integrals then are more general and are identified only with the beam *shape*, regardless of its *size*.

11.9 *Tabulated Results*

Design data for deflections of several arrangements of basic circular beams are provided.

Table 11.1 Complete elastic deflection coefficients for the circular cantilever for any central angle and for conventional intervals with concentrated end loading.

					ψ	$\frac{\pi}{4}$	$\frac{\pi}{2}$	π	$\frac{3\pi}{2}$	2π
α_x	$\frac{x_x}{Q_x}$				$\frac{3}{2}\psi - 2\sin\psi + \frac{1}{4}\sin 2\psi$	0.0139	0.3562	4.7124	9.0686	9.4248
α_{xy}	$\frac{x_y}{Q_y}$	α_{yx}	$\frac{y_x}{Q_x}$	$\frac{R^3}{EI}$	$\frac{1}{2} - \cos\psi + \frac{1}{2}\cos^2\psi$	0.0429	0.5000	2	0.5000	0
α_y	$\frac{y_y}{Q_y}$				$\frac{\psi}{2} - \frac{1}{4}\sin 2\psi$	0.1427	0.7854	1.5708	2.3562	3.1416
β_{zx}	$\frac{\theta_x}{Q_x}$	β_{xz}	$\frac{x_M}{M_z}$	$\frac{R^2}{EI}$	$\psi - \sin\psi$	0.0783	0.5708	3.1416	5.7124	6.2832
β_{zy}	$\frac{\theta_y}{Q_y}$	β_{yz}	$\frac{y_M}{M_z}$		$1 - \cos\psi$	0.2929	1	2	1	0
γ_z	$\frac{\theta}{M_z}$			$\frac{R}{EI}$	ψ	0.7854	1.5708	3.1416	4.7124	6.2832

		ψ	$\frac{\pi}{4}$	$\frac{\pi}{2}$	π	$\frac{3\pi}{2}$	2π
(a)	x_1 $\;\frac{\rho A R^4}{EI}$	$3\sin\psi - \frac{1}{2}\psi\sin^2\psi - \frac{3}{8}\sin 2\psi - \psi\cos\psi - \frac{5}{4}\psi$	0.0129	0.2511	-0.7854	-11.2467	-14.1372
	y_1	$\frac{3}{4}\sin^2\psi - \frac{1}{4}\psi\sin 2\psi + \cos\psi + \frac{1}{4}\psi^2 - 1$	0.0400	0.3669	0.4674	5.3017	9.8696
	θ_1 $\;\frac{\rho A R^3}{EI}$	$2\sin\psi - \psi\cos\psi - \psi$	0.0735	0.4292	0	-6.7124	-12.5664
(b)	x_1 $\;\frac{\rho A R^4}{EI}$	$\frac{1}{2}\psi\sin^2\psi + \frac{3}{8}\sin 2\psi - \frac{3}{4}\psi$	-0.0177	-0.3927	-2.3562	-1.1781	-4.7124
	y_1	$-\psi\sin\psi - \frac{1}{4}\psi\sin 2\psi - \frac{3}{4}\sin^2\psi - 2\cos\psi + \frac{1}{4}\psi^2 + 2$	-0.3867	0.2961	6.4674	11.5140	9.8696
	θ_1 $\;\frac{\rho A R^3}{EI}$	$-\psi\sin\psi - 2\cos\psi + 2$	0.0304	0.4292	4	6.7124	0

Table 11.2 Deflection of the constant planar circular cantilever with gravity loading. In (*a*) the free end is vertical and in (*b*) it is horizontal.

In Tables 11.1 and 11.2 we have free-fixed cantilever combinations, including end loads and distributed loads. The latter apply to the beam in a vertical plane subjected to gravity. Both tables include complete equations for the end-point deflection with any central angle.

Table 11.3 pertains to the pinned-supported statically determinate semicircular beam, with concentrated and distributed loads in the plane of the beam.

Table 11.3 Deflection of the statically determinate semicircular beam with concentrated and distributed weight loading.

		CONCENTRATED LOAD			DISTRIBUTED WEIGHT	
		a	b		c	d
x_1	$\frac{QR^3}{EI}$	0.2854	0.2500	$\frac{qR^4}{EI}$	0.3927	1.5708
y_1		0.2500	0.1781		0.2634	0.7348
x_2		0.7854	0.5000		0.7854	y_2 $\;$ 1.1190
θ_0		0.6427	-0.2854		0.4674	-0.7309
θ_1	$\frac{QR^2}{EI}$	-0.0719	0	$\frac{qR^3}{EI}$	0	-0.3669
θ_2		-0.3573	0.2854		-0.4674	2.0977

		a	b	c	d
x_{12}	$\dfrac{QR^3}{EI}$	9.4248	3.5708	1	0
x_{34}		3.5708	1.5708	0.5000	0
y_1		2	0.5000	0.7854	1.5708
y_3		-1.0708	-0.5000	0.1073	y_2 -1.5708

Table 11.4 The ungrounded complete open ring subjected to various separational loadings, including shear (d), has deflections as indicated.

The split, or piston ring, is analyzed in Table 11.4 for deflections caused by separational forces. All distributions are symmetrical except Table 11.4d, which is antisymmetrical. Table 11.5 also refers to this type of ring, but loading is provided by the weight of the ring with uniform internal or external uniform radial loading or pressure in Table 11.5d.

Tables 10.2 and 10.3 are useful for determining the proper dimensional combinations, particularly if the results are extended to the axially continuous shell. Calculation of bending stresses can proceed from Table 10.1, where moments are provided.

It should be noted that the tabular results apply to planar systems of any angular orientation. For instance in Table 11.1, the beam and the free end can be rotated as desired. In Table 11.3, the semicircle could be downward.

Table 11.5 Deflection of the complete open ring with weight loading combinations and uniform internal or external pressure.

		DISTRIBUTED WEIGHT			UNIFORM PRESSURE
		a	b	c	d
x_{12}	$\dfrac{qR^4}{EI}$	-1.5708	11.4680	14.1372	9.4248
x_{34}		-1.2146	5.1848	4.3562	3.5708
y_{12}		0	-1.5708	-9.8696	0
y_1		0.4674	0.7854	0	2
$y_3=y_4$ 0.5297			y_2 2.3562	y_5 9.4022	$y_3=y_4$ -1.0708

11.10   *The Piston Ring*

A piston ring is used for sealing purposes inside a cylindrical bore B (Fig. 11.4). The rectangular section involves a radial height h and a width b, with a mean ring radius R. As assembled, the outside of the ring is larger than the internal bore, and compression from the free shape to the circular bore results in radial pressure around the circumference. Ideally this pressure is distributed constantly around the entire contact surface in a radial sense, and it is generated by the distributed polar deflection of the circular beam.

If the developed contact pressure is to be constant, the required bending moment distribution in the compressed ring corresponding to constantly distributed radial loading is given in Table 10.1d.

$$M_p(\theta) = qR^2(1 - \cos\theta) = pbR^2(1 - \cos\theta)$$
$$= C_p(1 - \cos\theta) \tag{11.19a}$$

with θ ranging from 0 to 2π from a free end.

Additionally the distributed relative displacement pattern is obtained from pairs of opposed auxiliary loadings from

$$\delta_{Dp} = \int_0^\ell \frac{m}{FEI}\left[pbR^2(1 - \cos\theta)\right]R\,d\theta$$
$$= C_p\int_0^\ell m(1 - \cos\theta)\,d\theta \tag{11.19b}$$

where m is the auxiliary moment distribution due to the F loading applied diametrically; however, this relation is not particularly helpful in describing the required free shape of the ring. For assistance we turn to a related loading (Fig. 11.4a and Table 11.4a).

Figure 11.4   Deflection of the piston ring to the internal bore B is closely related to the tangential end loading (b).

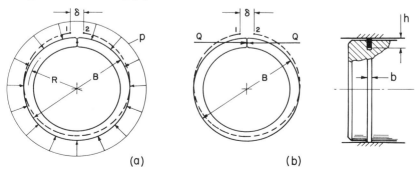

(a)   (b)

Taking this opposed tangential force loading at the ends, we have an actual bending moment distribution in the complete ring (Table 10.1b) of

$$M_Q(\theta) = QR(1 - \cos\theta) \qquad (11.20a)$$

As before, taking opposed F loadings between any two diametrically opposed points corresponding to those used to obtain the closure curve previously,

$$\delta_{DQ} = \int_0^\ell \frac{m}{FEI} [QR(1 - \cos\theta)R\,d\theta$$

$$= C_Q \int_0^\ell m(1 - \cos\theta)\,d\theta \qquad (11.20b)$$

Comparing Eqs. (11.19b) and (11.20b) we note that *the integral values are identical* in both cases. Further, by equating the relative deflections

$$C_p = \frac{pbR^3}{FEI} = C_Q = \frac{QR^2}{FEI}$$

$$p = \frac{Q}{bR} \qquad (11.21)$$

As a physical interpretation, assume a piston ring loaded with two closing forces Q (Fig. 11.4b) and that we:

1. Proceed to grind the outside to the shape of a true cylinder B, as constrained by the Q loading.
2. Release the ring and assemble it on a piston.
3. Compress the ring and insert it in the bore, thus deflecting it back to its position when ground.

Under these circumstances we have accomplished several important results.

1. The incremental distribution curve going from closed to free has been *identical* to that when going from free to closed.
2. We have achieved a *constantly distributed pressure* against the cylinder bore.
3. This pressure (Eq. 11.21) is directly proportional to Q and inversely proportional to b and R.

11.11 *Additional Ring Equations*

We have analytically satisfied the desired method of achieving constant pressure on the cylinder bore and have an equation for this resulting pressure. Obviously the force Q is difficult to measure physically, and it is much more convenient to index the pressure to the incremental tangential deflection at the ends, or the change in the gap dimension.

From Table 11.4*a*

$$\delta = x_{12} = \frac{3\pi Q R^3}{EI}$$

$$\frac{p}{E} = \frac{1}{36\pi}\left(\frac{h}{R}\right)^3\left(\frac{\delta}{R}\right) \tag{11.22}$$

with the ring pressure increasing directly as the gap deflection, as the cube of the radial height, and decreasing as the fourth power of the radius.

For cast iron rings there can be concern with the allowable bending stress in the ring, particularly with the tensile stress. Maximum bending stress during closure or opening of the ends occurs in the ring diametrically opposite the gap.

$$\sigma = \frac{M}{I/c} = \frac{2QR}{(bh^2/6)}$$

$$\frac{\sigma}{E} = \frac{1}{3\pi}\left(\frac{h}{R}\right)\left(\frac{\delta}{R}\right) \tag{11.23}$$

Equations (11.22) and (11.23) are in terms of dimensionless parameters. Both involve the ratios h/R and δ/R.

The proper physical shape of the ring is thus obtained by closing the ends of a rough ring and grinding it to a cylindrical shape. When released, the proper curve appears. This result is also achieved even if the ring does not have a constant radial dimension h. A given variation in I around the ring will have an identical effect upon the integrals in Eqs. (11.19b) and (11.20b) corresponding to the two loading cases (Fig. 11.4*a* and *b*). Deflection distributions will again be identical.

We have demonstrated a unique interplay between elastic curves and a potential engineering application of superposition of deflections to produce uniform piston ring pressure of a predictable magnitude.

Examples

11.1. A constant curved cantilever in the form of a complete circle carries a concentrated radial load Q at the free end. Determine:
   (a) The horizontal deflection of the load
   (b) The horizontal deflection of 3

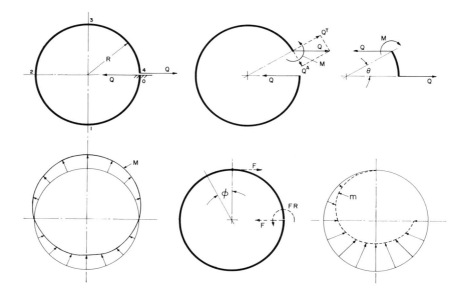

Solution:

(a) From the fundamental flexural energy equation

$$\frac{Qx_4}{2} = \int_0^\ell \frac{M^2}{2EI}\, ds = \frac{1}{2EI} \int_0^{2\pi} (QR \sin \theta)^2\, R\, d\theta$$

$$x_4 = \frac{QR^3}{EI} \int_0^{2\pi} \sin^2\theta\, d\theta = \frac{QR^3}{EI}\, \pi$$

(b) Using the auxiliary F loading

$$m = -FR(1 - \cos \phi)$$

This is negative, as compression on the outer fiber had previously been taken as positive bending moment:

$$Fx_3 = \int_0^\ell \frac{mM}{EI}\, ds = \frac{1}{EI} \int_0^{3/2\pi} - FR(1 - \cos \phi)(QR \sin \theta)\, R\, d\phi$$

Reducing to a common angular variable

$$\theta = \left(\phi + \frac{\pi}{2} \right) \qquad \sin \theta = \cos \phi$$

$$x_3 = \frac{QR^3}{EI} \int_0^{3/2\pi} (-\cos \phi + \cos^2\phi)\, d\phi$$

$$= \frac{QR^3}{EI} \left(1 + \frac{3}{4}\pi \right) = \frac{QR^3}{EI} (3.356)$$

or approximately 7 percent greater than x_4.

11.2. Two identical arcuate bars are pin connected and subjected to separational loading. Determine the relative deflection of the pins.

Solution: Because of symmetry, ends of both bars deflect equally in a horizontal sense, with no horizontal forces developed. External auxiliary loading is shown applied horizontally to the complete structure, resulting in $F/2$ applied at each end of the bars. Integration requires a common angular coordinate for both m and M. Using θ measured from the horizontal centerline, and the arms a, the moment expressions are

$$m = \frac{F}{2} R(\sin \theta - 0.707)$$

$$M = \frac{Q}{2} R(0.707 - \cos \theta)$$

The energy relation is

$$F\delta_{02} = \int_0^\ell \frac{mM}{EI}\, ds = \frac{4}{EI}\left(\frac{F}{2}\right)\left(\frac{Q}{2}\right)\int_{\pi/4}^{\pi/2} R^3(0.707 \sin \theta + 0.707 \cos \theta$$
$$- \sin \theta \cos \theta - 0.500)\, d\theta$$

$$\delta_{02} = \frac{QR^3}{EI}[0.500 + 0.2071 - 0.250 - (0.500)(\pi/4)]$$

$$= 0.0644 \frac{QR^3}{EI}$$

where definite integral values are taken from Table 10.3 by difference. Note the last constant term becomes the integral of $d\theta = \theta$.

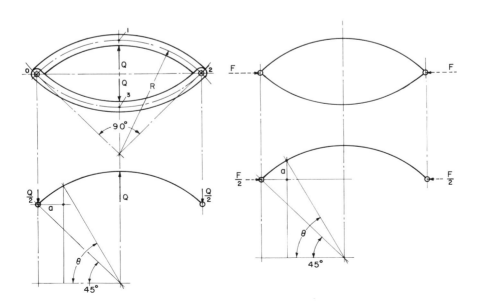

We could have preloaded only one of the bars, in which case we reduce to the same result. The factor before the integral then is 2, but the auxiliary force is F, not $F/2$.

11.3. For Example 11.2 use deflection factors from Table 11.1 to
(a) Verify the horizontal contraction
(b) Find the total vertical separation

Solution:
(a) A quadrant of the beam behaves as a curved cantilever beam. Taking the 0–1 section, a horizontal tangent is maintained at 1, with a fixed-end effect provided by the beam 1–2. Resolving the $Q/2$ end load into radial and tangential components, y and x, respectively,

$$Q_x = Q_y = \frac{Q}{2}(0.707) = 0.3536Q$$

Total tangential and radial deflections are

$$\delta_x = \alpha_x Q_x + \alpha_{xy} Q_y = \frac{0.3536QR^3}{EI}(0.0139 + 0.0429)$$

$$= 0.02008 \frac{QR^3}{EI}$$

$$\delta_y = \alpha_y Q_y + \alpha_{yx} Q_x = \frac{0.3536QR^3}{EI}(0.1427 + 0.0429)$$

$$= 0.06563 \frac{QR^3}{EI}$$

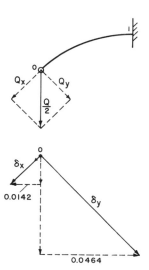

Summing the horizontal components,

$$x_0 = (-0.02008 + 0.06563)(0.707)\frac{QR^3}{EI} = 0.0322\frac{QR^3}{EI}$$

$$\delta_{02} = 2x_0 = 0.0644\frac{QR^3}{EI}$$

(b) Summing the previous components in the vertical direction,

$$y_0 = (0.0142 + 0.0464)\frac{QR^3}{EI} = 0.0606\frac{QR^3}{EI}$$

$$\delta_{13} = 2y_0 = 0.1212\frac{QR^3}{EI}$$

11.4. A curved spring is loaded by Q to close a gap δ_{23}. Find the expression for the gap-load relationship.

Solution: Application of Q causes both 2 and 3 to deflect downwardly. On this basis two independent solutions and F loadings are required at 2 and 3.

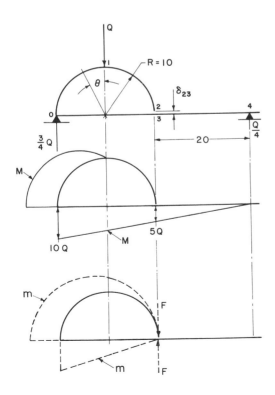

A simpler approach utilizes opposed F loads to obtain the relative displacement. Both m and M equations must be in terms of the common angle θ. For m, F_2 causes a shear F and a couple FR at 1. From 1 to 0

$$m = FR + FR \sin \theta = FR(1 + \sin \theta)$$
$$M = QR \sin \theta$$

In the straight beam, supports at 0 and 4 are neglected for auxiliary loading purposes, and $m = Fx_3$:

$$F\delta_{23} = \int_0^\ell \frac{mM}{EI} \, ds = \frac{1}{EI} \left[\int_0^{\pi/2} FR(1 + \sin \theta) QR \sin \theta R \, d\theta \right.$$
$$\left. + \int_0^{20} (Fx_3) M \, dx \right]$$

$$\delta_{23} = \frac{1}{EI} \left[QR^3 \int_0^{\pi/2} (\sin \theta + \sin^2 \theta) \, d\theta + \sum_0^{20} A \bar{x}_3 \right]$$

$$= \frac{1}{EI} \left[Q(10)^3 (1.7854) + \frac{10Q(20)^2}{3} + \frac{5Q(20)^2}{6} \right]$$

$$= 3452 \frac{Q}{EI}$$

$$\frac{\delta_{23}}{Q} = \frac{3452}{EI}$$

By taking F upward at 3, even though this absolute displacement is down, we involve the gap behavior by determining the *difference* between y_2 and y_3 in the calculation process.

11.5. The curved beam stands vertically, loaded by its own weight, $q = \rho A$, uniformly distributed. Determine the slope of the tangent to the elastic curve at 2.

Solution: Actual-moment distribution for the curved section corresponds to Table 10.1h, with $M = 0$ at 3. Auxiliary couples produce a constant auxiliary

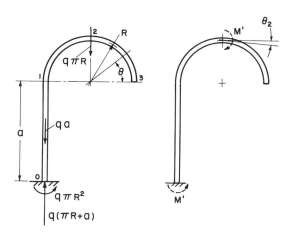

moment from 0 to 2, with no prestress in 2–3:

$$M'\theta_2 = \int_0^\ell \frac{M'M}{EI}\, ds = \frac{1}{EI}\left[\int_{\pi/2}^\pi M'qR^2(\sin\theta - \theta\cos\theta)\,R\,d\theta + \sum_0^1 A\right]$$

$$\theta_2 = \frac{1}{EI}\left[qR^3\{(2-1)-(-2-0.5708)\} + (q\pi R^2)a\right]$$

$$= \frac{qR^3}{EI}\left[3.5708 + \pi\left(\frac{a}{R}\right)\right]$$

with integral values from Table 10.3.

11.6. A steel U-bar rotates at constant speed about its central axis. Cross-sectional area is 2.50 cm², I is 0.50 cm⁴, and all dimensions are SI. Determine:
  (a) The maximum bending moment
  (b) The relative deflection of the ends δ_{24}

Solution:
  (a) Loading is composed of two parts, due to the straight and the curved portions. Considering the former first, total centrifugal force on 1–2 is

$$V_1 = (0.01)(\rho A\ell R\Omega^2) = (0.01)(0.008)(2.5)(15)(12)(3317)$$

$$= 119\ \text{N} \qquad q = \frac{Q}{\ell} = \frac{119}{15} = 7.96\ \text{N/cm}$$

$$M_1 = \frac{q\ell^2}{2} = 896\ \text{N}\cdot\text{cm}$$

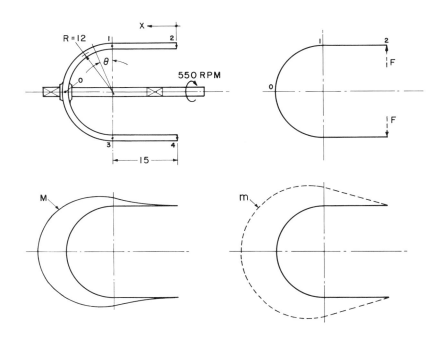

At 1 the curved section is loaded externally by an outward shear V_1 and a ccw moment M_1. Rotational effects on the curved section also produce a distributed moment (Tables 10.1j and 10.2a)

$$CR = (0.01)(\rho A R^2 \Omega^2) R = 119\frac{R^2}{\ell} = 1146 \text{ N} \cdot \text{cm}$$

$$M_c(\theta) = \left(\frac{1146}{2}\right) \sin^2\theta = 573 \sin^2\theta$$

Total moment distribution from the three components is

$$M(\theta) = 573 \sin^2\theta + 119(12)\sin\theta + 896$$

At $\theta = 90°$, $M_0 = 2900 \text{ N} \cdot \text{cm}$.

(b) For end deflection we apply opposite F loading. Auxiliary moments are

$$m_{21} = Fx_2 \qquad m_{10} = F(12 \sin\theta + 15)$$

Taking only half of the beam, the energy relation is

$$Fy_2 = \int_0^\ell \frac{Mm}{EI} ds = \frac{1}{EI}\left[\int_0^{15}(Fx)\left(\frac{7.98}{2}x^2\right) dx + \int_0^{\pi/2} F(12\sin\theta + 15)\right.$$

$$\left. \times (573\sin^2\theta + (119)(12)\sin\theta + 896) R \, d\theta\right]$$

$$y_2 = \frac{1}{EI}\int_0^{15} x^3 \, dx + \frac{(10)^6}{EI}\left[0.0825\int_0^{\pi/2}\sin^3\theta \, d\theta + 0.3095\int_0^{\pi/2}\sin^2\theta \, d\theta\right.$$

$$\left. + 0.3870\int_0^{\pi/2}\sin\theta \, d\theta + 0.1613\int_0^{\pi/2} d\theta\right]$$

$$= \frac{(10)^6}{EI}[0.050 + 0.055 + 0.243 + 0.387 + 0.253]$$

$$\delta_{24} = 2y_2 = \frac{0.988(10)^6}{0.50(20)(10)^6}(2) = 0.198 \text{ cm}$$

Note there are a total of seven contributing terms or effects producing a single deflection.

12

Indeterminate Circular Beams

DEFLECTIONS OF THE ARCUATE BEAM, or of ring elements, have now been explained and we turn our attention to more constrained systems requiring elasticity relations for their solution. These relate the manner in which these constraints participate in the load distribution. As in Chaps. 6 and 9, we apply auxiliary loads simulating constraints and equate the corresponding complementary strain energy to 0. When the system has been quantified in terms of external and internal loads, we finally determine selected deflections as required.

Ring and cylindrical structures of various types are common in design applications, and it is important to be able to analyze them properly. Information in Chaps. 10 and 11 will serve to expedite these solutions. As shown previously, we are restricted to planar cases that are usually two dimensional; however, methods are applicable also to those extending axially with constant section. The latter then are two dimensional in the sense that stress and deflection patterns are identical in any plane taken perpendicular to the longitudinal axis. These structures usually divide into two categories—those connected to and supported by ground and those that are independent of ground but indeterminate by virtue of being of a continuous or connected nature.

12.1 *Identifying Redundancies*

With excess structural constraints, we determine the extent of inde-terminancy by visualizing the elastic behavior under load as influenced by

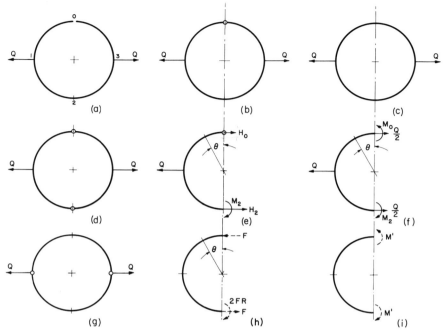

Figure 12.1 Statically determinate rings are shown in (a), (d), and (g) and once-redundant cases in (b) and (c). Free-body diagrams and auxiliary loadings are directly below the indeterminate cases.

the addition or removal of constraining elements. The arcuate beam is similar to the beam or bent in this respect, although there are some differences. For example, Fig. 12.1a, d, and g show statically determinate rings. In Fig. 12.1a there is a complete ring except for a small gap at the top, with loading on the horizontal diameter. Bending moment occurs only in the lower half, with a tensile force $Q_A = Q$ at 2.

Similarly, in Fig. 12.1d, with pins at 0 and 2, statics equations will determine beam loading at any point, and the maximum moment occurs at the load points 1 and 3.

In Fig. 12.1g the upper and lower halves share the load equally; that is, a shear of $Q/2$ is applied to each, producing maximum bending moment at 0 and 2 of $QR/2$.

With pin connection at 0 (Fig. 12.1b), a tensile force is developed at this point that carries a portion of the load. The problem is once redundant because of the introduction of a single constraint. Theoretically, the pin can carry vertical shear loads from one half to the other; however, no shear is possible because of symmetry.

Further redundancy is introduced if the ring has continuous stiffness at 0 (Fig. 12.1c), with both H_0 and M_0 connecting the left and right halves.

From the bisymmetrical nature of the ring, we note that $H_0 = H_2$ and $M_0 = M_2$ (Fig. 12.1f). Further $H_0 = H_2 = Q/2$, and the horizontal forces are determined. Thus although Fig. 12.1c is potentially twice redundant, only the couple remains to be found, and we need only one equation from elasticity and energy.

These illustrations are not completely general, as the pins and loads are located on perpendicular axes, but they do introduce the interplay between loading, constraints, and symmetry. In some cases lacking symmetry, joining of the split ring will create a maximum of three redundancies—axial and transverse forces and the internal moment. Similarly, joining the end of a curved beam to ground can introduce the same three components of beam reaction at the connected point.

12.2   *Indeterminate Energy Solution*

Solving the once-pinned case (Fig. 12.1b) we have an actual loading (Fig. 12.1e) that is symmetrical about the vertical, but not the horizontal axis. We therefore apply auxiliary loading to the complete decoupled ring (Fig. 12.1a) by forcing it apart at 0 with F. Induced preload is seen in Fig. 12.1h, from which the auxiliary moment is given by

$$m = FR(1 - \cos\theta) \tag{12.1}$$

Actual moment consists of two terms. The first is

$$M_H = -H_0 R(1 - \cos\theta) \tag{12.2a}$$

continuous from 0 to π. Due to Q we have

$$M_Q = QR \sin\left(\theta - \frac{\pi}{2}\right)$$
$$= -QR \cos\theta \tag{12.2b}$$

valid from $\theta = \pi/2$ to $\theta = \pi$.

Although we have isolated only one half of the ring, to argue that there is no work done by the internal F loads we should sum the complementary energy in the entire ring to 0:

$$0 = \int_0^\ell \frac{mM}{EI}\,ds = 2\int_0^{\ell/2} \frac{mM}{EI}\,ds = -2FH_0 R^3 \int_0^\pi (1 - \cos\theta)^2\,d\theta$$
$$- 2FQR^3 \int_{\pi/2}^\pi (\cos\theta)(1 - \cos\theta)\,d\theta$$

$$0 = -H_0(4.7124) - Q(-1.5708 - 0.2146)$$
$$H_0 = 0.3789Q \tag{12.3}$$

taking integral values from Table 10.3. From this, 38 percent of the Q load is taken at the pin. Continuing with statics,

$$H_2 = 0.6211Q \qquad M_2 = 0.2433QR \qquad (12.4)$$

12.3 Verification by Deflection Restoration

Although the results just obtained follow a simple procedure and the computations are not involved, possibilities for error always exist. If we can verify a result independently, the confidence level increases by an order of magnitude. In fact, a check virtually certifies an answer and is usually worth the effort. In this case, referring to Table 11.4a and b, we allow the disconnected ring to separate (Fig. 12.1a) and then bring the ends together to insert the pin by providing the H_0 loading at closure:

$$x_0 = 3.5708 \frac{QR^3}{EI} - 9.4248 \frac{H_0 R^3}{EI} = 0$$

$$H_0 = 0.3789Q \qquad (12.5)$$

With tabulated coefficients available, this is obviously a direct and rapid result; however, when the system does not conform to the tabular cases, it is impossible. One can still derive the deflection coefficients using energy methods, if a check is desired. This is not really an independent check, as the deflection solution is closely related to the auxiliary loading procedure used in Sec. 12.2.

12.4 The Continuous Ring

Taking the problem in Fig. 12.c, we have symmetrical ring loading (Fig. 12.1f and i), permitting us to integrate only from 0 to $\pi/2$. Complementary energy sums to 0 in each quadrant, in each half, and in the entire ring. Cancelling M', EI, and R^2, the energy equation for the redundant solution reduces to

$$0 = \frac{-QR}{2} \int_0^{\pi/2} (1 - \cos\theta)\, d\theta + M_0 \int_0^{\pi/2} d\theta$$

$$M_0 = \frac{0.2854}{1.5708} QR = 0.1817QR \qquad (12.6)$$

Verification of this result is found in Table 12.1 at M_2 with $N = 2$.
From statics we find M_1 using the 0–1 quadrant:

$$M = 0.1817QR - \frac{Q}{2}R - M_1 = 0$$

$$M_1 = -0.3183QR \qquad (12.7)$$

M_1 is assumed clockwise, corresponding to positive bending moment, or compression in the outer fibers. The negative result indicates a reverse direction, or compression at the inner fibers. This is again confirmed in Table 12.1.

12.5 Deflection Solution

In Fig. 12.1*b* we now determine the change in the horizontal diameter x_{13}. This requires diametral loading by auxiliary forces (Fig. 12.2). As in Eq. (12.2b) for Q,

$$m = -FR \cos \theta \tag{12.8}$$

The actual moment function is

$$M(\theta) = -(0.3789)QR(1 - \cos \theta) - QR \cos \theta \tag{12.9}$$

And for the complementary energy in the entire ring,

$$Fx_{13} = \int_0^\ell \frac{mM}{EI}\, ds = \frac{FQR^3}{EI} 2 \left[\int_{\pi/2}^{\pi} 0.3789 \cos \theta (1 - \cos \theta)\, d\theta \right.$$

$$\left. + \int_{\pi/2}^{\pi} \cos^2\theta\, d\theta \right]$$

$$x_{13} = \frac{2QR^3}{EI} \left[-0.3789(-1.5708 - 0.2146) + (1.5708 - 0.7854) \right]$$

$$= 0.2178 \frac{Qr^3}{EI} \tag{12.10}$$

By superposition (Table 11.4), we have the same result:

$$x_{13} = 1.5708 \frac{QR^3}{EI} - (0.3789Q)(3.5708) \frac{R^3}{EI} = 0.2178 \frac{QR^3}{EI}$$

Figure 12.2 Decoupling options for calculating the deflection of the horizontal diameter of the continuous ring (Fig. 12.1*c*).

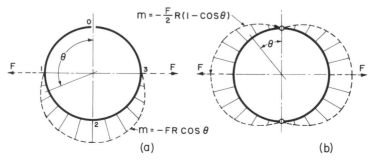

$$m = -\frac{F}{2} R(1 - \cos\theta)$$

$$m = -FR \cos \theta$$

(a)   (b)

Decoupling can also be effected by introducing pivotal flexibility at 0 and 2 (Fig. 12.2b). The m distribution has two components, one continuous from 0 to 2, and one from 1 to 2:

$$m = -\frac{F}{2}R(1 - \cos\theta)\Big|_0^\pi - Fr\cos\theta\Big|_{\pi/2}^\pi \qquad (12.11)$$

Note that although the auxiliary diagram is bisymmetrical, we cannot integrate only from 0 to $\pi/2$. This is because the M diagram is not bisymmetrical. The simpler option is Fig. 12.2a, but both lead to identical results.

With the continuous ring (Fig. 12.1c), we obtain x_{13} similarly using the Fig. 12.2 decoupling and auxiliary loading. The result, also found in Table 12.1, is

$$x_{13} = \frac{2(0.2976)}{4EI}QR^3 = 0.1488\frac{QR^3}{EI} \qquad (12.12)$$

The upper pin is seen to increase the diametral flexibility as it replaces the continuous ring at 0, and the factor is $0.2178/0.1488 = 1.46$.

Obviously, many other deflections can be calculated with appropriate auxiliary loading and decoupling, including the relative displacement in the vertical direction and slopes to the elastic curve. The M' couples for the latter would usually be applied in symmetrical pairs.

12.6 Multiple Redundancies

A more challenging indeterminate beam is shown in Fig. 12.3a, with distributed, or weight loading, and with both ends fixed. There is no symmetry to utilize, and we must determine a total of six unknown external reactions. Three equations are available from planar static equilibrium, and three equations must be derived from the elastic behavior using energy relations. These, in turn, follow from three auxiliary loadings (Fig. 12.3b–d) with respective auxiliary moments of

$$m_y = FR(1 - \cos\theta) \qquad (12.13a)$$

$$m_x = FR\sin\theta \qquad (12.13b)$$

$$m_\theta = M' \qquad (12.13c)$$

Note that the end at 1 is entirely decoupled from the support plane before the independent auxiliary loads are introduced.

Actual moment distribution is obtained by including the weight loading from Table 10.1h, with 1 as a free end:

$$M(\theta) = V_1 R(1 - \cos\theta) + H_1 R\sin\theta + M_1 - qR^2(\sin\theta - \theta\cos\theta) \qquad (12.14)$$

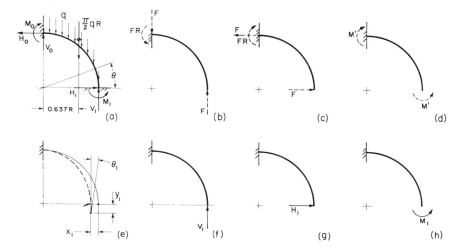

Figure 12.3 The fixed–fixed beam, three times redundant, requires three auxiliary loadings [(b), (c), and (d)] for the energy solution. Alternate solution by deflection superposition (e) relates to the three supporting reaction components at 1 [(f), (g), and (h)].

From energy

$$0 = \frac{1}{EI} \int_0^{\pi/2} FR(1 - \cos\theta) M(\theta) R\,d\theta \qquad (12.15a)$$

$$0 = \frac{1}{EI} \int_0^{\pi/2} FR(\sin\theta) M(\theta) R\,d\theta \qquad (12.15b)$$

$$0 = \frac{1}{EI} \int_0^{\pi/2} M'M(\theta)\,d\theta \qquad (12.15c)$$

These reduce to

$$0 = \int_0^{\pi/2} (1 - \cos\theta) M(\theta)\,d\theta \qquad (12.16a)$$

$$0 = \int_0^{\pi/2} (\sin\theta) M(\theta) R\,d\theta \qquad (12.16b)$$

$$0 = \int_0^{\pi/2} M(\theta)\,d\theta \qquad (12.16c)$$

Substituting for the definite integrals, the final simultaneous linear equations are

$$\begin{cases} 0.3562 V_1 + 0.5000 H_1 + 0.5708 M_1/R = 0.2961 qR & (12.17a) \\ 0.5000 V_1 + 0.7854 H_1 + \quad\quad M_1/R = 0.3927 qR & (12.17b) \\ 0.5708 V_1 + \quad\quad H_1 + 1.5708 M_1/R = 0.4292 qR & (12.17c) \end{cases}$$

with the following numerical results:

$$V_1 = 1.3048qR \qquad H_1 = -0.3950qR \qquad M_1 = 0.0506qR^2$$

Both V_1 and M_1 are directed as assumed (Fig. 12.3a), but H_1 is opposite or to the left.

From static equilibrium we now solve for the reactions at 0, including the resultant weight as shown at the centroid of the distribution. These are

$$V_0 = 0.2660qR \qquad H_0 = 0.3950qR \qquad M_0 = -0.0390qR^2$$

It is usually easiest to assume positive bending moment originally. The negative result for M_0 indicates compression at the inner fibers.

12.7   Deflection Restoration Analogy

Figure 12.3e indicates the deflection components due to the distributed load at 1 if unsupported. These can be derived or taken from Table 11.2b:

$$y_1 = 0.2961\frac{qR^4}{EI} \qquad x_1 = 0.3927\frac{qR^4}{EI} \qquad \theta_1 = 0.4292\frac{qR^3}{EI}$$

From Table 11.1 we consider all nine direct and coupled deflection components at the end of the cantilevered beam produced by the V_1, H_1, and M_1 end loadings. Summing all effects and equating the resultant deflections at 1 to 0,

$$
\left\{
\begin{aligned}
\sum y_1 &= 0.3562\frac{V_1 R^3}{EI} + 0.5000\frac{H_1 R^3}{EI} + 0.5708\frac{M_1 R^2}{EI} - 0.2961\frac{qR^4}{EI} = 0 \\
&\hspace{9.5cm}\text{(12.18a)}\\
\sum x_1 &= 0.5000\frac{V_1 R^3}{EI} + 0.7854\frac{H_1 R^3}{EI} + \frac{M_1 R^2}{EI} - 0.3927\frac{qR^4}{EI} = 0 \\
&\hspace{9.5cm}\text{(12.18b)}\\
\sum \theta_1 &= 0.5708\frac{V_1 R^3}{EI} + \frac{H_1 R^3}{EI} + 1.5708\frac{M_1 R^2}{EI} - 0.4292\frac{qR^4}{EI} = 0 \\
&\hspace{9.5cm}\text{(12.18c)}
\end{aligned}
\right.
$$

We note that these three simultaneous equations in V_1, H_1, and M_1 are *identical to those obtained from the fundamental approach* using auxiliary loading and integration [Eqs. (12.17)]. Solution will yield identical results.

This exercise illustrates how the complementary-energy solution relates the several reactions to the elastic behavior of the beam by incorporating the deflection parameters. Equation (12.15a), for example, based upon the vertical F loading (Fig. 12.3b), derives information related to the zero vertical condition at 1, which is an end condition. Thus although the two methods of solving the redundancy appear distinct, they are in fact interrelated.

12.8 *Selective Reduction of Redundancies*

The beam (Fig. 12.3a) is shown in Table 12.6d with all reactions summarized and indicates various degrees of constraint at the lower end. In Fig. 12.3c there is no slope constraint and M_1 is 0. Relating this case to the displacement solution [Eqs. (12.18)], we can solve for the pinned-end condition by deleting the proper terms. Now $M_1 = 0$ and $\theta_1 \neq 0$, and Eqs. (12.18a) and (12.18b) become

$$\begin{cases} y_1 = 0.3562 \dfrac{V_1 R^3}{EI} + 0.5000 \dfrac{H_1 R^3}{EI} - 0.2961 \dfrac{qR^4}{EI} = 0 & (12.19a) \\[4mm] x_1 = 0.5000 \dfrac{V_1 R^3}{EI} + 0.7854 \dfrac{H_1 R^3}{EI} - 0.3927 \dfrac{qR^4}{EI} = 0 & (12.19b) \end{cases}$$

Similarly in Table 12.6b, we have only horizontal support. Then $V_1 = M_1 = 0$, and θ_1 and y_1 are not 0. Only Eq. (12.18b) applies and it degenerates to

$$x_1 = 0.7854 \frac{H_1 R^3}{EI} - 0.3927 \frac{qR^4}{EI} = 0 \qquad (12.20)$$

Finally, in Table 12.6a we have from Eq. (12.18a)

$$y_1 = 0.3562 \frac{V_1 R^3}{EI} - 0.2961 \frac{qR^4}{EI} = 0 \qquad (12.21)$$

which determines V_1.

Note the load coefficients involving q are unchanged, as they depend only upon the deflections of the unsupported beam under distributed load. Further combinations could include moment constraint ($\theta_1 = 0$) with horizontal or vertical displacement permitted; however, slope constraints per se are rather rare.

12.9 *Tabulated Values*

Tables are included for representative indeterminate circular beams. They are by no means complete, as there are an endless number of beam and loading combinations. The listed values of reactions, moments, and deflections result from energy solutions predicated on bending effects only.

		N	2	3	4	6	8
		ψ	$\frac{\pi}{2}$	$\frac{\pi}{3}$	$\frac{\pi}{4}$	$\frac{\pi}{6}$	$\frac{\pi}{8}$
$\dfrac{M_1}{QR}$		$\frac{1}{2}\left[\dfrac{\cos\psi}{\sin\psi}-\dfrac{1}{\psi}\right]$	-0.3183	-0.1888	-0.1366	-0.0889	-0.0661
$\dfrac{M_2}{QR}$		$\frac{1}{2}\left[\dfrac{1}{\sin\psi}-\dfrac{1}{\psi}\right]$	0.1817	0.0999	0.0705	0.0451	0.0333
δ_1	$\dfrac{QR^3}{4EI}$	$\dfrac{\psi}{\sin^2\psi}+\dfrac{\cos\psi}{\sin\psi}-\dfrac{2}{\psi}$	0.2976	0.0638	0.0243	0.0067	0.0028
δ_2		$\dfrac{\psi\cos\psi}{\sin^2\psi}+\dfrac{1}{\sin\psi}-\dfrac{2}{\psi}$	-0.2732	-0.0570	-0.0215	-0.0059	-0.0024

Table 12.1 The uniform circular ring is loaded by symmetrically spaced radial loads in its plane. Deflections are radial, positive outward.

Table 12.2 The continuous obround beam in (a) and (b). As a continuous shell it is subjected to internal pressure (c).

		LOADING ON AXES			PRESSURE	
		(a)	(b)		(c)	
$M(\theta)$		$A+\frac{1}{2}\sin\theta$	$C-\frac{1}{2}[1-\cos\theta]$		$D+\cos\theta$	
M_0	QR	A	C	$pR^2\gamma$	$D+1$	
M_1		$A+\frac{1}{2}$	$C-\frac{1}{2}$		D	
M_2		$A+\frac{1}{2}$	$C-\frac{1}{2}[1+\gamma]$		$D-\frac{1}{2}\gamma$	
x_{04}		$2A[1+\gamma]+0.7854+\gamma$	$-[AB+\frac{1}{2}+\gamma+\gamma^2]$		$2D[1+\gamma]+1-\frac{2}{3}\gamma^2$	
y_{26}	$\dfrac{QR^3}{EI}$	$-[AB+\frac{1}{2}+\gamma+\frac{1}{2}\gamma^2]$	$-CB+0.3562+\gamma+\gamma^2+\frac{1}{3}\gamma^3$	$\dfrac{pR^4\gamma}{EI}$	$-D[1.1416+2\gamma+\gamma^2]+D_1$	
y_{17}		$2A+\frac{1}{2}$	$2C-0.2146$		$2D+1.5708$	
$y_{17}=y_{35}$		$A=\left[\dfrac{-(1+\gamma)}{\pi+2\gamma}\right]$	$C=\left[\dfrac{0.2854+\frac{1}{2}\gamma+\frac{1}{4}\gamma^2}{1.5708+\gamma}\right]$		$D=\left[\dfrac{-1+\frac{1}{3}\gamma^2}{1.5708+\gamma}\right]$	
	$\gamma=\dfrac{a}{R}$		$B=[1.1416+2\gamma+\gamma^2]$		$D_1=[-0.4292+\frac{2}{3}\gamma^2+0.4166\gamma^3]$	

	HYDROSTATIC		WEIGHT		CENTRIFUGAL	
	a		b		c	
$M(\theta)$	$1 - \theta \sin \theta - \frac{1}{2}\cos\theta$				$-\frac{1}{2} + \cos^2\theta$	
M_0	$\rho \frac{R^3}{2}$	$\frac{1}{2}$		$q R^2$	$\frac{\mu R^3 \Omega^2}{2}$	$\frac{1}{2}$
M_2		$\frac{3}{2}$				$\frac{1}{2}$
H_0	ρR^2	$\frac{3}{4}$	$\frac{1}{2}$	qR	$\mu R^2 \Omega^2$	1
H_2		$\frac{5}{4}$	$-\frac{1}{2}$			1
y_0	$\dfrac{\rho R^5}{2EI}$	0.4674		$\dfrac{qR^4}{EI}$	$\dfrac{\mu R^5 \Omega^2}{EI}$	0.5000
y_1		0.2797				0.2500
x_{13}		0.4292				0.1667
θ_1	$\dfrac{\rho R^4}{2EI}$	0.0708		$\dfrac{qR^3}{EI}$	—	0

Table 12.3 Equations for bending moment in the constant ring are identical for hydrostatic and weight loadings with base support. Centrifugal loading with rotation about the diameter results in a bisymmetrical system.

Table 12.4 The 90° circular segment is loaded by a central concentrated force, with various symmetrical end conditions.

		a	b	c	d
V_0		0.5000	0.5000	0.5000	0.5000
Q_A	Q	—	0.7071	0.9967	1.1492
Q_T		—	0	0.2896	0.4421
M_0	QR	0	0	0	0.0503
M_1		0.3536	0.2071	0.0872	0.0743
y_1	$\dfrac{QR^3}{EI}$	0.0606	0.0139	0.0020	0.0013
δ_{02}		0.0644	δ_0 −0.0290	0	0
θ_0	$\dfrac{QR^2}{EI}$	0.1313	0.0554	−0.0136	0

		a	b	c	d
V_0		0.7854	0.7854	0.7854	0.7854
Q_A	qR	—	1.1107	1.1777	1.1967
Q_T		—	0	0.0669	0.0860
M_0	qR^2	0	0	0	0.0063
M_1		0.2625	0.0324	0.0047	0.0031
y_1	$\dfrac{qR^4}{EI}$	0.0569	0.0030	0.0002	0.00015
δ_{02}		0.0623	δ_0 −0.0048	0	0
θ_0	$\dfrac{qR^3}{EI}$	0.1328	0.0136	−0.0008	0

Table 12.5 Similar to Table 12.4, but with distributed weight loading in the plane.

In Table 12.1 we have the complete ring with symmetrically disposed equal radial loads. Bending moments and radial deflections are given at the loads, and at midpoints between the loads.

Table 12.2 indicates the complexities arising as the circular beam is elongated. With pressure loading, the shell has bending moments not present in the cylindrical shell, and these tend to produce the governing stresses.

Table 12.6 The cantilevered quadrant, loaded by its own weight, has various types of end support.

		a	b	c	d
V_0		0.7395	1.5708	0.3541	0.2660
V_1	qR	0.8313	0	1.2167	1.3048
H_1		0	0.5000	0.2745	0.3950
M_0	qR^2	0.1687	0.5000	0.0578	0.0390
M_1		0	0	0	0.0506
x_1	$\dfrac{qR^4}{EI}$	0.0230	0	0	0
y_1		0	0.0461	0	0
θ_1	$\dfrac{qR^3}{EI}$	−0.0453	−0.0708	0.0092	0

		a	b	c	d
V_o	qR	1.1036	1.5708	1.3981	1.3006
V_1		0.4672	0	0.1727	0.2702
H		0	0.7049	0.4625	0.3913
M_o	qR^2	0.1036	0.1341	0.0644	0.0497
M_1		0	0	0	0.0410
x_1	$\dfrac{qR^4}{EI}$	-0.0175	0	0	0
y_1		0	0.5976	0	0
θ_1	$\dfrac{qR^3}{EI}$	-0.0380	0.0284	-0.0075	0

Table 12.7 Similar to Table 12.6, but with the beam inverted.

Table 12.8 Reactions and moments for the pinned–pinned semicircular beam with different center loads. Deflections are also given for the center point.

		a	b		c
$\dfrac{V}{Q}$		0.5000	0.5000	$\dfrac{V}{M_z/R}$	0.5000
$\dfrac{H}{Q}$		0.5000	0.3183	H	0
$\dfrac{M}{QR}$		0	0	$\dfrac{M}{M_z}$	0
$\dfrac{M_1}{QR}$		0	0.1817	$\dfrac{M_1}{M_z}$	±0.5000
x_1	$\dfrac{QR^3}{EI}$	0.0708	0	$\dfrac{M_zR^2}{EI}$	-0.0719
y_1		0	0.0190		0
θ_1	$\dfrac{QR^2}{EI}$	-0.0719	0	$\dfrac{M_zR}{EI}$	0.1781

	(a)	(b)		(c)
$\dfrac{V_0}{Q}$	0.2272	0.6211	$\dfrac{V_0}{M_z/R}$	0.5455
$\dfrac{V_2}{Q}$	0.2272	0.3789	$\dfrac{V_2}{M_z/R}$	0.5455
$\dfrac{M_0}{QR}$	0.5455	0.2423	$\dfrac{M_0}{M_z}$	0.0911
$\dfrac{M_1}{QR}$	0.2272	0.3789	$\dfrac{M_1}{M_z}$	-0.4545 , 0.5455
x_1	0.1129	0.0943		-0.0134
y_1 $\left(\dfrac{QR^3}{EI}\right)$	0.0943	0.1090	$\dfrac{M_z R^2}{EI}$	0.0260
x_2	0.2400	0.2577		0.0911
θ_1	-0.0134	0.0260		0.1683
θ_2 $\left(\dfrac{QR^2}{EI}\right)$	-0.1431	-0.1902	$\dfrac{M_z R}{EI}$	-0.1431

Table 12.9 Similar to Table 12.8, but for the fixed–supported beam.

Basic distributed loadings are shown in Table 12.3a and b, with a cylindrical tube full of liquid resting on a surface, and a ring or tube supporting its own weight. As seen, these cases have much in common. In the rotational case (Table 12.3c), distributed centrifugal loading develops as the ring revolves about its diametral axis. We note the deflections increase as the *fifth* power of the radius.

Tables 12.4 and 12.5 provide design data for the symmetrical 90° arc with central and weight loading and various end supports.

In Table 12.6 the circular quadrant beam is subjected to weight loading and clamped at the upper end. Various constraints are present at the lower end. Table 12.7 parallels Table 12.6, but the center of the beam is reversed.

Tables 12.8–12.11 pertain to the semicircular beam with different end conditions. Horizontal, vertical, and couple loads are applied at the midpoint.

Numerical values contain some round-off error, but are sufficiently accurate for most purposes and generally exact to at least three decimal places.

	(a)	(b)		(c)
$\dfrac{V}{Q}$	0.3682	0.4696 0.5304	$\dfrac{V}{M_z/R}$	0.5991
$\dfrac{H}{Q}$	0.6678 0.3322	0.3760	$\dfrac{H}{M_z/R}$	0.1262
$\dfrac{M_0}{QR}$	0.2635	0.0608	$\dfrac{M_0}{M_z}$	0.1982
$\dfrac{M_1}{QR}$	0.0360	0.1544	$\dfrac{M_1}{M_z}$	-0.5271 , 0.4792
x_1 $\left[\dfrac{QR^3}{EI}\right]$	0.0332	0.0087	$\dfrac{M_zR^2}{EI}$	-0.0437
y_1	0.0087	0.0170		-0.0065
θ_1 $\left[\dfrac{QR^2}{EI}\right]$	-0.0437	-0.0065	$\dfrac{M_zR}{EI}$	0.1568
θ_2	0.0783	0.0477		-0.0590

Table 12.10 Similar to Table 12.8, but for the fixed–pinned case.

Table 12.11 Similar to Table 12.8, but with fixed–fixed ends.

	(a)	(b)		(c)
$\dfrac{V}{Q}$	0.3183	0.5000	$\dfrac{V}{M_z/R}$	0.6366
$\dfrac{H}{Q}$	0.5000	0.4591	H	0
$\dfrac{M}{QR}$	± 0.1817	0.1106	$\dfrac{M}{M_z}$	± 0.1366
$\dfrac{M_1}{QR}$	0	0.1515	$\dfrac{M_1}{M_z}$	± 0.5000
x_1 $\left[\dfrac{QR^3}{EI}\right]$	0.0189	0	$\dfrac{M_zR^2}{EI}$	0.0329
y_1	0	0.0117		0
θ_1 $\left[\dfrac{QR^2}{EI}\right]$	0.0329	0	$\dfrac{M_zR}{EI}$	0.1488

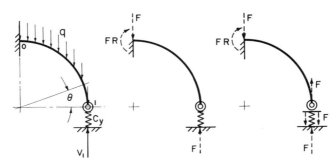

Figure 12.4 Vertical flexibility at 1 can involve preload from base or a separational displacement at 1.

12.10 *Flexible Supports*

Since we have examined the Table 12.6 beam in some detail with respect to the various support possibilities, we now include the effects of compliance replacing the assumed rigid support elements. Although there cannot be a completely rigid support physically, these are usually sufficiently rigid so that beam flexure is the only significant parameter. In effect, elastic energy in bending greatly exceeds that created in the supports. Any energy in the supports, however, does alter both the indeterminate and deflection solutions.

For instance, if we have a known vertical compliance C_y in the fixed-supported beam (Table 12.6a), we have as a model Fig. 12.4. An F preload can be applied from ground through the support, or as a separational force between the beam and the spring. In both cases, there is no work done by the preload or its reactions during the application of the weight loading. We have auxiliary effects ($m = FR \sin \theta$) in the beam and F in the spring. Actual load in the spring is V_1, and we have for the complementary energy

$$0 = \int_0^\ell \frac{mM}{EI}\, ds + \sum CFQ \tag{12.22a}$$

$$= \frac{1}{EI} \int_0^{\pi/2} (FR \sin \theta)\, M(\theta)\, R\, d\theta + C_y FV_1 \tag{12.22b}$$

$$= \int_0^{\pi/2} \sin \theta M(\theta)\, d\theta + \left(\frac{C_y EI}{R^2} \right) V_1 \tag{12.22c}$$

from which we solve for V_1 that also appears in the $M(\theta)$ term. Because of the spring energy term, we cannot cancel EI nor R. The coefficient $(C_y EI / R^2)$ is dimensionally a length. It serves to compare the beam and spring energies and becomes 0 for a rigid support.

By a similar procedure (Chap. 2), we can also account for horizontal or angular flexibilities at either support. As explained in Sec. 11.5, strain energy due to direct stress in the beam can be included by integration over the length of the beam in terms of the preload and actual load distributions.

12.11 *Degenerate Solutions*

In special cases involving bending, it is not possible to solve a redundancy using the basic (mM) integrals. For instance, a semicircular shell with uniform internal pressure (Fig. 12.5) has an actual-moment expression for a unit strip of

$$M(\theta) = -qR^2(1 - \cos\theta) + qR^2(1 - \cos\theta) - H_2 R \sin\theta \quad (12.23)$$

where the first term from the vertical support and the second term from the distributed loading (Table 10.1*d*) cancel, thereby eliminating the load factor q from the solution. This obviously precludes an evaluation of H_2 since this redundant force must be proportional to q. Since Fig. 12.5 represents a real problem, there must be a solution.

This can be obtained by involving the strain energy from the direct stress [Eq. (11.12)] and distributions from Table 10.1*a*, *b*, and *d*:

$$m = FR \sin\theta \qquad F_A = F \sin\theta$$

$$0 = \frac{2}{EI} \int_0^{\pi/2} (FR \sin\theta)(-H_2 R \sin\theta) R \, d\theta$$

$$+ \frac{2}{EA} \int_0^{\pi/2} (F \sin\theta) qR \cos\theta - H_2 \sin\theta + qR(1 - \cos\theta) R \, d\theta$$

$$0 = \frac{-H_2 R^3}{I} \int_0^{\pi/2} \sin^2\theta \, d\theta - \frac{HR}{A} \int_0^{\pi/2} \sin^2\theta \, d\theta + \frac{qR^2}{A} \int_0^{\pi/2} \sin\theta \, d\theta$$

$$0.7854 H_2 R \left[\frac{R}{I} + \frac{1}{A} \right] = \frac{qR^2}{A}$$

$$H_2 = 1.2732 qR \left[\cfrac{1}{1 + \cfrac{AR^2}{I}} \right] \qquad (12.24a)$$

For constant wall, or a rectangular unit strip,

$$\frac{AR^2}{I} = \frac{tR^2}{t^3/12} = 12\left(\frac{R}{t}\right)^2$$

$$H_2 = 1.2732 qR \left[\cfrac{1}{1 + 12\left(\cfrac{R}{t}\right)^2} \right] \qquad (12.24b)$$

And the actual bending moment is $-H_2 R \sin\theta$ from Eq. (12.23).

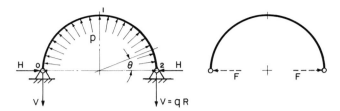

Figure 12.5 The semicircular beam, once redundant with respect to H, requires consideration of direct strain energy for solution.

This will be 0 at 2 and a maximum of $H_2 R$ at the center of the shell. With our slender beam assumption, however, t must be considerably less than R. H_2 will be very small, but not 0. In spite of the horizontal support, the shell is substantially in simple hoop tension with negligible bending.

12.12 *Simple Thermal Deformation*

As indicated in Sec. 2.8, misalignment at assembly or differential temperatures can lead to internal loading in statically indeterminate structures. Solutions for these loads are based upon the load-deflection characteristics, and the stiffer a system, the more the loads developed by a given displacement error.

If a circular ring is heated uniformly (Fig. 12.6), it expands radially and circumferentially. Basically, there is a change in the length of the bar circumferentially:

$$\delta_C = 2\pi R\lambda(\Delta T) = 2\pi R\epsilon_T \tag{12.25a}$$

where $\lambda =$ coefficient of thermal expansion for the material
  $\Delta T =$ increase of temperature from ambient
  $\epsilon_T = \lambda(\Delta T) =$ dimensionless unit elongation due to temperature

Figure 12.6 Free thermal expansion of a ring develops circumferential and radial deflections.

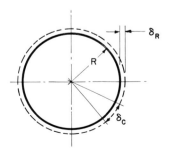

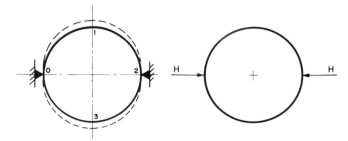

Figure 12.7 With thermal expansion, rigid horizontal constraint of a ring develops horizontal reactions at the supports.

Also diametral and radial increments are:

$$\delta_D = \frac{\delta_C}{\pi} = 2R\epsilon_T \qquad (12.25b)$$

$$\delta_R = \frac{\delta_D}{2} = R\epsilon_T \qquad (12.25c)$$

Note the diametral increase is the same as for a straight bar of length $2R$, visualized as in position on the ring diameter.

The circular ring is *unstressed* by the expansion, as the growth is unresisted. In other words, with this type of deformation, the ring behaves as if statically determinate. This would also be true if the ring contained diametral spokes of the same material at the same temperature.

12.13   *Thermal Stresses*

With the simple ring expanding, but constrained between fixed supports (Fig. 12.7), the expansion is prevented at the two points. Contact reactions are developed creating bending stresses and direct stresses. Physically, we can view the restraining forces as superimposed after the free expansion has occurred. The required reactions are related to the compliance of the ring when diametrically loaded. Ring stiffness can be determined by energy methods for deflection after a redundant solution, but with Table 12.1 available, we have simply

$$x_{02} = \delta_D - \frac{2(0.2976)}{4}\frac{HR^3}{EI} = 0$$

$$H = 13.44\left(\frac{EI}{R^2}\right)\epsilon_T \qquad (12.26)$$

The horizontal reactions are directly proportional to the beam-section factor EI, to the impressed thermal unit elongation, and inversely proportional to R^2.

From Table 12.1 we can now determine the corresponding bending stresses due to H, maximum at the contact points:

$$M_0 = M_2 = 0.3183HR = 4.278\left(\frac{EI}{R}\right)\epsilon_T \qquad (12.27a)$$

$$M_1 = M_3 = -0.1817HR = -2.442\left(\frac{EI}{R}\right)\epsilon_T \qquad (12.27b)$$

Signs of the moment factors are reversed because the loading is inward rather than outward. Further converting the moments to bending stresses with (I/c),

$$\sigma_0 = \sigma_2 = 4.278\left(\frac{Ec}{R}\right)\epsilon_T \qquad (12.28a)$$

$$\sigma_1 = \sigma_3 = 2.442\left(\frac{Ec}{R}\right)\epsilon_T \qquad (12.28b)$$

where the stresses are independent of I.

12.14 Thermal Deflections

Again in the simple case (Fig. 12.7), we consider the vertical radial deflection y_1. This has two components:

1. The unconstrained thermal displacement δ_R (Fig. 12.6).
2. The superimposed elastic displacement due to H (Table 12.1):

$$y_1 = R\epsilon_T + 0.2732\left[13.44\left(\frac{EI}{R^2}\right)\epsilon_T\right]\frac{R^3}{4EI}$$

$$= (1 + 0.9180)R\epsilon_T = 1.918R\epsilon_T \qquad (12.29)$$

where this deflection is independent of both E and I.

12.15 Thermal Loads in a Grounded Beam

A somewhat more comprehensive structure is now analyzed (Fig. 12.8). The semicircular beam is clamped-pinned, as in Table 12.10; however, the tabular data are not applicable since we do not have center loading. Instead the latent thermal expansion, shown dashed, is resisted by the supports at 0 and 2. With the pin removed the expanded beam is still circular, and the vertical tangent maintained at 0 by the fixed end. Thus the expanded points

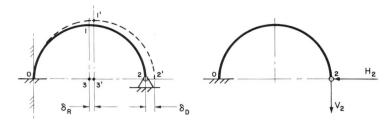

Figure 12.8 Resistance to thermal expansion of the fixed–pinned semicircle is evaluated by superposition of H_2 and V_2 to produce δ_{02} horizontally.

$3'$ and $2'$ lie on the original centerline. We then force the end point $2'$ back to its original position, and the reinserted pin carries a single vector load by the forcing phase.

Taking perpendicular components of this force and applying Table 11.1 factors for the circular cantilever, we have by superposition,

$$\begin{cases} \sum y_2 = 4.7124\dfrac{V_2 R^3}{EI} + 2\dfrac{H_2 R^3}{EI} = 0 & (12.30a) \\[4mm] \sum x_2 = 2\dfrac{V_2 R^3}{EI} + 1.5708\dfrac{H_2 R^3}{EI} - \delta_D = 0 & (12.30b) \end{cases}$$

With V_2 and H_2 in the assumed directions shown, all influence coefficients are positive. It is important to use consistent signs when interpreting the table. Solving the equations,

$$V_2 = 0.5878\left(\frac{EI}{R^3}\right)\delta_D \quad \text{and} \quad H_2 = -2.770\left(\frac{EI}{R^3}\right)\delta_D \quad (12.31a)$$

If δ_D is an assembly error, this dimension is substituted directly, or if due to expansion we use $2R\epsilon_T$, and

$$V_2 = 1.176\left(\frac{EI}{R^2}\right)\epsilon_T \quad \text{and} \quad H_2 = -5.540\left(\frac{EI}{R^2}\right)\epsilon_T \quad (12.31b)$$

Contrary to intuitive conclusions, H_2 is to the right because of coupling interplay. Indeed, in Eq. (12.30a) with the sum 0, one term must be negative.

Having the reaction components at 2, we can determine thermal stresses. For instance, the shear developed in the pin is the vector sum of V_2 and H_2. Bending moment at 0 is obtained from statics, or $\sum M_0 = 0$.

As in Sec. 12.14, we can further obtain final deflections. To find θ_2, we use Table 11.1 and superimpose the effects of V_2 and H_2 on the cantilever that have produced the horizontal displacement from 2' to 2:

$$\theta_2 = \beta_{zx}V_2 + \beta_{zy}H_2 = -1.845\epsilon_T \tag{12.32}$$

This dimensionless slope increment is also absolute because the expanded slope of the tangent at 2' is vertical. It is *independent of EI and R*, depends only on the unit thermal expansion, and is counterclockwise.

12.16   *Hoop Stress*

The long cylindrical vessel with internal pressure has been mentioned, and it is an elementary problem. As indicated (Fig. 10.2), the shell wall is under uniform circumferential tension. The problem is not indeterminate and does not involve bending. Taking a unit strip, the tensile hoop stress is

$$\sigma_A = \frac{Q_A}{A} = \frac{qR}{t} = \frac{pR}{t} \tag{12.33}$$

where q is the constantly distributed circumferential loading per unit axial length, numerically equal to p. Circumferential elongation of the bar results in an enlarged circle (Fig. 12.6), as with thermal expansion but now with stress present. The dimensional increases are

$$\delta_C = 2\pi R\left(\frac{\sigma A}{E}\right) = 2\pi\frac{pR^2}{Et} \tag{12.34a}$$

$$\delta_R = \frac{\delta C}{2\pi} = \frac{pR^2}{Et} \tag{12.34b}$$

With constraints (Fig. 12.7), redundancy and bending moments are introduced, and solutions for reactions, moments, and deflections are patterned after those for thermally produced expansion. Reversed, or external pressure, is then analogous to contractions caused by lowered temperatures.

Examples

12.1. The semicircular beam, pin-connected to a straight beam of the same *EI* factor, carries a central concentrated load and is supported at the bottom. Find:
  (a) The tensile force in the straight bar
  (b) The vertical deflection of 0

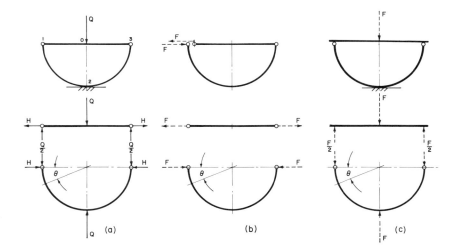

Solution:

(a) We determine the structure to be once redundant, as there would be static stability with the circular beam carrying the vertical loads minus the tension provided by the horizontal bar. The horizontal tension H is unknown, and in simulation of its function, we apply the auxiliary preload F at 1. As seen in part *b* of the figure, m is 0 in the straight section, though M exists:

$$0 = 2 \int_1^2 \frac{mM}{EI} \, ds$$

$$= \frac{2}{EI} \int_0^{\pi/2} (-FR \sin \theta) \left[\frac{Q}{2} R(1 - \cos \theta) - HR \sin \theta \right] R \, d\theta$$

$$0 = \int_0^{\pi/2} -\frac{Q}{2} \sin \theta (1 - \cos \theta) + H \sin^2 \theta \, d\theta$$

With integrals from Table 10.3, $H = 0.3183Q$.

(b) For y_0 we decouple at the pins (part *c* of the figure), eliminating H. In the curved beam $m = (FR/2)(1 - \cos \theta)$, and the deflection contribution of the straight beam is well known:

$$y_0 = \int_0^\ell \frac{mM}{FEI} \, ds + \frac{Q\ell^3}{48EI}$$

$$= \frac{QR^3}{EI} 2\left(\frac{1}{2} \right) \left[\frac{1}{2}(0.3562) - (0.3183)(0.5000) \right] + \frac{QR^3}{6EI}$$

$$= 0.1856 \frac{QR^3}{EI}$$

12.2. The geometry of Example 12.1 is now formed as a continuous beam. Determine:
(a) Bending moment at 1
(b) Slope developed at 1

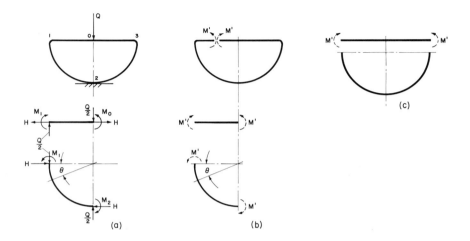

Solution:
(a) With the pins replaced by fixed corners, moments are transmitted at 1 and 3. Considering symmetry, we solve for the two redundancies H and M_1. Applying first the auxiliary loads F (Example 12.1b), and assuming M_1 positive, the first elasticity equation is

$$0 = \int_0^\ell \frac{mM}{EI} \, ds$$

$$= \frac{2}{EI} \int_0^{\pi/2} (-FR \sin\theta) \left[\frac{Q}{2} R(1 - \cos\theta) - HR \sin\theta + M_1 \right] R \, d\theta$$

$$= \frac{-QR}{4} + 0.7854 HR - M_1$$

$$0.7854 HR - M_1 = 0.2500 QR \qquad (1)$$

For the second equation we apply *internal* loading using M' (part b of the figure),

$$0 = \int_0^\ell \frac{mM}{EI} \, ds = \frac{2}{EI} \int_0^{\pi/2} M'M(\theta) R \, d\theta + \frac{2M'}{EI} \sum A$$

$$= \left[\frac{QR}{2}(0.5708) - HR(1) + M_1(1.5708) \right] R + \left[M_1 R + \frac{Q}{2} \frac{R^2}{2} \right]$$

$$HR - 2.5708 M_1 = 0.5354 QR \qquad (2)$$

Solving (1) and (2) simultaneously, $H = 0.1053Q$ and $M_1 = -0.1673QR$.

(b) For rotation of the corner, we decouple at the corners and apply *external* M' moments symmetrically (part c of the figure). Known M for the straight beam from 1 to 0, as part of the complete structure with Q loading is

$$M = -0.1673QR + \frac{Q}{2}x_1$$

Considering complementary energy in 1-0-3 *only*,

$$2M'\theta_1 = \frac{2M'}{EI}\sum_1^0 A$$

$$\theta_1 = \frac{1}{EI}\left[-0.1673QR^2 + \frac{Q}{2}\left(\frac{R^2}{2}\right)\right] \quad \text{(Table 5.1)}$$

$$= 0.0827\frac{QR^2}{EI} \quad \text{(clockwise as assumed)}$$

12.3. A continuous ring consists of two halves, each of constant but different sectional moment of inertia. In terms of the relative inertia ratio, I_a/I_b, find:
   (a) The ratio of the load in a to the load in b
   (b) The bending moment at the junctures

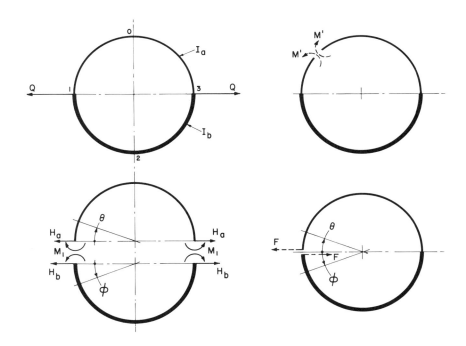

Solution:

(a) There are effectively three unknowns to be determined from elasticity—the shears H_a and H_b and the connecting moments M_1. No axial, or tangential, loads occur at 1 or 3 because of symmetry about the 0–2 axis. If there were a tensile load at 1 in I_a, for equilibrium there would have to be a corresponding compressive load at 3, but this violates symmetry. For the elasticity equations, we apply M' couples and shear forces F at 1. Then

$$0 = 2 \int_0^{\pi/2} \frac{M(\theta)}{EI_a} \, d\theta + 2 \int_0^{\pi/2} \frac{M(\phi)}{EI_b} \, d\phi$$

$$0 = 2 \int_0^{\pi/2} \frac{(\sin\theta) M(\theta)}{EI_a} \, d\theta + 2 \int_0^{\pi/2} \frac{(-\sin\phi) M(\phi)}{EI_b} \, d\phi$$

Reducing and substituting,

$$0 = \frac{1}{I_a} \int_0^{\pi/2} (H_a R \sin\theta + M_1) \, d\theta + \frac{1}{I_b} \int_0^{\pi/2} (H_b R \sin\phi + M_1) \, d\phi$$

$$0 = \frac{1}{I_a} \int_0^{\pi/2} (\sin\theta)(H_a R \sin\theta + M_1) \, d\theta$$

$$+ \frac{1}{I_b} \int_0^{\pi/2} (-\sin\phi)(H_b R \sin\phi + M_1) \, d\phi$$

With integral values and $\alpha = (I_a/I_b)$,

$$H_a + \alpha H_b + 1.5708(1 + \alpha)\left(\frac{M_1}{R}\right) = 0$$

$$0.7854 H_a - 0.7854 \alpha H_b + (1 - \alpha)\left(\frac{M_1}{R}\right) = 0$$

And with $H_a + H_b = Q$

$$\frac{H_a}{H_b} = \frac{\alpha(\alpha + 9.556)}{(1 + 9.556\alpha)}$$

For equal stiffness, $\alpha = 1$ and $H_a = H_b$.

(b) From the simultaneous elasticity relations we now solve for the moments:

$$M_1 = M_3 = \left[-\frac{6.720\alpha}{1 + 19.112\alpha + \alpha^2}\right] QR$$

For $\alpha = 1$,

$$M_1 = M_3 = -0.3183 QR = \frac{-1}{\pi} QR \quad \text{(Table 12.1)}$$

Numerical values for M_0 and M_2 can be determined for a given α ratio applying equilibrium equations to the quadrant beams. M_2 is greater then M_0 and

$$\frac{M_0}{M_2} = \frac{\alpha(\alpha + 2.836)}{1 + 2.836\alpha}$$

12.4. A cylindrical tube half full of liquid is suspended from the top. Determine:
  (a) Moment and shear at the horizontal diameter
  (b) Change in the horizontal diameter

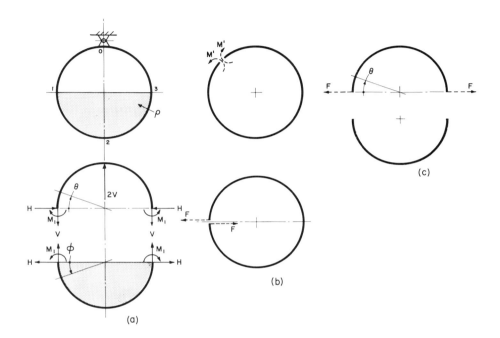

Solution:

(a) For the indeterminate solution, we require the actual-moment distributions for the upper and lower halves in terms of the horizontal shear, vertical tension, bending moment, and moment due to the hydrostatic pressure, breaking the structure at 1, and using Table 10.1*f*,

$$M(\theta) = -HR\sin\theta - VR(1 - \cos\theta) + M_1$$

$$M(\phi) = +HR\sin\phi - VR(1 - \cos\phi) + M_1 + \tfrac{1}{2}\rho R^3(\sin\phi - \phi\cos\phi)$$

With the auxiliary couple loading (part *b* of the figure), and $V = \rho(\pi R^2/4)$,

$$0 = -HR - (0.5708)VR + (1.5708)M_1$$
$$+ HR - (0.5708)VR + (1.5708)M_1 + \tfrac{1}{2}\rho R^3(1 - 0.5708)$$

and

$$-1.1416VR + M_1 = -0.2146R^3 \qquad M_1 = 0.2171\rho R^3$$

Note we have obtained M_1 without simultaneously finding H because of the antisymmetrical condition of the shears. Next, to find H we apply the related auxiliary F loading:

$$m(\theta) = FR \sin \theta$$
$$m(\phi) = -FR \sin \phi$$

The energy summation reduces to

$$0 = -(0.7854) HR - (0.5000) VR + M_1 - (0.7854) HR + (0.5000) VR$$
$$- M_1 - \tfrac{1}{2}\rho R^3 (0.7854 - 0.3927)$$

$$H = -0.1250\rho R^2$$

Shears are directed oppositely to the assumed position, and now with antisymmetrical m, the V terms and the M_1 terms cancel, with a direct solution for H.

(b) For the deflection we decouple the two halves and apply the pair of F loads (part c of the figure). Then $m(\theta) = FR \sin \theta$ and with

$$M(\theta) = +HR \sin \theta - VR(1 - \cos \theta) + M_1$$

we need only consider the top half, avoiding completely the ϕ expressions and the hydrostatic fluid loading:

$$x_{13} = \frac{2R}{EI} \int_0^{\pi/2} \sin \theta \, M(\theta) \, R \, d\theta$$

$$= \frac{2\rho R^5}{EI} (0.1250)(0.7854) - (0.7854)(0.5000) + (0.2171)(1)$$

$$= -0.1549 \frac{\rho R^5}{EI} \quad \text{(contraction)}$$

As in all shells, forces, moments, and moment of inertia are understood to be per unit length (Table 10.2).

12.5. An obround tube carries constant internal pressure (Table 12.2c). If $R = 24$, $a = 10$, and the wall thickness is 0.30, find the expression for the maximum bending stress.

Solution: Substituting values, $\gamma = 0.417$ and $D = -0.474$

$$M_0 = (D + 1) pR^2 \gamma = (1 - 0.474)(576)(0.417) p = 126.3p$$
$$M_1 = DpR^2 \gamma = -113.8p$$
$$M_2 = \left(D - \frac{0.417}{2} \right) p(576)(0.417) = -163.9p$$

Taking the largest, and from Table 10.2 the bending stress is

$$\sigma_2 = \frac{6M_2}{t^2} = \frac{6(163.9p)}{(0.30)^2} = 10{,}930p$$

13

Torsional and Transverse Shear

UNTIL NOW, WE HAVE CONSIDERED only the direct and bending stresses in structures. These are directed parallel to the axes of the members and can be superimposed algebraically. Respective deflections, however, are normal to each other. We now introduce the topic of shear stress and the related energy utilized to effect deflection and redundant solutions.

First to be discussed is torsional shear, present in prismatic elements with twist about the central longitudinal axis due to torque. This subject is restricted to purely torsional cases, but is of particular importance when incorporated with bending in later chapters. Secondly, we investigate transverse shear effects in beams, and their potential contribution to the elastic energy evaluation is considered. This can occur in beams with significant depth, and methods are given to quantify this effect.

Torsional elements, as with direct stress elements, however, are *independent of the slenderness restriction*, provided the length is not so short that end distortions preclude the development of a substantially classical torsional stress pattern.

13.1 *Vector Notation*

As we introduce torsional loads and combine them with bending moments, and as we analyze more complex geometries, it is necessary to define couples vectorially. An extremely useful notation is the double-headed vector (Fig.

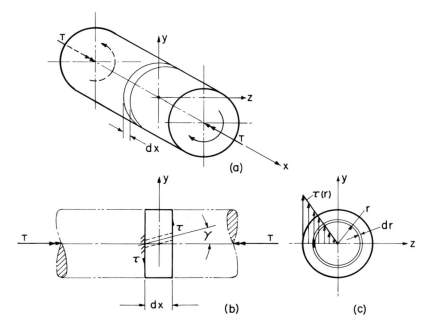

Figure 13.1 Axial element of an elastic cylinder in torsion carries face shear stresses varying linearly from 0 to maximum.

13.1). Significance of this vector is as follows:

1. Length indicates the magnitude of the couple.
2. The vector is coincident with the axis about which the couple acts.
3. The double head indicates the sense of the couple, *clockwise* when viewed from the tail of the arrow, similar to the advancement of a right-hand thread in the direction of arrow flight when so rotated.
4. Either the head or the tail of the arrow can coincide with the plane perpendicular to the vector in which the couple acts.

When the vector appears as a point in a projected view, this convention can further include a circle or a cross to indicate a view of the head or tail of the vector, respectively.

Torsional and moment couples have space orientation relative to a structure. As a vector quantity the double-headed vector can be resolved into components or combined into resultants. These components and resultants, in turn, maintain all the characteristics of the double-headed vector with respect to specifying magnitude, direction, and location of a couple.

13.2   *Stresses and Deflections in Simple Torsion*

The cylindrical shaft subjected to twist about its longitudinal axis is indeed a common element in mechanical devices. Torque can be carried statically, but usually the shaft is rotating and transmitting power. Although in a state of three-dimensional stress, there is a relatively simple shear stress distribution on the circular surfaces intercepted by any plane perpendicular to the axis. This stress increases linearly from 0 at the center to maximum at the surface (Fig. 13.1*c*):

$$\tau(r) = \frac{Tr}{J} \tag{13.1}$$

where T = the torsional load
   r = radius to a fiber
   $J = \pi D^4/32$ = polar moment of inertia of the sectional area
   D = outside diameter of the shaft

Angular deformation of the cylindrical bar is

$$\theta = \frac{T\ell}{GJ} \tag{13.2}$$

where θ = total angle of twist (radians)
   ℓ = axial length over which the twist occurs
   G = shear modulus of elasticity

If subjected to pure torque, there will be only angular twist, whether the shaft is constrained radially by bearings or not. As with simple tension–compression members, the flexibility of the torsional element can be expressed as either angular compliance or spring rate:

$$C = \frac{\theta}{T} = \frac{1}{K}$$

$$K = \frac{T}{\theta} = \frac{1}{C}$$

13.3   *Noncircular Sections*

Frequently, the cylinder of Fig. 13.1 is hollow or tubular. If so, the same relations for stress and deflection are valid, but J is reduced by removal of the core material. For a given torque and outside diameter, both stresses and deflections are increased, with

$$J = \frac{\pi}{32}(D^4 - d^4) \quad \text{and} \quad \frac{J}{r} = \frac{\pi(D^4 - d^4)}{16D} \tag{13.3}$$

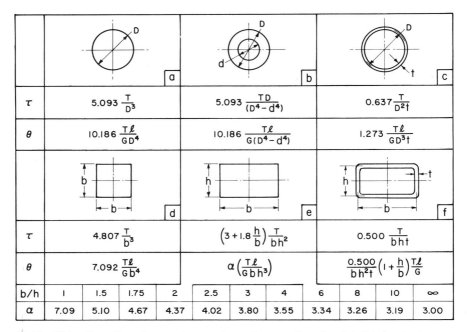

Table 13.1 Equations for maximum shear stress and angle of twist due to torque load T on prismatic sections. G is the shear modulus of elasticity and ℓ the length of the torsional element.

Note the J/r value is not the difference between the outer and inner sectional moduli.

Further prismatic sections that are sometimes used to carry torque include regular polygons, solid rectangles, rectangular tubes, angles, channels, and I-beams. In these the stresses do not vary simply as in Eq. (13.1). Formulas are available for most possibilities. We are primarily concerned with equivalent stiffness factors, J_e or K_e, which also can be found in the literature (Table 13.1):

$$\theta = \frac{T\ell}{GJ_e} \tag{13.4}$$

Although similar in purpose, J_e is *not* a polar moment of inertia, rather only a stiffness term. It will generally be allowable to substitute J_e for J in all the energy relations and procedures that follow in which it is necessary to define a noncircular torsional elastic element.

Stress and deflection relations are summarized in Table 13.1 for torsional loading of bars with various cross sections. Bending stress equations for the same cross sections are given in Table 13.2.

	a	b	c
σ	$10.186\,\dfrac{M}{D^3}$	$10.186\,\dfrac{MD}{(D^4-d^4)}$	$1.273\,\dfrac{M}{D^2 t}$
$I_x = I_y$	$0.0491\,D^4$	$0.0491\,(D^4 - d^4)$	$0.3926\,D^3 t$

	d	e	f
σ_x	$6\,\dfrac{M}{b^3}$	$\dfrac{1}{6}bh^2$	$\dfrac{h^2 t}{3}\left[1+3\left(\dfrac{b}{h}\right)\right]$
σ_y		$\dfrac{1}{6}b^2 h$	$\dfrac{b^2 t}{3}\left[1+3\left(\dfrac{h}{b}\right)\right]$
I_x	$\dfrac{1}{12}b^4$	$\dfrac{1}{12}bh^3$	$\dfrac{h^3 t}{6}\left[1+3\left(\dfrac{b}{h}\right)\right]$
I_y		$\dfrac{1}{12}b^3 h$	$\dfrac{b^3 t}{6}\left[1+3\left(\dfrac{h}{b}\right)\right]$

Table 13.2 Equations for maximum bending stress and for moment of inertia about the principal axes x and y. In (c) and (f) thin walls are assumed.

13.4 *Elastic Energy in Torsional Shear*

Returning to the fundamental cylindrical geometry (Fig. 13.1), we now determine the total strain energy contained within a torsionally stressed cylindrical volume. The procedure closely parallels that for bending (Sec. 3.5) and the results are strikingly similar.

We take first an elementary length dx, and within this consider an elementary annular tube of radius r and thickness dr (Fig. 13.1c). The tube in turn contains angular subelements as shown. Under loading there are angular shear distortions γ, and by definition

$$\gamma = \frac{\tau}{G} = \frac{Tr}{GJ} \tag{13.5}$$

The angular deformation varies as the radius, and the annular area $(2\pi r\,dr)$ displaces torsionally a linear tangential distance of

$$\gamma\,dx = \frac{Tr}{GJ}\,dx \tag{13.6}$$

Total tangential shear force carried by the annular element is $\tau(2\pi r\,dr) = (T/J)2\pi r^2\,dr$. During deformation the energy stored in the tubular element is then

$$\frac{1}{2}\left[\frac{2\pi r^2 T}{J}\,dr\right]\left[\frac{Tr}{GJ}\,dx\right] \tag{13.7}$$

Summing over the entire surface we obtain the elemental strain energy in a cylinder of length dx:

$$dU_s = \frac{1}{2}\frac{T^2}{GJ^2}\int_0^R (2\pi r^3\,dr)\,dx = \frac{T^2}{2GJ}\,dx \tag{13.8}$$

This expression is analogous to Eq. (3.9). And for a length ℓ, the total elastic energy stored is

$$U_s = \int_0^\ell \frac{T^2}{2GJ}\,dx \tag{13.9}$$

13.5   *Basic Torsional Deflection*

Torsional problems are relatively more simple than bending in that a constant torque usually exists over definite spans. Then the energy integral reduces to

$$U_s = \frac{T^2}{2G}\int_0^\ell \frac{1}{J}\,dx \tag{13.10a}$$

And for constant diameter,

$$U_s = \frac{T^2}{2GJ}\int_0^\ell dx = \frac{T^2 \ell}{2GJ} \tag{13.10b}$$

Figure 13.2   The simple bar in torsion has a rectangular torque diagram and stores strain energy corresponding to the shaded triangular area.

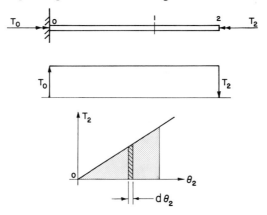

Applying Eq. (13.10b) to the torsional cantilever (Fig. 13.2), we equate external and internal energies:

$$U_e = \frac{T_2\theta_2}{2} = U_s = \frac{T_2^2}{2GJ} \tag{13.11}$$

which reduces to the classic relation [Eq. (13.2)]. Work done by the gradually applied torque corresponds to the shaded triangular area under the torque–deflection curve (Fig. 13.2). This procedure can obviously be extended to include multiple spans of different diameters and geared combinations by summing strain energies in all spans.

13.6 Variable Sections

The basic deflection equation in torsion is then

$$\theta = \frac{T}{G} \int_0^\ell \frac{dx}{J(x)} \tag{13.12}$$

where any variation in J must be included in the integration process. If the variation is arbitrary, we can use numerical integration to determine $1/J(x)$ from specific coordinates and the area under this curve.

The tapered section, or frustum of a cone, can be treated as explained in Sec. 4.9 for the cantilever. For the torsional equivalent (Fig. 13.3),

$$\frac{\theta_1}{\theta_{TC}} = \frac{1}{3(1-r)}\left[\frac{1}{r^3} - 1\right] \tag{13.13}$$

where θ_1 = angle of twist at the small end
$\quad \theta_{TC} = T_1\ell/GJ_0$ = reference twist of the large enclosing cylinder
$\quad r = d_1/d_0$ = ratio of smaller to larger conical diameters

This result follows from the geometry and

$$\frac{J(x)}{J(0)} = \left[\frac{d(x)}{d_0}\right]^4 = \left[r + \frac{x}{\ell}(1-r)\right]^4 \tag{13.14}$$

Figure 13.3 The frustum of a cone carries an axial torsional load.

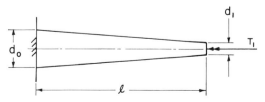

13.7   *Torsional Deflection at Any Point*

Although it is usually not difficult to determine the angular displacement at any point on a shaft using proportionality, we can again develop the concept of auxiliary loading and complementary energy for this purpose. The procedure is similar to that for bending (Sec. 4.1), leading to Eq. (4.4c). Now, however, we are concerned with actual and auxiliary torques and angular displacements. In rotational terms, Eq. (4.4c) becomes

$$T'\theta_1 = \int_0^\ell \frac{tT}{GJ}\,dx \tag{13.15}$$

where $T' = $ an auxiliary couple applied at the deflection point
$\quad\quad \theta_1 = $ angular displacement at T' under actual loading
$\quad\quad t = t(x) = $ auxiliary torque distribution
$\quad\quad T = T(x) = $ actual torque distribution

For instance, in Fig. 13.2 we obtain the deflection at 1 with T' at 1 and a product integral from 0 to 1.

13.8   *Torsional Indeterminacies*

Fixed–fixed torsional elements are once redundant (Fig. 13.4). This case can be solved by treating the two springs in parallel with a common angular displacement at 1; however, using the complementary energy approach [Eqs. (2.3) and (6.1)] and focusing on the shaded rectangular area (Fig.

Figure 13.4   A fixed–fixed torsional shaft is solved for the redundancy by means of auxiliary couple loading T'.

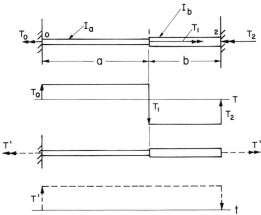

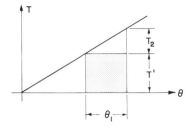

Figure 13.5 Complementary energy in angular coordinates corresponds to the area of the shaded rectangle.

13.5), Eq. (13.15) becomes, with $\theta_2 = 0$,

$$\int_0^\ell \frac{tT}{GJ} \, dx = 0 \qquad (13.16)$$

With T' applied at 2, we have an actual torque distribution resembling a shear diagram and a constant t diagram (Fig. 13.4)

$$0 = \frac{T'}{GJ_a} \int_0^a T_0 \, dx - \frac{T'}{GJ_b} \int_0^b T_2 \, dx$$

$$\frac{T_0}{T_2} = \frac{b}{a} \left(\frac{J_a}{J_b} \right) \qquad (13.17)$$

And summing external torques to 0, we have

$$\frac{T_2}{T_1} = \frac{1}{1 + \dfrac{b}{a} \left(\dfrac{J_a}{J_b} \right)} \qquad (13.18a)$$

$$\frac{T_0}{T_1} = \frac{1}{1 + \dfrac{a}{b} \left(\dfrac{J_b}{J_a} \right)} \qquad (13.18b)$$

Note the importance of the sign of the energy integrals as determined by the sense of the t and T torques, which is not as readily decided as for bending. This is usually done by establishing the sign from the similarity or the dissimilarity of the directions of the respective vectors.

13.9 *Longitudinal Shear Stress*

As explained in texts on strength of materials, a beam carrying vertical shear loading does not have a corresponding uniform shear stress over the beam section. On the contrary, the transverse shear stresses are 0 at the extreme fibers and maximum at the neutral axis (Fig. 13.6). The equation

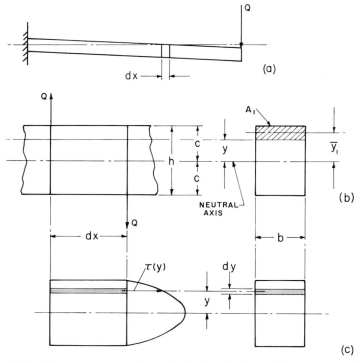

Figure 13.6 Transverse shear is parabolically distributed in a rectangular beam, with strain energy summed over the elemental volumes.

for the shear stress is

$$\tau(y) = \frac{A_1 y_1 V}{Ib} \tag{13.19}$$

where A_1 = sectional area beyond y
y_1 = centroidal distance of A_1 from the neutral axis
V = vertical shear load
I = sectional moment of inertia about neutral axis
b = width of section at y

Although this calculation can become involved for unusual geometries, we illustrate with a rectangular section for which

$$A_1 = b\left(\frac{h}{2} - y\right) = b(c - y)$$

$$y_1 = \frac{1}{2}\left(\frac{h}{2} + y\right) = \frac{1}{2}(c + y)$$

And

$$\tau(y) = \frac{b}{2}\frac{(c^2 - y^2)V}{Ib} = \frac{c^2V}{2I}\left[1 - \left(\frac{y}{c}\right)^2\right]$$ (13.20a)

Substituting for I,

$$\tau(y) = \frac{3}{2}\frac{V}{A}\left[1 - \left(\frac{y}{c}\right)^2\right]$$ (13.20b)

with the dimensionless quantity in brackets governing the parabolic variation, maximum at the neutral axis.

13.10 *Elastic Energy in Longitudinal Shear*

We now sum the energy stored in the entire volume of a beam related to the shear distortions. Taking a horizontal element of area $(b\,dy)$ of an axial beam element dx (Fig. 13.6), the elemental strain energy is

$$dU_s = \frac{1}{2}\int_{-c}^{c}(\tau b\,dy)\left(\frac{\tau}{G}\,dx\right) = \frac{b}{G}\int_{0}^{c}\tau^2\,dy\,dx$$

$$= \frac{9}{4}\left(\frac{c}{h}\right)\frac{V^2}{GA}\left[1 - 2\left(\frac{y}{c}\right)^2 + \left(\frac{y}{c}\right)^4\right]$$

$$= \frac{3}{5}\left(\frac{V^2}{GA}\right)dx = \frac{1}{2}\left(\frac{V^2}{GA_s}\right)dx$$ (13.21)

where $A_s = \frac{5}{6}A = 0.8333(bh) =$ a modified shear area for a *rectangular* section.

A similar integration over a *solid round* section yields

$$A_s = \frac{9}{10}A = 0.7069d^2$$

Both are somewhat less than the true areas. Then for the entire beam length

$$U_s = \int_{0}^{\ell}\frac{V^2}{2GA_s}\,dx$$ (13.22)

analogous to $\int_0^\ell(M^2/2EI)\,dx$ in bending [Eq. (3.10)].

13.11 *Shear Deflection at a Load*

For the cantilever (Fig. 13.6a), we equate the external energy from Q to the internal strain energy stored in the volume of the beam:

$$U_e = \frac{Qy_s}{2} = U_s = \int_{0}^{\ell}\frac{V^2}{2GA_s}\,dx$$ (13.23)

v_s = end deflection of the beam attributed to shear deformation throughout the beam

V = the vertical shear force developed by Q

Solving for the cantilever of rectangular section,

$$y_s = \frac{Q\ell}{GA_s} = \frac{6}{5}\left(\frac{Q\ell}{GA}\right) = \left(\frac{6}{5}\frac{\ell}{GA}\right)Q = C_sQ \qquad (13.24)$$

This expression is analogous to $(\ell/EA)Q = CQ$, [Eq. (1.3)] for tension or compression.

13.12   Shear Deflection at Any Point

Following previously developed concepts, we now introduce auxiliary loading and the related complementary energy. With identical arguments to those in Sec. 4.1, we have (Fig. 13.7)

$$U_s = \int_0^\ell \frac{(v + V)^2}{2GA_s}\,dx = \int_0^\ell \frac{v^2}{2GA_s}\,dx + \int_0^\ell \frac{V^2}{2GA_s}\,dx$$

$$+ \int_0^\ell \frac{vV}{GA_s}\,dx \qquad (13.25)$$

Figure 13.7   Shear deflection at a point on a beam involves an auxiliary force and shear diagram.

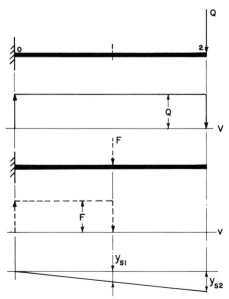

where the first two integrals are the primary energies associated with the independently applied loads, F and V, respectively. With Q applied after the preload F, the cross product or complementary-energy integral is the important third term. For the midpoint (Fig 13.7)

$$Fy_{s1} = \int_0^{\ell/2} \frac{FQ}{GA_s} \, dx$$

$$y_{s1} = \frac{1}{2} \frac{Q\ell}{GA_s} \tag{13.26}$$

As the center deflection is $\frac{1}{2}$ of the end deflection, we have verified the simple nature of the shear-deflection distribution in a constant beam with constant shear. It is a straight-line variation. Slope due to shear will often also be a simple function. For the cantilever with end loading, it is constant along the length of the beam (Fig. 13.6a)

$$\theta_s = \frac{y_s}{\ell} = \frac{Q}{GA_s} \tag{13.27}$$

We have assumed that the cantilever loading is along one of the principal axes of the rectangular section, but not which one. Unlike bend-

Table 13.3 Deflection characteristics of basic beam types due to transverse shear involve the shear modulus and the equivalent shear area A_s.

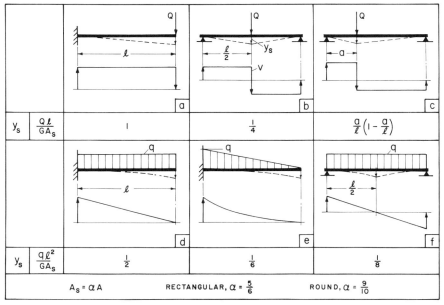

y_s	$\frac{Q\ell}{GA_s}$		1		$\frac{1}{4}$		$\frac{a}{\ell}\left(1 - \frac{a}{\ell}\right)$

y_s	$\frac{q\ell^2}{GA_s}$		$\frac{1}{2}$		$\frac{1}{6}$		$\frac{1}{8}$

$$A_s = \alpha A \qquad\qquad \text{RECTANGULAR, } \alpha = \frac{5}{6} \qquad\qquad \text{ROUND, } \alpha = \frac{9}{10}$$

ing, the same shear deflections result regardless of whether the beam is loaded in the edgewise or flatwise direction.

Table 13.3 indicates the shear diagrams and the maximum beam deflections due to shear for six types of basic beam loadings.

13.13 *Summation of Strain Energy Components*

Elastic energy distribution in a complete physical structure can be a very complex matter indeed. Stresses can vary throughout a volume and as many as six stress components may apply at a given point or elementary cube. But in the energy solutions proposed in this book, we assume generally elements that are, or can be modeled as slender straight or curved beams, or combinations thereof. These are prismatic, have constant defined central and neutral axes, and are subjected to classically distributed direct, bending, and shear stresses.

Fortunately, with these restrictions combined stresses within these structures *need not be considered for strain energy purposes*; that is, the loads and associated stresses are treated completely independently with respect to direct, bending, and shear loads. We can treat them as uncoupled for purposes of energies and deflections. With shear effects now outlined, we are able to account for all significant types of energy within the limitations indicated.

In complete form, including possible variations in cross section, we have finally for actual loading only,

$$U_s = \int_0^\ell \frac{M^2}{2EI}\,ds + \int_0^\ell \frac{Q^2}{2EA}\,ds + \int_0^\ell \frac{T^2}{2GJ}\,ds + \int_0^\ell \frac{V^2}{2GA_s}\,ds \quad (13.28)$$

where the integrals account for the strain energy due to bending, direct, torsional shear, and longitudinal shear stresses.

If auxiliary loading precedes the external loading, the total complementary strain energy is

$$U_{sc} = \int_0^\ell \frac{mM}{EI}\,ds + \int_0^\ell \frac{fQ}{EA}\,ds + \int_0^\ell \frac{tT}{GJ}\,ds + \int_0^\ell \frac{vV}{GA_s}\,ds \quad (13.29)$$

We note the similarity of the forms and the simple algebraic summation of the component energies. In Eq. (13.28), because of the squares, *all integrals will be positive* signifying that any type of induced stress due to a load contributes in an additive sense to all others. In Eq. (13.29) the sense of the factors can be opposite and the integral values negative, indicating the possibility of subtractive loads and stresses.

Eqs. (13.28) and (13.29) can be truncated to the extent that some effects are negligible relative to others. This decision will depend upon loading and geometric considerations, and it usually is not difficult to determine the significant terms from brief calculations.

Examples

13.1. An input of 60 hp is provided to a steel shaft 0–3 rotating at 450 rpm. Loads of 10 and 15 hp are applied at 1 and 2, respectively, with the balance taken to a geared output at 3. Determine the torsional deflection of the shaft.

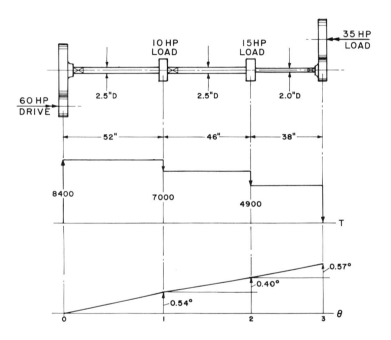

Solution: Calculation of shaft torques leads to the torque diagram:

$$T_{01} = \frac{(HP)(63{,}024)}{RPM} = \frac{(60)(63{,}024)}{450} = 8400 \text{ lb in.}$$

$$T_{12} = \frac{(50)(63{,}024)}{450} = 7000 \text{ lb in.}$$

$$T_{23} = \frac{(35)(63{,}024)}{450} = 4900 \text{ lb in.}$$

Applying opposed T' torsional couples to the two ends 0 and 3, a constant t diagram results, assumed positive. From Eq. (13.15) we calculate the total deflection, with $J = \pi d^4/32$:

$$T'\theta_{03} = \int_0^{\ell} \frac{tT}{GJ}\, dx = \frac{T'}{G} \int_0^1 \frac{8400}{3.835}\, dx + \frac{T'}{G} \int_1^2 \frac{7000}{3.835}\, dx$$

$$+ \frac{T'}{G} \int_2^3 \frac{4900}{1.571}\, dx$$

with $G = 12(10)^6$ lb/in.^{2} and T' cancelled,

$$\theta_{03} = 0.0095 + 0.0070 + 0.0099 = 0.0264 \text{ rad}$$
$$= 0.54° + 0.40° + 0.57° = 1.51°$$

The successive terms represent contributions of the strain energies in each span leading to the contributions to the total deflection. In the deflection plot, these are additive.

13.2. The shaft of Example 13.1 is grounded at the ends and opposite torques applied statically at 1 and 2. Find the torque distribution.

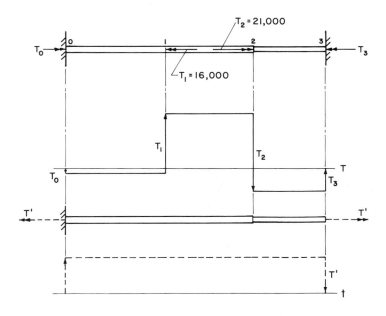

Solution: With two end torque reactions there is one redundancy, requiring a preload couple T' at 3 producing a constant positive t diagram. Assuming directions of T_0 and T_3 as shown, the energy relation [Eq. (13.16)] is

$$0 = \int_0^{\ell} \frac{tT}{GJ}\, dx = \frac{T'}{G} \int_0^1 \frac{-T_0}{3.835}\, dx + \frac{T'}{G} \int_1^2 \frac{(-T_0 + 16,000)}{3.835}\, dx$$

$$+ \frac{T'}{G} \int_2^3 \frac{(-T_0 - 5,000)}{1.571}\, dx$$

$$T_0 = T_{01} = 1430 \text{ lb in.} \quad \text{(in the direction assumed)}$$

Then by torque summation

$$T_{12} = -1430 + 16,000 = 14,570 \text{ lb in.}$$
$$T_{23} = -1430 + 16,000 - 21,000 = -6430 \text{ lb in.}$$

13.3. A horizontal simply supported steel beam of round cross section carries a uniformly distributed load q over its length. Find:
(a) The center deflection including the shear component
(b) The d/ℓ ratio for which the shear effect on the deflection will be 10 percent of the bending

Solution:
(a) From Table 5.3a, the bending deflection is

$$y = \frac{5}{16}\left(\frac{1}{24}\right)\frac{q\ell^4}{EI} = \frac{5}{384}\frac{q\ell^4}{EI}$$

and from Table 13.1f

$$y_s = \frac{1}{8}\frac{q\ell^2}{GA_s}$$

Taking the summation,

$$\Sigma y = \frac{5}{384}\frac{q\ell^4}{EI}\left[1 + 1.67\left(\frac{d}{\ell}\right)^2\right]$$

having substituted $E/G = \frac{30}{12}$ and $\alpha = \frac{9}{10}$
(b) In the Σy result we take the ratio of the second term to the first as $\frac{1}{10}$:

$$\frac{1.67\left(\dfrac{d}{\ell}\right)^2}{1} = 0.10$$

$$\left(\frac{d}{\ell}\right) = 0.25$$

Or the length will be only four diameters.

13.4. Taking the complete circular cantilever of Example 11.1 find the horizontal deflection x_4 attributable to
(a) Direct stress
(b) Transverse shear

Solution:
(a) From Table 10.1a, $Q_A = Q \sin \theta$

$$\frac{Qx_4}{2} = \int_0^\ell \frac{(Q \sin \theta)^2}{2EA} ds$$

$$x_4 = \frac{Q}{EA}\int_0^{2\pi} \sin^2\theta R\, d\theta = 0$$

This effect cancels over the complete ring.

(b) From Table 10.1b, $Q_T = Q \cos \theta$

$$\frac{Q x_4}{2} = \int_0^{\ell} \frac{(Q \cos \theta)^2}{2 G A_s} ds$$

$$x_{4s} = \frac{Q}{G A_s} \int_0^{\pi/2} \cos^2 \theta R \, d\theta = \pi \frac{QR}{G A_s}$$

Total horizontal deflection including bending is

$$x_4 = \pi Q R \left[\frac{R^2}{EI} + \frac{1}{G A_s} \right]$$

13.5. Repeat Example 13.4 for the x_3 deflection.

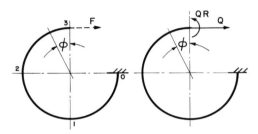

Solution:
(a) The effect of Q_4 at 3 is a tensile force Q and a couple QR at 3. We disregard the couple in obtaining the required deflections. With auxiliary force F at 3

$$F x_3 = \int_0^{\ell} \frac{fQ}{EA} ds$$

$$x_3 = \frac{Q}{EA} \int_0^{3/2\pi} \cos^2 \phi R \, d\phi = 2.356 \frac{QR}{EA}$$

(b) For shear

$$F x_{s3} = \int_0^{\ell} \frac{vV}{G A_s} ds$$

$$x_{s3} = \frac{Q}{G A_s} \int_0^{3/2\pi} \sin^2 \phi R \, d\phi = 2.356 \frac{QR}{G A_s}$$

Total deflection including bending is

$$x_3 = QR \left[3.356 \frac{R^2}{EI} + 2.356 \left(\frac{1}{EA} + \frac{1}{G A_s} \right) \right]$$

14

Deflections with Bending and Torsion

UNTIL NOW, WE HAVE ARBITRARILY LIMITED ALL STRUCTURES to those lying substantially in a single plane, the only exceptions being the continuous tubular geometry. The tubular cases, however, are essentially planar, with identical conditions prevailing in any plane taken perpendicular to the longitudinal axis. The torsional cases just discussed are not strictly planar, but are effectively one dimensional with respect to a central axis. This restriction has been intentional, to fully outline procedural concepts in the various categories of elementary elastic structures.

A prismatic elastic element can be subjected to axial loading, to bending loading, and to torque loading, each producing related stress and energy distributions. We now consider the particular combination involving bending and torsion—specifically, the corresponding deflections resulting in determinate structures. Component elements are again modeled primarily as prismatic in nature, although nonuniform cross sections can be accommodated within the energy integrals if necessary.

14.1   *Deflections of the Cantilevered Bent*

The planar bent (Fig. 14.1a), loaded perpendicularly to its plane, carries bending moment throughout its length. Simultaneously, it is subjected to a constant torque in *a*. For small deflections 2 moves substantially vertically, with the displacement consisting of three components:

1. Cantilever bending of *b*.
2. Cantilever bending of *a*.
3. Rotation of *a* about the 0–1 axis.

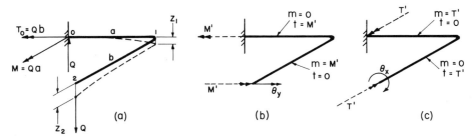

Figure 14.1 The simple bent with an end load perpendicular to its plane develops one translational and two angular deflection components. Auxiliary couples for determining the slopes are shown in (b) and (c).

We add the several contributions from conventional equations:

$$z_2 = \frac{Qa^3}{3EI} + \frac{Qb^3}{3EI} + b\left[\frac{(Qb)a}{GJ}\right] \qquad (14.1)$$

where the last term is the linear vertical displacement of 2 resulting from the product of the angle of twist and the radius b, and

$$EI = \text{flexural rigidity}$$
$$GJ = \text{torsional rigidity}$$

Because of bending in vertical planes, I is understood to be about the appropriate horizontal neutral axis and could be different for each leg. The torsional section factor is independent of the polar position of the major or minor axes.

We can also arrive at the expression for z_2 by equating energies

$$\frac{Qz_2}{2} = \int_0^\ell \frac{M^2}{2EI}\, ds + \int_0^\ell \frac{T^2}{2GJ}\, ds$$

$$= \frac{1}{2EI}\int_0^b (Qx_2)^2\, dx_2 + \frac{1}{2EI}\int_0^a (Qx_1)^2\, dx_1 + \frac{1}{2GJ}\int_0^a (Qb)^2\, dx_1$$

$$z_2 = \frac{Q}{3EI}(a^3 + b^3) + \frac{Q}{GJ}(ab^2) \qquad (14.2)$$

There are advantages to the energy approach. We do not have to locate the individual deflection formulas, nor visualize the deflection curve, nor convert the angular displacement to linear. It is necessary to visualize the moment and torque loadings; however, these are required in any case for purposes of strength analysis.

As explained in Sec. 13.13, and as now proven, bending and torsional energies in a are uncoupled and determined completely independently.

Except for the second-order deflection of the arcs, 2 only moves vertically, with no displacement in the horizontal plane. There is zero bending moment in this plane, and therefore no possible strain energy in this sense. Shear deflections due to Q exist over the entire length but are negligible.

14.2   *Relative Stiffness Parameters*

Equation (14.2) indicates a deflection dependent upon the a and b dimensions, the load, and the material–geometric properties of the beam, EI and GJ. These are termed the *flexural rigidity* and the *torsional rigidity* of the elements, respectively. The equation becomes more general if rewritten,

$$z_2 = \frac{Qb^3}{3EI}\left\{\left[1 + \left(\frac{a}{b}\right)^3\right] + 3\lambda\left(\frac{a}{b}\right)\right\} \tag{14.3}$$

where $Qb^3/3EI$ = a reference cantilever deflection of b
$\quad\quad\quad a/b$ = a dimensionless geometric ratio
$\quad\quad\quad\quad \lambda = EI/GJ$ = the dimensionless ratio of flexural to torsional rigidity

With both bending and torsion present, we will invariably have the λ ratio in the solution. By treating the bending terms as reference, torsional effects are related through λ. In this factor we have the ratio of the material moduli, in turn connected by Poisson's ratio ν by the equation

$$\frac{E}{G} = 2(1 + \nu) \tag{14.4}$$

If ν varies from about 0.25 to 0.33 for most common alloys, the E/G ratio then varies from 2.50 to 2.66, with 2.5 a reasonable value for usual accuracy requirements.

The second portion of λ, or I/J, is quantified by the geometry of the cross section. For round sections $(I/J) = \frac{1}{2}$ and $\lambda = 1.25$. With rectangular sections, an equivalent torsional J factor is necessary (Table 13.1), and the corresponding behavior of the total factor is given in Fig. 14.2. For instance, if in Fig. 14.1, the structure is a thin-walled square tube, $\lambda = 1.67$ and $3\lambda = 5$ in Eq. (14.3). Note that this term being relative is independent of the absolute dimensions and wall thickness. Absolute information regarding size will appear in the value of I in the first factor and in the absolute reference deflection.

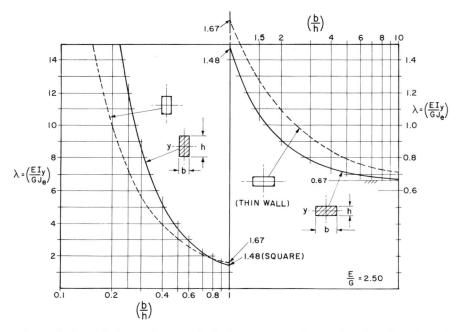

Figure 14.2 Relative stiffness ratio λ for a rectangular section is a function of (b/h). Solid curves apply to a solid section and the dashed to a thin-walled rectangular tube.

14.3 Coupled Deflections

As an example of the use of auxiliary loading, we find the slope developed at 2 by applying M' (Fig. 14.1b). Then

$$M'\theta_y = \int_0^\ell \frac{mM}{EI}\, ds + \int_0^\ell \frac{tT}{GJ}\, ds$$

$$\theta_y = \frac{1}{EI}\int_0^b (Qx_2)\, dx_2 + \frac{1}{GJ}\int_0^a (Qa)\, dx_0$$

where $t = 0$ in b and $m = 0$ in a. Finally

$$\theta_y = \frac{Qb^2}{2EI}\left[1 + 2\lambda\frac{a}{b}\right] \tag{14.5}$$

Applying an auxiliary couple about the b axis (Fig. 14.1c),

$$T'\theta_x = \int_0^a \frac{T'M}{EI}\, ds$$

$$\theta_x = \frac{1}{EI}\sum_0^a A = \frac{1}{EI}\left(\frac{Qa^2}{2}\right) \tag{14.6}$$

					$+M_y$, $+Q_z$, $+M_x$, b, a (cantilevered bent)	b, a, a, $+Q_z$, $+M_x$, $+M_y$ (U-bar)
				$\lambda = \dfrac{EI}{GJ}$		
α_z	$\dfrac{z}{Q_z}$			$\dfrac{b^3}{3EI}$	$1 + \left(\dfrac{a}{b}\right)^3 + 3\lambda\left(\dfrac{a}{b}\right)$	$1 + 2\left(\dfrac{a}{b}\right)^3 + 3\left(\dfrac{a}{b}\right)\lambda\left[1 + \dfrac{a}{b}\right]$
β_{xz}	$\dfrac{\theta_x}{Q_z}$	β_{zx}	$\dfrac{z}{M_x}$	$\dfrac{b^2}{2EI}$	$\left(\dfrac{a}{b}\right)^2$	$-2\left(\dfrac{a}{b}\right)\left[\dfrac{a}{b} + \lambda\right]$
β_{yz}	$\dfrac{\theta_y}{Q_z}$	β_{zy}	$\dfrac{z}{M_y}$		$1 + 2\lambda\left(\dfrac{a}{b}\right)$	
γ_x	$\dfrac{\theta_x}{M_x}$			$\dfrac{b}{EI}$	$\left(\dfrac{a}{b}\right) + \lambda$	$2\left(\dfrac{a}{b}\right) + \lambda$
γ_{xy}	$\dfrac{\theta_x}{M_y}$	γ_{yx}	$\dfrac{\theta_y}{M_x}$		0	0
γ_y	$\dfrac{\theta_y}{M_y}$				$1 + \lambda\left(\dfrac{a}{b}\right)$	$1 + 2\lambda\left(\dfrac{a}{b}\right)$

Table 14.1 End deflection characteristics of the cantilevered bent and U-bar for out-of-plane flexure.

These results are included in Table 14.1 with the complete response values for force and moment loads at the end of the bent. Including reciprocal terms there are nine out-of-plane influence factors. There were similarly nine such factors for the three coordinates involved with in-plane deflections (Table 8.1). The mutual independence, or orthogonality, of these two cases provides the separate uncoupled results. Note also that torsional components only occur in Table 14.1.

14.4 *Orthogonality of Bending Components*

As we consider three-dimensional problems, it is important to predict bending behavior in several directions. Except for special simple cross-sectional cases, such as the circular, bending is influenced by the orientation of the principal axes of the section. Taking a basic rectangular cantilever (Fig. 14.3a) with an end load directed at an angle θ from the horizontal, we do not obtain simple cantilever deflection in the Q direction. The beam has maximum compliance vertically and minimum horizontally, involving I_y and I_z.

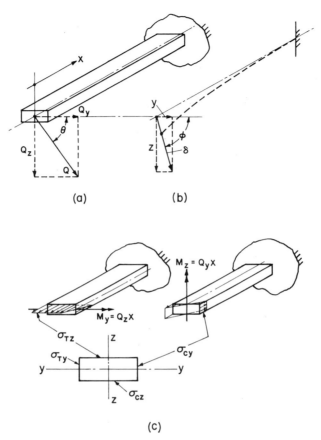

(a)   (b)

(c)

Figure 14.3 With an end load directed obliquely to the neutral axes, deflection components must be obtained using load components along these axes. Bending stress patterns are mutually perpendicular (c).

We must resolve Q into Q_y and Q_z components that in turn produce uncoupled y and z deflections:

$$y = \frac{Q_y \ell^3}{3EI_z} = \left(\frac{Q\ell^3}{3EI_z} \right) \sin \theta \qquad (14.7a)$$

$$z = \frac{Q_y \ell^3}{3EI_y} = \left(\frac{Q\ell^3}{3EI_y} \right) \cos \theta \qquad (14.7b)$$

The resultant deflection δ is the vector sum of the components in the directions of the principal axes and *will* not coincide with the spatial direction of Q. Rather the total deflection is the sum of the perpendicular components and

$$\phi = \tan^{-1}\left(\frac{I_z}{I_y} \cot \theta \right) \tag{14.8}$$

This argument is closely related to the discussion of principal axes (Sec. 1.13), except that in Fig. 14.3 the principal axial directions are established immediately as edgewise and flatwise relative to the rectangular section.

14.5 Orthogonality of Bending Energy

The perpendicular load components (Q_y and Q_z), in turn, cause perpendicular bending stress patterns with respect to the perpendicular axes z and y (Fig. 14.3c). Moment ($Q_z x$) produces tension in the top fibers and compression in the bottom but *no stresses* on the vertical short edges. Similarly, ($Q_y x$) causes no stresses on the horizontal long edges. As shown, the stress patterns are mutually independent by virtue of the perpendicular orientation of the principal (neutral) axes.

It follows that the strain energy integrals in the sense of M_y and M_z are also mutually independent. A further aspect of this feature is in the complementary integral of the product of the auxiliary and actual moments. If the m and M moments are aligned with the respective principal axes, as is usually the case, *the product is* 0 and decoupling is verified.

If an auxiliary loading develops zero stress on the extreme fiber of a beam at which an actual moment loading develops maximum bending stress, the product and complementary energy must be 0. No coupling exists. This is a very useful test in three-dimensional structures.

In torsion this complication does not exist. If auxiliary and actual torques occur in an element, a (tT) product exists as does the complementary-energy integral; however, we must be careful to identify this energy as negative if the senses of the two torques are opposite.

14.6 Orthogonal Cantilevered Bent

Although cantilevered structures are not well supported and tend to have large stresses and deflections, they do occur in practice. Since they are statically determinate, it is now in order to consider a more comprehensive solution. In Fig. 14.4 a three-leg bent carrying an end load is assumed to be of constant section. It is in equilibrium with end support as indicated.

We separate the three beam elements and place each in equilibrium with equal and opposite reactions at the junctures. There is bending moment in each leg and a constant torque in *b*. Corresponding moment diagrams are

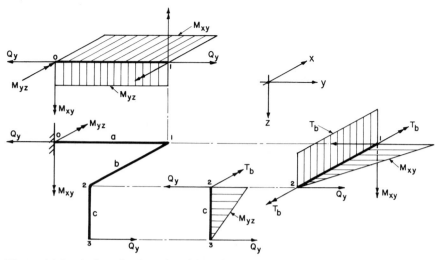

Figure 14.4   A three-leg bent is subjected to a combination of bending and torque distributions.

shown, with a subjected to constant bending moments in perpendicular planes.

Summing strain energies for the deflection y_3, and omitting the subscript for Q_y,

$$\frac{Qy_3}{2} = \frac{1}{2EI}\left[\int_3^2 (Qz)^2\, dz + \int_2^1 (Qx)^2\, dx + \int_0^1 (cQ)^2\, dy \right.$$

$$\left. + \int_0^1 (bQ)^2\, dy\right] + \frac{1}{2GJ}\int_2^1 (cQ)^2\, dx$$

$$y_3 = \frac{Q}{EI}\left[\frac{c^3}{3} + \frac{b^3}{3} + c^2 a + b^2 a\right] + \frac{Q}{GJ}[c^2 b] \tag{14.9}$$

where in the four bending terms the first two represent simple cantilever deflections and the second two relate to the cantilever a with perpendicular end moments. Both end moments contribute to y_3 even though the displacements at 1 are perpendicular to each other. On this point, if we assume a round cross section, the bending energy in a is

$$U_s = \int_0^1 \frac{\left(M_{xy}^2 + M_{zy}^2\right)}{2EI}\, dy \tag{14.10}$$

Since the resultant moment is constant in a,

$$M_R^2 = M_{xy}^2 + M_{zy}^2$$

Equation (14.10) becomes

$$U_s = \int_0^1 \frac{M_R^2}{2EI}\, dy \tag{14.11}$$

where a is bending in the resultant oblique plane.

14.7 Arbitrary Cross Sections

The bent (Fig. 14.4) has several bending moment distributions relating to the various legs, and they are in defined vertical and horizontal planes. Bending occurs relative to the respective neutral axes normal to the plane of the moments. Should the beam section have directional properties, each integration must contain the appropriate I value. In a with perpendicular moments we would have to maintain the orthogonal components and could not use the resultant moment. Each beam could also have a different absolute value of I. A more complete form of solution than Eq. (14.9) is then

$$\frac{Qy_3}{2} = \int_3^2 \frac{(Qz)^2}{2EI_{cx}}\, dz + \int_2^1 \frac{(Qx)^2}{2EI_{by}}\, dx + \int_0^1 \frac{(cQ)^2}{2EI_{ax}}\, dy$$

$$+ \int_0^1 \frac{(bQ)^2}{2EI_{az}}\, dy + \int_2^1 \frac{(cQ)^2}{2GJ_e}\, dx \tag{14.12}$$

14.8 Coupled Deflections

For the bent in Fig. 14.4, there are six potential displacement components at 3 due to Q_y, including three perpendicular deflections and three slopes about the mutually perpendicular reference axes. Taking for illustration the x component, we apply F_x (Fig. 14.5a). By inspection the only coincident stresses for both auxiliary and actual loadings occur as bending in the horizontal plane in a, and the required auxiliary diagrams are therefore reduced to the m diagram shown:

$$F_x x_y = \int_0^1 \frac{(Fy_1)M}{EI_{az}}\, dy_1$$

$$x_y = \frac{1}{EI_{az}} \sum A\bar{y} = \frac{1}{EI_{az}} (bQ_y) \frac{a^2}{2} \tag{14.13}$$

Physically, this single coupled effect is caused by the rearward x deflection of the cantilever a with an end moment (bQ_y) transmitted from 3 to 1.

To find the vertical component we apply F_z (Fig. 14.5b) and similarly determine only one interactive situation, or the product of the triangular m

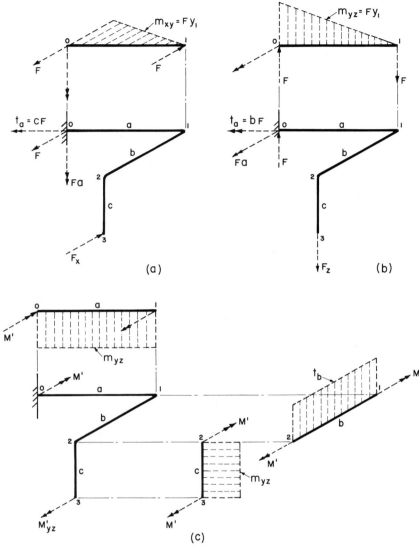

Figure 14.5 Auxiliary loadings are applied to the bent of Figure 14.4 for coupled deflection solutions.

diagram and the rectangular M diagram in a. The respective stresses have opposite sense:

$$F_z z_y = \int_0^1 \frac{(Fy_1)M}{EI_{ax}} \, dy_1$$

$$z_y = \frac{-1}{EI_{ax}} \sum A\bar{y} = \frac{-1}{EI_{ax}} (cQ_y) \frac{a^2}{2} \qquad (14.14)$$

Again this is a single cantilever effect in a transposed unaltered to 3 as z_y. Incidentally, the primary deflections x_3 and z_3 caused by external loads in these directions can be obtained with actual loads Q_x and Q_y replacing F_x and F_y by completing the abbreviated diagrams (Fig. 14.5a and b) and using

$$\int_0^\ell \frac{M^2}{2EI}\,ds$$

Slopes at 3 are determined taking auxiliary couples at 3. For θ_x we apply M'_{yz} (Fig. 14.5c) and proceed to identify the compatible (mM) and (tT) products throughout the structure. There are now two such cases—a moment effect in c and a torque effect in b. Evaluation of these integrals provides the required deflection component.

If deflections other than at the end are of interest, the method requires the shifting of the actual or auxiliary load to the points and in the directions of the problem as specified.

14.9 *The Rectangular Bent*

An unclosed rectangularly arranged beam can be loaded by pairs of opposite forces without reacting to ground (Table 14.2). Two basic gap

Table 14.2 Deflection due to separational loading of a split rectangular planar bar—also maximum bending moments and torques.

		a	b	c
δ_{01}	$\frac{Qb^3}{EI}$	$\left(\frac{a}{b}\right)^2[1+\frac{2}{3}\left(\frac{a}{b}\right)]$	$\frac{1}{6}+\frac{1}{2}\left(\frac{a}{b}\right)$	$[\frac{1}{6}+\frac{2}{3}\left(\frac{a}{b}\right)^3]+\lambda\left(\frac{a}{b}\right)[\frac{1}{2}+\frac{a}{b}]$
M	Q	a	$\frac{b}{2}$	$a,\frac{b}{2}$
T		0	0	$a,\frac{b}{2}$

		d	e	f
δ_{01}	$\frac{Qb^3}{EI}$	$\left(\frac{a}{b}\right)[1+\frac{2}{3}\left(\frac{a}{b}\right)]$	$\frac{2}{3}+\frac{a}{b}$	$\frac{2}{3}[1+\left(\frac{a}{b}\right)]+\lambda\left(\frac{a}{b}\right)[1+\frac{a}{b}]$
M	Q	a	b	a,b
T		0	0	a,b

positions are analyzed, with the gap at the center of a side and at a corner. Shear and planar separational loads (Q_y and Q_x) involve no torsion, but the Q_z loads are out-of-plane and involve the λ factor. The former can apply to long rectangular tubes, but the latter (Table 14.2c and f) are of finite section I_y and J_e.

14.10 *Distributed Out-of-Plane Loading*

Distributed loads on a bent create no torque in the planar case with the bent vertical, but weight loading causes torsion in a horizontal bent. Table 14.3 indicates end-deflection behavior with gravitational loading in a uniform beam. In Table 14.3a the resultant of the load on b produces a torque in a of $qb^2/2$. With an auxiliary load F in the z direction, the torsional contribution to the z deflection is

$$\int_0^\ell \frac{tT}{GJ}\,ds = \int_0^a \frac{(Fb)}{GJ}\left(\frac{qb^2}{2}\right)ds = \frac{Fq}{GJ}\left(\frac{ab^3}{2}\right) \tag{14.15}$$

Table 14.3 End deflection tabulation for rectangular bents with uniformly distributed out-of-plane loading, including supporting moments and torques.

		a	b
z	$\dfrac{qb^4}{EI}$	$[\frac{1}{8} + \frac{1}{3}(\frac{a}{b})^3 + \frac{1}{8}(\frac{a}{b})^4] + \lambda\frac{1}{2}(\frac{a}{b})$	$[\frac{1}{8} + \frac{1}{3}(\frac{a}{b}) - \frac{1}{6}(\frac{a}{b})^3 + \frac{1}{6}(\frac{a}{b})^4]$ $+\lambda(\frac{a}{b})[\frac{1}{2} + (\frac{a}{b}) + \frac{1}{2}(\frac{a}{b})^2]$
θ_x	$\dfrac{qb^3}{EI}$	$(\frac{a}{b})^2[\frac{1}{2} + \frac{1}{6}(\frac{a}{b})]$	$\frac{1}{2}(\frac{a}{b})^2[1 - \lambda]$
θ_y		$\frac{1}{6} + \lambda\frac{1}{2}(\frac{a}{b})$	$[\frac{1}{2}(\frac{a}{b}) + \frac{1}{6}] + \lambda(\frac{a}{b})[\frac{1}{2} + \frac{a}{b}]$
M_0	qb^2	$(\frac{a}{b})[1 + \frac{1}{2}(\frac{a}{b})]$	$(\frac{a}{b})[1 + \frac{a}{b}]$
T_0		$\frac{1}{2}$	$\frac{1}{2} + \frac{a}{b}$

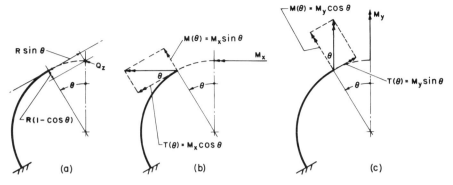

Figure 14.6 Concentrated out-of-plane loading of the circular beam causes distributed moments and torques.

as indicated by the $\lambda\frac{1}{2}(a/b)$ term in Table 14.1a. Bending terms follow as described in Chap. 8 and are superimposed.

The cantilevered symmetrical U-bar is summarized in Table 14.3b. Distributed loading develops no torque in the outboard leg.

14.11 Bending and Torsion in Circular Beams

An end load perpendicular to the plane of a circular beam develops both bending and torsional stresses (Fig. 14.6a) and deflection components. The distributed internal loadings are (Table 14.4),

$$M(\theta) = Q_z R \sin \theta \qquad (14.16a)$$

$$T(\theta) = Q_z R(1 - \cos \theta) \qquad (14.16b)$$

where $M(\theta)$ = bending moment about the radial neutral axis in the central plane of the beam
$T(\theta)$ = torque about the central tangential axis

Transverse deflection of the load is

$$\frac{Q_z z}{2} = \int_0^\ell \frac{M^2}{2EI} \, ds + \int_0^\ell \frac{T^2}{2GJ} \, ds$$

with the result in Table 14.5.

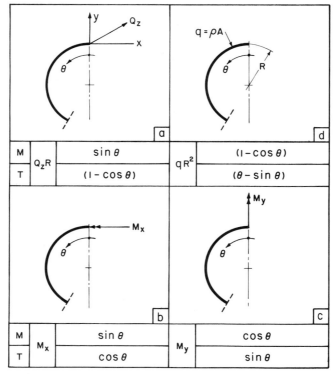

Table 14.4 Reference expressions for moment and torque distributions with various loadings, proceeding from the free end.

If the end load is an axial or transverse couple (Fig. 14.6*b* and *c*), these are resolved into moment and torque components as shown. Complete deflection characteristics for all loading and response combinations are in Table 14.5.

14.12 *Distributed Loading of Circular Beams*

The horizontal circular beam with weight loading normal to its plane can be visualized as carrying a resultant overhung load acting at the center of gravity of the distributed pattern (Fig. 14.7*a*). The centroid of the circular

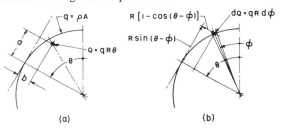

					ψ	$\frac{\pi}{4}$	$\frac{\pi}{2}$	π	$\frac{3\pi}{2}$	2π
α_z	$\dfrac{Z}{Q_z}$			$\dfrac{R^3}{EI}$	$\frac{\psi}{2}-\frac{1}{4}\sin 2\psi$	0.1427	0.7854	1.5708	2.3562	3.1416
					$+\lambda$ $\frac{3}{2}\psi-2\sin\psi+\frac{1}{4}\sin 2\psi$	0.0139	0.3562	4.7124	9.0686	9.4248
β_{xz} $\dfrac{\theta_x}{Q_z}$	β_{zx} $\dfrac{Z}{M_x}$			$\dfrac{R^2}{EI}$	$\frac{\psi}{2}-\frac{1}{4}\sin 2\psi$	0.1427	0.7854	1.5708	2.3562	3.1416
					$+\lambda$ $\frac{\psi}{2}-\sin\psi+\frac{1}{4}\sin 2\psi$	-0.0644	-0.2146	1.5708	3.3562	3.1416
β_{yz} $\dfrac{\theta_y}{Q_z}$	β_{zy} $\dfrac{Z}{M_y}$				$\frac{1}{2}\sin^2\psi$	0.2500	0.5000	0	0.5000	0
					$+\lambda$ $1-\cos\psi-\frac{1}{2}\sin^2\psi$	0.0429	0.5000	2	0.5000	0
γ_x	$\dfrac{\theta_x}{M_x}$			$\dfrac{R}{EI}$	$\frac{\psi}{2}-\frac{1}{4}\sin 2\psi$	0.1427	0.7854	1.5708	2.3562	3.1416
					$+\lambda$ $\frac{\psi}{2}+\frac{1}{4}\sin 2\psi$	0.6427	0.7854	1.5708	2.3562	3.1416
γ_{xy} $\dfrac{\theta_x}{M_y}$	γ_{yx} $\dfrac{\theta_y}{M_x}$				$\frac{1}{2}\sin^2\psi$	0.2500	0.5000	0	0.5000	0
					$+\lambda$ $-\frac{1}{2}\sin^2\psi$	-0.2500	-0.5000	0	-0.5000	0
γ_y	$\dfrac{\theta_y}{M_y}$				$\frac{\psi}{2}+\frac{1}{4}\sin 2\psi$	0.6427	0.7854	1.5708	2.3562	3.1416
					$+\lambda$ $\frac{\psi}{2}-\frac{1}{4}\sin 2\psi$	0.1427	0.7854	1.5708	2.3562	3.1416

Table 14.5 Complete deflection coefficients for the circular cantilever with out-of-plane concentrated loading.

arc is located on the radial bisector, with

$$\bar{x} = R\left(\frac{2}{\theta}\right)\sin\left(\frac{\theta}{2}\right) \qquad (14.17a)$$

and the resultant load at this point is

$$Q = \int_0^\theta \rho A\, ds = qR\theta \qquad (14.17b)$$

Figure 14.7 Uniformly distributed loading causing moment and torque can be solved by centroidal or integrational procedures.

Taking this moment about the radial axis the bending moment is

$$M(\theta) = Qa = (qR\theta)\bar{x}\sin\left(\frac{\theta}{2}\right) = qR^2(1 - \cos\theta) \qquad (14.18a)$$

and for torsion about the tangential axis

$$T(\theta) = Qb = (qR\theta)\left(R - \bar{x}\cos\frac{\theta}{2}\right) = qR^2(\theta - \sin\theta) \qquad (14.18b)$$

Alternatively, we can obtain these results by direct integration as in Sec. 10.3. Using an intermediate ranging variable ϕ, and treating the limit θ as a constant (Fig. 14.7b),

$$dM = [R\sin(\theta - \phi)]\,dQ$$
$$= qR^2[\sin\theta\cos\phi - \cos\theta\sin\phi]\,d\phi$$
$$M = qR^2\left[\sin\theta\int_0^\theta\cos\phi\,d\phi - \cos\theta\int_0^\theta\sin\phi\,d\phi\right]$$
$$= qR^2[\sin\theta\sin\phi + \cos\theta\cos\phi]_0^\theta$$
$$= qR^2(1 - \cos\theta) \qquad (14.19a)$$

Table 14.6 End deflection characteristics of the circular cantilever with out-of-plane distributed loading.

$$\lambda = \frac{EI_y}{GJ_e}$$

				ψ	$\frac{\pi}{4}$	$\frac{\pi}{2}$	π	$\frac{3\pi}{2}$	2π
z	$\dfrac{qR^4}{EI}$		$1 - \cos\psi - \tfrac{1}{2}\sin^2\psi$		0.0429	0.5000	2	0.5000	0
		$+\lambda$	$\tfrac{1}{2}[\psi - \sin\psi]^2$		0.0031	0.1629	4.9348	16.3157	19.7392
θ_x	$\dfrac{qR^3}{EI}$		$1 - \cos\psi - \tfrac{1}{2}\sin^2\psi$		0.0429	0.5000	2	0.5000	0
		$+\lambda$	$1 - \cos\psi - \psi\sin\psi + \tfrac{1}{2}\sin^2\psi$		-0.0125	-0.0708	2	6.2124	0
θ_y			$-\tfrac{1}{2}\psi + \sin\psi - \tfrac{1}{4}\sin2\psi$		0.0644	0.2146	-1.5708	-3.3562	-3.1416
		$+\lambda$	$-\tfrac{1}{2}\psi - \psi\cos\psi + \sin\psi + \tfrac{1}{4}\sin2\psi$		0.0091	0.2146	1.5708	-3.3562	-9.4244
M_o	qR^2		$1 - \cos\psi$		0.2929	1	2	1	0
T_o			$\psi - \sin\psi$		0.0783	0.5708	3.1416	5.7124	6.2832

And for the torque,

$$dT = \left[R\{1 - \cos(\theta - \phi)\} \right] dQ$$
$$= qR^2 \left[1 - \cos\theta\cos\phi - \sin\theta\sin\phi \right] d\phi$$
$$T = qR^2 \left[\int_0^\theta d\phi - \cos\theta \int_0^\theta \cos\phi \, d\phi - \sin\theta \int_0^\theta \sin\phi \, d\phi \right]$$
$$= qR^2 \left[\phi - \cos\theta\sin\phi + \sin\theta\cos\phi \right]_0^\theta$$
$$= qR^2(\theta - \sin\theta) \tag{14.19b}$$

Deflections of the end of the beam, derived using these moment and torque equations and appropriate auxiliary loading (Table 14.5), are given in Table 14.6.

14.13 Circular–Straight Combinations

Circular beams are often joined to straight elements, but such structures present no particular problems when determining deflections by the energy

Table 14.7 Separational compliances of the split planar obround beam loaded in various orthogonal directions.

		a	b	c
δ_{01} $\dfrac{QR^3}{EI}$		$3\pi + 8\gamma$	$\pi + 8\gamma + 2\pi\gamma^2 + \frac{4}{3}\gamma^3$	$\left[\pi + \frac{4}{3}\gamma^3\right] + \lambda\left[3\pi + 8\gamma\right]$
M	QR	2	$1+\gamma$	γ
T		0	0	2

		d	e	f
δ_{01} $\dfrac{QR^3}{EI}$		$3\pi + A$	$\pi + 4\gamma$	$[10.283 + A]$ $+\lambda[3\pi + 18.283\gamma + 2\pi\gamma^2]$
M	QR	$2(1+\gamma)$	1	$1+2\gamma$
T		0	0	$2(1+\gamma)$

$\gamma = \dfrac{a}{R}$	$\lambda = EI/GJ$	$A = 24.566\,\gamma + 20.566\,\gamma^2 + 5.333\,\gamma^3$	

method. For instance, Table 14.7 illustrates the separational deflections of various obround configurations. It is understood that only the primary deflection components in the direction of the loads are given. The nonplanar cases involving torsion are shown in Table 14.7c and f.

In Table 14.7c the results follow, with Q_z perpendicular to the beam, from the relation

$$\frac{Qz}{2} = 4\int_0^a \frac{(Qx)^2}{2EI} dx + 4\int_0^{\pi/2} \frac{(QR\sin\theta)^2}{2EI} R\,d\theta$$

$$+ 2\int_0^\pi \frac{[QR(1 - \cos\theta)]^2}{2GJ} R\,d\theta + \int_0^{2a} \frac{(2QR)^2}{2GJ} R\,d\theta \quad (14.20)$$

from which we reduce to the final deflection result in Table 14.7c, consisting of the component bending and torsional terms.

The M and T values in Table 14.7 indicate the maximum loads in the structure for stress purposes.

14.14 *Arbitrary Curvature by Numerical Integration*

Although a curved beam can be nonplanar and of nonuniform section, we illustrate the numerical integration problem with a constant planar canti-lever, avoiding these complications. An end load produces distributed bending and torsion when applied perpendicular to the plane (Fig. 14.8). It is necessary to determine these functions about the central axes of the beam at multiple discrete points to sum the distributed bending and torsional elastic energies.

Solving for the vertical deflection,

$$\frac{Qz}{2} = \int_0^\ell \frac{M^2}{2EI} ds + \int_0^\ell \frac{T^2}{2GJ} ds$$

$$= \frac{1}{2EI} \int_0^\ell (Qa)^2 ds + \frac{1}{2GJ} \int_0^\ell (Qb)^2 ds \quad (14.21a)$$

$$z = \frac{Q}{EI} \int_0^\ell a^2 ds + \frac{Q}{GJ} \int_0^\ell b^2 ds$$

$$= \frac{Q}{EI} [A_1 + \lambda A_2] \quad (14.21b)$$

where $\lambda = EI/GJ$ and the integrals have the dimensions of distance cubed, defining a reference $\ell^3/3$ for an equivalent simple cantilever. They also represent full-scale areas under the a^2 and b^2 functional distributions plotted against the arc length (Fig. 14.8).

To obtain the component of slope of the end in the plane perpendicular to the reference axis, we apply T' in this sense and evaluate the auxiliary

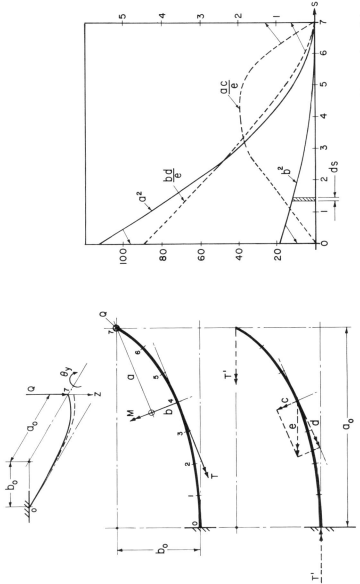

Figure 14.8 Deflection solution with an end load and a generally curved beam requires numerical integration of the geometric products indicated.

distributions:

$$m = \left(\frac{c}{e}\right)T' \quad \text{and} \quad t = \left(\frac{d}{e}\right)T' \tag{14.22}$$

by taking components of T' normal to and along the central axis (tangential) at selected points. Then

$$T'\theta_y = \int_0^\ell \frac{mM}{EI}\,ds + \int_0^\ell \frac{tT}{GJ}\,ds \tag{14.23a}$$

$$\theta_y = \frac{1}{EI}\int_0^\ell \left(\frac{c}{e}\right)(Qa)\,ds + \frac{1}{GJ}\int_0^\ell \left(\frac{d}{e}\right)(Qb)\,ds$$

$$= \frac{Q}{EI}\int_0^\ell \left(\frac{ac}{e}\right)\,ds + \frac{Q}{GJ}\int_0^\ell \left(\frac{bd}{e}\right)\,ds$$

$$= \frac{Q}{EI}[A_3 + \lambda A_4] \tag{14.23b}$$

where now the integral areas are in actual area dimensions or distance squared.

14.15 Out-of-Plane Distributed Loading

The previous solution for the beam of general curvature can be adapted to distributed loading by a series of summations about discretely spaced intervals. Taking the weight of each element as acting normal to the plane of the beam, we add the contributing moments (Fig. 14.9). Strictly speaking,

Figure 14.9 Distributed transverse load on the planar curved beam requires a different set of dimensional summations for each interval of the beam.

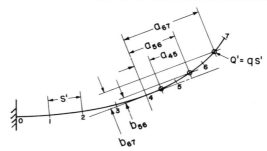

the centroid of a curved element is neither on the beam axis nor on the chord, but there is usually little difference. Assuming equal intervals to 7, and, for example, finding the moment and torque at 4,

$$M_4 = qs'[a_{67} + a_{56} + a_{45}] \tag{14.24a}$$

$$T_4 = qs'[b_{67} + b_{56} + b_{45}] \tag{14.24b}$$

where $qs' = Q' =$ total distributed load within an interval
a and $b =$ perpendicular distances from an interval point to the centroids of the elements

Note that as we shift from one point to the next, a completely different set of a and b coordinate distances must be determined, with the total number of terms increasing as we move from the end.

To obtain the transverse deflection z_7, we require an auxiliary load F at 7 in the transverse direction. As in Fig. 14.8, these are, at any point,

$$m = Fa \quad \text{and} \quad t = Fb$$

These must be summed cumulatively as in Eqs. (14.24) to provide the m and t distributions.

Finally, having tabulated m, t, M, and T at each point,

$$Fz_7 = \int_0^\ell \frac{(Fa)M(s)}{EI} \, ds + \int_0^\ell \frac{(Fb)T(s)}{GJ} \, ds$$

$$z_7 = \frac{1}{EI} \left[\int_0^\ell aM(s) \, ds + \lambda \int_0^\ell bT(s) \, ds \right] \tag{14.25}$$

Similarly for θ_y, we can refer to Fig. 14.8 and the T' loading which produces the necessary (c/e) and (d/e) values for m and t.

All the integrations are interpreted as areas under curves over the length of the beam to evaluate the required numerical factors. These functional distributions, in turn, reflect in various combinations the effect of the beam curvature upon the physical behavior of the beam under load. Actual stresses from the moments and torques are obtainable from the M and T values, tending to maximum at the base.

Examples

14.1. A bent offset planar bar has a transverse end load Q. Find all linear deflection components at 3.

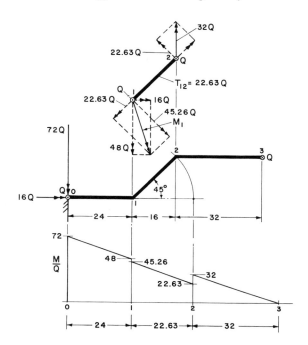

Solution: The entire beam is in equilibrium with end moment, torque, and shear at 0. Isolating 1–2 having a length of 22.63, there is a constant torque of $32 \cos 45° = 22.63Q$. At the corner 1, the torque and moment produce a resultant couple M_1, not oriented to a beam axis, but a maximum value. In turn, M_1 is resolved into torque and moment components of $48Q$ and $16Q$, respectively, aligned with the axis 0–1. By this vector process, we transfer from one leg to the next.

It is often convenient to rectify the moment diagram to the total beam length. As shown, the (M/Q) ratios are numerically equal to the distance from 3 in 0–1 and 2–3. All sections have the same slope of unity, but there are discontinuities as 1 and 2.

For the transverse deflection

$$\frac{Qz_3}{2} = \int_0^{32} \frac{(Qx_3)^2}{2EI} \, dx_3 + \int_0^{22.63} \frac{Q(22.63 + x_2)^2}{2EI} \, dx_2$$

$$+ \int_0^{24} \frac{Q(48 + x_1)^2}{2EI} \, dx_1$$

$$+ \int_0^{22.63} \frac{(22.63Q)^2}{2GJ} \, dx_2 + \int_0^{24} \frac{(16Q)^2}{2GJ} \, dx_1$$

$$z_3 = \frac{Q}{EI}(125,500 + \lambda 17,700)$$

Any F loads in the plane of the bar (xy) develop auxiliary moments and stresses maximum in the horizontal plane shown. Since no bending stresses occur in this sense due to Q, there are not possible deflections of 3 in the plane due to bending. Similarly, there are no torques caused by F loads in the plane, and the z_3 component is the total linear deflection.

14.2. Repeat Example 14.1 for possible slopes at 3.

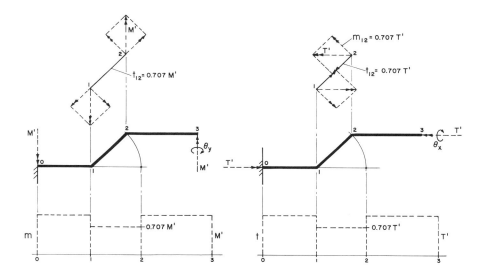

Solution: Applying M' at 3, we first solve for θ_y and have auxiliary torque only in 1–2:

$$M'\theta_y = \int_3^2 \frac{M'M}{EI}\,dx_3 + \int_2^1 \frac{(0.707M')\,M}{EI}\,dx_2$$

$$+ \int_1^0 \frac{M'M}{EI}\,dx_1 + \int_2^1 \frac{(0.707M')\,T}{GJ}\,dx_2$$

$$\theta_y = \frac{1}{EI}\left[\sum_3^2 A + 0.707\sum_2^1 A + \sum_1^0 A + 0.707\lambda T_{12}\ell_{12}\right]$$

$$= \frac{Q}{EI}[2495 + \lambda 362]$$

With T' at 3 for θ_x, there is auxiliary torque in the entire bar varying stepwise as shown, but auxiliary moment only in 1–2:

$$T'\theta_x = \int_2^1 \frac{(0.707T')\,M}{EI}\,dx_2 + \int_2^1 \frac{(0.707T')\,T_{12}}{GJ}\,dx_2 + \int_1^0 \frac{T'T_{01}}{GJ}\,dx_1$$

$$\theta_x = \frac{Q}{EI}[362 + \lambda 746]$$

Note the predominant effect of the torsional terms in the torsional deflection, although it is not independent of bending. Applying an auxiliary couple about the z axis produces auxiliary stresses orthogonal to those of the Q loading; therefore, there is no slope in the horizontal plane.

14.3. The bar of Example 14.1 carries a uniform unit load of $q = 0.30$ transverse to its plane. Calculate the end deflection z_3.

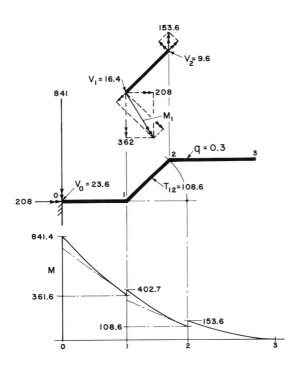

Solution: The distributed load causes parabolic actual moment components in the several legs and shears at 0, 1, and 2. Transition effects in 1–2 are indicated, and the total moment diagram. Auxiliary moments are identical to the actual in Example 14.1, with F replacing Q, and the product terms are

$$Fz_3 = \int_3^2 \frac{(Fx_3)\,M}{EI}\,dx_3 + \int_2^1 \frac{F(22.63 + x_3)\,M}{EI}\,dx_2 + \int_1^0 \frac{F(48 + x_1)\,M}{EI}\,dx_1$$

$$+ \int_2^1 \frac{(22.63F)(108.6)}{GJ}\,dx_2 + \int_1^0 \frac{(16F)(208)}{EI}\,dx_1$$

With seven component actual moment terms and four auxiliary moment terms between 0 and 2, there are 13 evaluations involving bending; however, there are only two torsional terms:

$$z_3 = \frac{(10)^6}{EI}[0.039 + 0.056 + 0.056 + 0.013 + 0.028 + 0.037$$

$$+ 0.0098 + 0.417 + 0.227 + 0.033 + 0.104$$

$$+ 0.076 + 0.012] + \frac{(10)^6}{GJ}[0.056 + 0.080]$$

$$= \frac{(10)^6}{EI}[1.107 + \lambda 0.135]$$

14.4. A cantilevered crankshaft carries a vertical transverse load at 4 on the central axis. The cheeks 1–2 and 3–4 are assumed rigid, and the round sections all have the same diameter. Find:

(a) The vertical deflection z_4

(b) Same for z_5

(c) Same for z_3

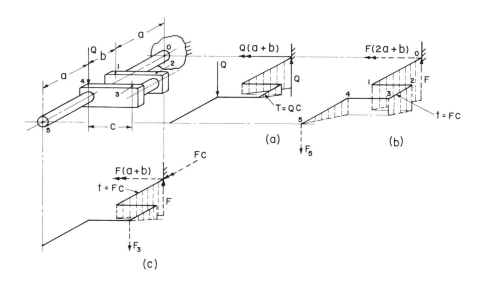

(a)

(b)

(c)

Solution:

(a)
$$\frac{Qz_4}{2} = \int_0^\ell \frac{M^2}{2EI}\,ds + \int_0^\ell \frac{T^2}{2GJ}\,ds$$

$$z_4 = \frac{1}{EI}\int_3^2 (Qx_3)^2\,dx_3 + \int_1^0 Q^2(b + x_1)^2\,dx_1 + \frac{1}{GJ}\int_3^2 (Qc)^2\,dx_3$$

$$= \frac{Q}{EI}\left[\frac{(a + b)^3}{3} + \lambda bc^2\right]$$

(b)
$$Fz_5 = \frac{F}{EI} \sum_5^0 A\bar{x} + \frac{1}{GJ} \int_0^b (Fc)(Qc)\, ds$$

$$z_5 = \frac{1}{EI}\left[\frac{Q(a+b)^2}{2}\left\{ a + \frac{2}{3}(a+b) \right\} \right] + \frac{Qc^2 b}{GJ}$$

$$= \frac{Q}{EI}\left[\frac{(a+b)^2}{6}(5a + 2b) + \lambda bc^2 \right]$$

(c)
$$Fz_3 = \int_0^\ell \frac{mM}{EI}\, ds + \int_0^\ell \frac{tT}{GJ}\, ds$$

$$= \frac{F}{EI} \int_0^b x_3 M\, dx_3 + \int_0^a (b + x_1) M\, dx_1$$

$$z_3 = \frac{1}{EI} \sum_0^b A\bar{x}_3 + b\sum_0^a A + \sum_0^a A\bar{x}_1$$

$$= \frac{Q}{3EI}(a+b)^3$$

In (c) there is no torsional contribution, and the bending effects are identical with (a).

14.5. The crank of Example 14.4 carries an end torque T_5. Determine the corresponding torsional deflection θ_5 about the central axis.

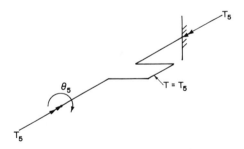

Solution: Without any supporting bearings, there can be no shear at the support 0 and a constant torque over the lengths of the cylindrical sections. There is bending moment in the cheeks, but as they are stiff, there is no deflection due to bending. Summing over the length,

$$\theta_5 = \frac{T_5}{GJ}(2a + b)$$

14.6. A split planar ring is forced apart at the split by Q loads. Find:
(a) The out-of-plane compliance
(b) The total relative angle of twist about the tangential axis

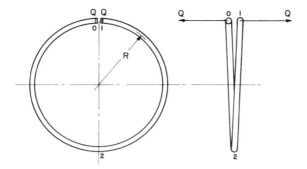

Solution:

(a) This bar is equivalent to two semicircles cantilevered at 2, or it can be considered as a complete 360° ring cantilevered at one end. Taking the latter option and Table 14.4,

$$\delta_{01} = z = Q_z \alpha_z = \frac{QR^3 \pi}{EI}[1 + 3\lambda]$$

(b) Also, Table 14.4 provides the angular deflection:

$$\theta_{01} = Q_z \beta_{xz} = \frac{QR^2 \pi}{EI}[1 + \lambda]$$

14.7. An obround split bar is similarly subjected to separational out-of-plane loading. Calculate:

(a) The relative deflection

(b) The maximum primary bending and shear stresses

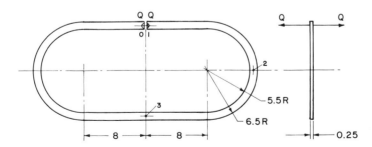

Solution:

(a) Applying Table 14.7c, with $\gamma = a/R = \frac{8}{6} = 1.333$,

$$\delta_{01} = \frac{QR^3}{EI}[(\pi + 3.160) + \lambda(3\pi + 10.666)]$$

Also $I = \frac{1}{12}bh^3 = \frac{1}{12}(1)(0.25)^3 = 0.0013$

From Fig. 14.2 and $b/h = 4$, $\lambda = 0.74$

$$\delta_{01} = \frac{Q(6)^3}{E(0.0013)}[6.302 + (0.74)(20.09)] = 3.5(10)^6\frac{Q}{E}$$

(b) Maximum bending stress (Table 14.7c) with $I/c = bh^2/6$ is

$$\sigma = \frac{\gamma QR}{I/c} = \frac{(1.333)Q6}{(0.0104)} = 770Q$$

and it occurs at 2.

Maximum torque is $2QR$ at 3, giving a maximum shear stress in the rectangular section (Table 13.1) of

$$\tau = \left[3 + (1.8)\left(\frac{0.25}{1.00}\right)\right]\frac{2Q6}{(1)(0.25)^2} = 660Q$$

On the rectangular section at 2, maximum tensile and compressive stresses exist at the top and bottom of the flat surfaces. At 3 the maximum shear stress occurs at the surface at the center of the longer sides. There is no bending at 3, but there is torsion at 2. If maximum combined stresses are important, it is necessary to search the structure at selected points on the perimeter.

14.8. For the end load on a 270° cantilever, calculate

(a) z_2

(b) z_1

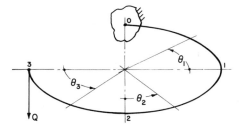

Solution:

(a)
$$M = QR\sin\left(\theta_2 + \frac{\pi}{2}\right) = QR\cos\theta_2$$

$$T = QR\left[1 - \cos\left(\theta_2 + \frac{\pi}{2}\right)\right] = QR(1 + \sin\theta_2)$$

$$m = FR\sin\theta_2$$

$$t = FR(1 - \cos\theta_2)$$

$$Fz_2 = \int_0^\ell \frac{mM}{EI}\,ds + \int_0^\ell \frac{tT}{GJ}\,ds$$

$$z_2 = (\pi + 2)\frac{QR^3}{GJ}$$

(b) With the auxiliary load at 1,

$$M = QR \sin(\theta_1 + \pi) = -QR \sin \theta_1$$
$$T = QR[1 - \cos(\theta_1 + \pi)] = QR(1 + \cos \theta_1)$$
$$m = FR \sin \theta_1$$
$$t = FR(1 - \cos \theta_1)$$
$$z_1 = \frac{\pi}{4}[-1 + \lambda]\frac{QR^3}{EI}$$

14.9. Repeat Example 14.8 as check using superposition and factors from Table 14.4.

Solution:
(a) At 2, we have $V = Q$, $M = QR$, and $T = -QR$

$$z_2 = \frac{QR^3}{EI}\left[\frac{\pi}{2} + \lambda\frac{3\pi}{2}\right] - \frac{(QR)R^2}{EI}\left[\frac{\pi}{2} + \lambda\frac{\pi}{2}\right]$$
$$+ \frac{(QR)R^2}{EI}(0 + \lambda 2) = 5.1416\frac{QR^3}{GJ}$$

(b) At 1, Q transmits to this section, $V = Q$, $M = 0$, $T = -2QR$.

$$Z_1 = \frac{QR^3}{EI}\left[\frac{\pi}{4} + 0.3562\right] - \frac{(2QR)R^2}{EI}\left[\frac{\pi}{4} - 0.2146\lambda\right]$$
$$= \frac{QR^3}{EI}\left[\frac{\pi}{4}(-1 + \lambda)\right]$$

These results verify the previous complete solutions. Note the importance of maintaining the proper sign convention when relating to the tabular coordinates.

Indeterminate Systems with Bending and Torsion

LOADED THREE-DIMENSIONAL STRUCTURES TYPICALLY ARE in a state of complex stress. In those that can be modeled as connected beam or bar geometric elements, both bending and torsion are apt to be present. Also, in the more general cases, the structure will involve redundant constraints. We now apply energy solutions to these situations, specifically equating the combined complementary energy to 0 under restraint-type auxiliary preload.

As in the previously outlined methods, it is also possible to solve redundancies by means of displacement restoration. This procedure is especially useful in the more involved bending and torsional problems and also aids in identifying redundancies and symmetry. When used in conjunction with the tabulated deflection data of Chap. 14, such solutions become a valuable tool to verify results obtained by energy techniques. If deflection characteristics are not available, they must be derived but provide an independent means of cross-checking results. It is strongly recommended that all solutions based upon energy be confirmed by a second procedure if possible. Although the method is infallible, opportunities abound for errors in sign, calculation, and in the free-body analysis.

15.1 *The Once-Redundant Bent*

The simple bent, fixed at both ends with a central load (Fig. 15.1*a*), appears to be multiply indeterminate. By applying statics and recognizing its symmetrical nature, however, we find that shears and bending moments are immediately evaluated at the supports. The supports in turn prevent rotation of the two inboard legs a, and torques T_0 are developed by the load (Fig. 15.1b), but of equal magnitude from symmetry and of opposite sense. Thus the problem is considerably simplified by this preliminary analysis and reduces to one energy equation, with the actual loading indicated (Fig. 15.1c).

We can further visualize the situation in a direct front view (Fig. 15.1e and f). Without the T_0 constraints the deflection curve is as shown dashed, but the external torques effectively superimpose the slope $\Delta\theta_3$ for a final center deflection of y_2. This exercise further establishes the sense of the T_0 couples.

The obvious and preferred auxiliary loading involves a pair of equal and opposite couples T' that simulate the indeterminate feature (Fig. 15.2a). These create auxiliary torque in a and constant bending moment in b. We then have

$$0 = \int_0^\ell \frac{mM}{EI}\, ds + \int_0^\ell \frac{tT}{GJ}\, ds$$

$$= \frac{T'}{EI} \sum_0^b A + \frac{2T'}{GJ} \int_0^a T_0\, ds$$

$$= \left[T_0 b - \frac{Qb^2}{8} \right] + \lambda 2 T_0 a$$

$$T_0 = \frac{Qb}{8} \left[\frac{1}{1 + \lambda 2 \dfrac{a}{b}} \right] \tag{15.1}$$

as in Table 15.4b.

We note the contributions of the $\lambda = EI/GJ$ factor to the torsional reaction, with I taken about a horizontal axis on the cross section. As J becomes significantly greater than I, this particular term will become negligible in its effect upon T_0.

Alternately, we can introduce an internal shear at 3 (Fig. 15.2b), with auxiliary torque t_a then only at 0. This approach, although somewhat more complicated, will yield an identical result.

A similar attempt to introduce the shear preload at 2, however, will be unsuccessful. Auxiliary-moment and torque distributions are then *antisymmetrical*, and, when combined with the actual *symmetrical* moments and torques produce, a degenerate energy equation. This particular preload tends to violate a physical feature of the problem; namely, there is no internally reactive shear at 2 where the load is applied.

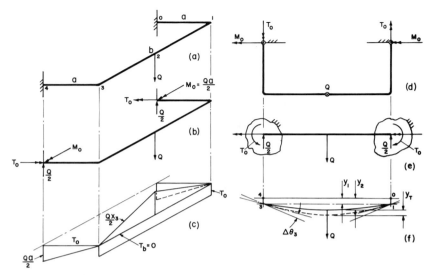

Figure 15.1 The symmetrical bent carrying a central load has one redundancy T_0. In (f) T_0 is shown restraining the displacement of Q.

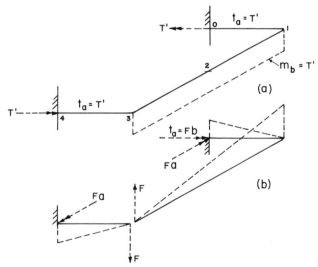

Figure 15.2 Two options are available for auxiliary loading for solution of the indeterminancy in Fig. 15.1.

15.2 *Displacement Restoration*

As an alternative solution, we can decouple the bar at 2 and apply first half of the actual load, or $Q/2$ to each of the cantilevered bents (Fig. 15.3a). There are then opposed slopes θ_2 as shown that can be evaluated readily for

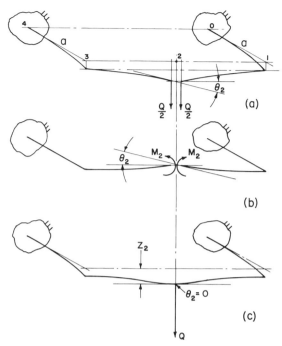

Figure 15.3 Redundancy can also be solved by superimposing the internal moment M_2 to return the slope to 0 after decoupling.

the determinate bents. For the original structure, there must obviously be zero slope at 2 (Fig. 15.3*c*), and this condition is achieved by superimposing opposed moments M_2 corresponding to the actual bending moment present at 2 (Fig. 15.3*b*).

Adding displacements, with θ_2 from Table 14.1,

$$\beta yz\left(\frac{Q}{2}\right) - \gamma_y M_2 = 0 \tag{15.2}$$

where

$$\beta_{yz} = \frac{(b/2)^2}{2EI}\left[1 + 2\lambda\frac{a}{(b/2)}\right]$$

$$\gamma_y = \frac{(b/2)}{EI}\left[1 + \lambda\frac{a}{(b/2)}\right]$$

Solving,

$$M_2 = \frac{Qb}{8}\left[\frac{1 + 4\lambda\frac{a}{b}}{1 + 2\lambda\frac{a}{b}}\right] \tag{15.3}$$

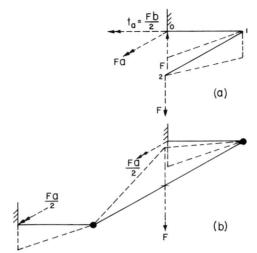

Figure 15.4 The decoupled structure is preloaded for the deflection solution with several possibilities. In (*b*) ball swivels provide decoupling at 1 and 3 to eliminate auxiliary torque.

This result agrees with Table 15.3*b*, and from statics

$$\sum M_{01} = T_0 + M_2 - \frac{Q}{2}\left(\frac{b}{2}\right) = 0$$

$$T_0 = \frac{Qb}{8}\left[\frac{1}{1 + 2\lambda\frac{a}{b}}\right] \qquad (15.4)$$

A further option corresponds to decoupling the torque constraints, applying Q, and determining the axial rotations at 0 and 4. Then the compliance of the two ends with respect to opposed couples T_0 and T_4 establishes the restoration relationship.

15.3   Displacement at a Load

For the Fig. 15.1*a* bent, having determined the reactions, we proceed to find displacements, in particular z_2 (Fig. 15.4*a*). After decoupling at 2, we preload one half, obtaining auxiliary bending in 2–1 and 1–0 and torque in *a.* With actual loading from Fig. 15.1*c*,

$$Fz_2 = \int_0^\ell \frac{mM}{EI}\, ds + \int_0^\ell \frac{tT}{GJ}\, ds$$

$$= \frac{F}{EI} \sum_0^1 A\bar{x}_2 + \frac{F}{EI} \sum_1^0 A\bar{x}_1 + \frac{Fb}{2GJ} \int_0^a T_0\, ds$$

$$z_2 = \frac{1}{EI}\left[-\frac{Q}{2}\left(\frac{b}{2}\right)^3\frac{1}{6} + T_0\left(\frac{b}{2}\right)^2\frac{1}{2} + \frac{Q}{2}\left(\frac{a^3}{3}\right)\right] + \frac{bT_0 a}{2GJ} \quad (15.5)$$

and this expression can be reduced to the deflection equation shown in Table 15.3*b*.

The bent can also be decoupled for deflection preload purposes by visualizing ball joints at 1 and 3 (Fig. 15.4*b*). This prevents *F* from inducing auxiliary torque in the *a* members. Preload consists of *b* as a simple beam with a center load and the *a* legs as simple end-loaded cantilevers. Then with torsion eliminated,

$$Fz_2 = \frac{2}{EI}\left[\frac{F}{2}\sum_0^{b/2} A\bar{x}_1 + \frac{F}{2}\sum_0^a A\bar{x}_1\right]$$

$$z_2 = \frac{1}{EI}\left[\sum A\bar{x}_1 + \sum A\bar{x}_1\right] \quad (15.6)$$

which reduces to the indicated tabular result.

15.4 *Displacement by Superposition*

It is convenient to arrive at the z_2 displacement using the philosophy indicated in Sec. 15.2 by which the actual condition is preceded by the decoupled and restoration phases (Fig. 15.3). Subtracting the upward displacement due to M_2 (Fig. 15.3*b*), from the decoupled (Fig. 15.3*a*), and using Table 14.1*a* for deflections

$$z_2 = \alpha_z\left(\frac{Q}{2}\right) - \beta_{zy} M_2 \quad (15.7)$$

Data are provided in Table 15.1 for double concentrated loading of a continuous rectangular frame of constant section. Tables 15.2 to 15.5 pertain to various fixed–fixed bent geometries with both concentrated and distributed loading. In some cases the results are explicit, but in unsymmetrical bents coefficients must be entered from deflection data in Chapter 14, as just indicated.

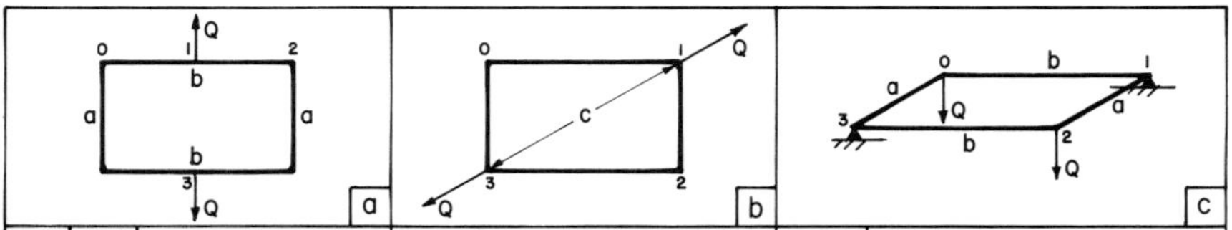

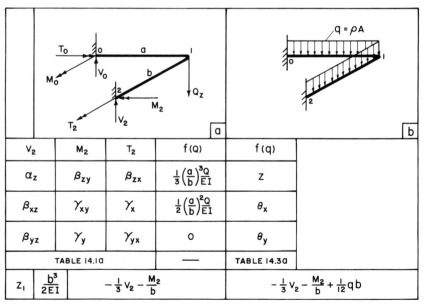

δ_{13}	$\frac{Qb^3}{EI}$	$\frac{1}{96A}\left[\frac{1}{4}+\frac{a}{b}\right]$	$\frac{1}{24}A\left(\frac{a}{c}\right)^2$	$Z_0 = Z_2$	$\frac{1}{48}\left[1+\left(\frac{a}{b}\right)^3\right]+\frac{1}{16}\lambda\left(\frac{a}{b}\right)A$	
M_0	Qb	$\frac{1}{8A}$	$\frac{1}{4}\left(\frac{a}{c}\right)$	$M_{0b} = -M_{1b}$	Qb	$\frac{1}{4}$
M_1		$-\frac{1}{8A}\left[1+2\left(\frac{a}{b}\right)\right]$	$-\frac{1}{4}\left(\frac{a}{c}\right)$	$M_{0a} = -M_{1a}$		$\frac{1}{4}\left(\frac{a}{b}\right)$
$\lambda = \frac{EI}{GJ}$			$A = \left[1+\left(\frac{a}{b}\right)\right]$	T_a		$\frac{1}{4}$
				T_b		$\frac{1}{4}\left(\frac{a}{b}\right)$

Table 15.1 The rectangular planar bar carries separating forces in the plane (a) and (b) and out-of-plane loads (c).

Table 15.2 The fixed–fixed bent with unequal legs has a transverse corner load (a) and distributed load (b).

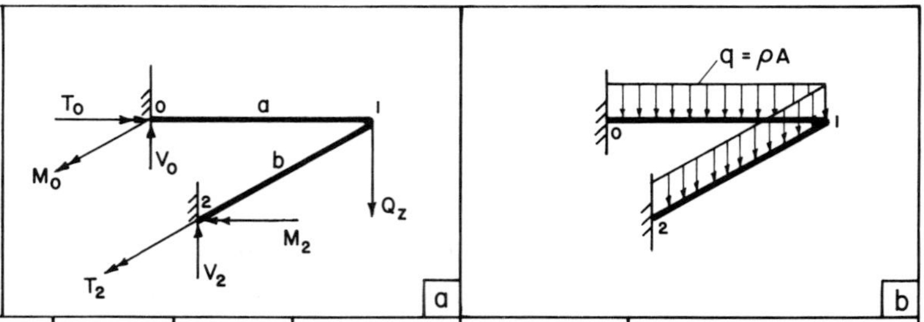

V_2	M_2	T_2	f(Q)	f(q)
α_z	β_{zy}	β_{zx}	$\frac{1}{3}\left(\frac{a}{b}\right)^3\frac{Q}{EI}$	Z
β_{xz}	γ_{xy}	γ_x	$\frac{1}{2}\left(\frac{a}{b}\right)^2\frac{Q}{EI}$	θ_x
β_{yz}	γ_y	γ_{yx}	0	θ_y
TABLE 14.1a			—	TABLE 14.3a

Z_1	$\frac{b^3}{2EI}$	$-\frac{1}{3}V_2 - \frac{M_2}{b}$	$-\frac{1}{3}V_2 - \frac{M_2}{b} + \frac{1}{12}qb$

Table 15.3 The cantilevered bent carries symmetrical concentrated loading.

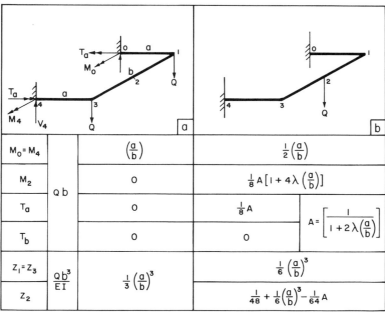

		(a)	(b)
$M_0 = M_4$	Qb	$\left(\dfrac{a}{b}\right)$	$\dfrac{1}{2}\left(\dfrac{a}{b}\right)$
M_2		0	$\dfrac{1}{8}A\left[1 + 4\lambda\left(\dfrac{a}{b}\right)\right]$
T_a		0	$\dfrac{1}{8}A$
T_b		0	0
$Z_1 = Z_3$	$\dfrac{Qb^3}{EI}$	$\dfrac{1}{3}\left(\dfrac{a}{b}\right)^3$	$\dfrac{1}{6}\left(\dfrac{a}{b}\right)^3$
Z_2			$\dfrac{1}{48} + \dfrac{1}{6}\left(\dfrac{a}{b}\right)^3 - \dfrac{1}{64}A$

$$A = \left[\dfrac{1}{1 + 2\lambda\left(\dfrac{a}{b}\right)}\right]$$

Table 15.4 Similar to Table 15.3, but with symmetrical distributed loading.

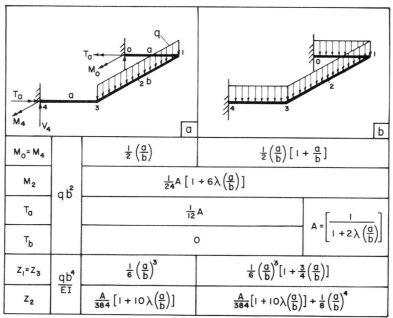

		(a)	(b)
$M_0 = M_4$	qb^2	$\dfrac{1}{2}\left(\dfrac{a}{b}\right)$	$\dfrac{1}{2}\left(\dfrac{a}{b}\right)\left[1 + \dfrac{a}{b}\right]$
M_2		$\dfrac{1}{24}A\left[1 + 6\lambda\left(\dfrac{a}{b}\right)\right]$	
T_a		$\dfrac{1}{12}A$	
T_b		0	
$Z_1 = Z_3$	$\dfrac{qb^4}{EI}$	$\dfrac{1}{6}\left(\dfrac{a}{b}\right)^3$	$\dfrac{1}{6}\left(\dfrac{a}{b}\right)^3\left[1 + \dfrac{3}{4}\left(\dfrac{a}{b}\right)\right]$
Z_2		$\dfrac{A}{384}\left[1 + 10\lambda\left(\dfrac{a}{b}\right)\right]$	$\dfrac{A}{384}\left[1 + 10\lambda\left(\dfrac{a}{b}\right)\right] + \dfrac{1}{8}\left(\dfrac{a}{b}\right)^4$

$$A = \left[\dfrac{1}{1 + 2\lambda\left(\dfrac{a}{b}\right)}\right]$$

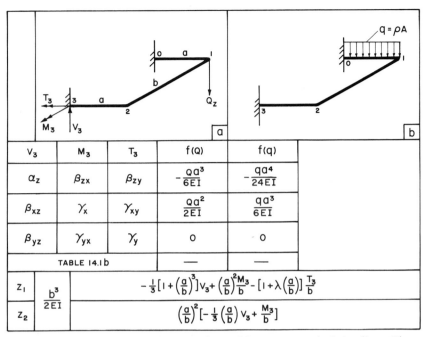

V₃	M₃	T₃	f(Q)	f(q)
α_z	β_{zx}	β_{zy}	$-\dfrac{Qa^3}{6EI}$	$-\dfrac{qa^4}{24EI}$
β_{xz}	γ_x	γ_{xy}	$\dfrac{Qa^2}{2EI}$	$\dfrac{qa^3}{6EI}$
β_{yz}	γ_{yx}	γ_y	0	0
	TABLE 14.1b		—	—

z_1	$\dfrac{b^3}{2EI}$	$-\frac{1}{3}\left[1+\left(\frac{a}{b}\right)^3\right]V_3 + \left(\frac{a}{b}\right)^2\frac{M_3}{b} - \left[1+\lambda\left(\frac{a}{b}\right)\right]\frac{T_3}{b}$
z_2		$\left(\frac{a}{b}\right)^2\left[-\frac{1}{3}\left(\frac{a}{b}\right)V_3 + \frac{M_3}{b}\right]$

Table 15.5 Similar to Table 15.3, but with unsymmetrical loading. Three simultaneous equations are indicated, with coefficients of V_3, M_3, and T_3 from Table 14.1b.

15.5 *Bent with Multiple Redundancies*

In Fig. 15.5a a constant bar in the shape of a planar tee carries a transverse load Q. Ends of the crossbar are completely fixed at 1 and 2, with a simple support at 4. If the supports at 1 and 2 are also simple, the structure is not indeterminate; however, as constrained, there are additionally bending moments and torques at 1 and 2 that alter the division of shear support.

From symmetry, $M_1 = M_2$ and $T_1 = T_2$, and the structure is therefore twice redundant. Constructing the actual moment diagrams toward the central juncture 0, we have Fig. 15.5b due to Q. The torque diagram is discontinuous at 0.

To employ the required auxiliary conditions, we first introduce T' (Fig. 15.5c), and there are two possibilities. We can twist 1–2 with *opposed* auxiliary torques at the ends, or we can have T' loads of the *same sense* reacting at 4 with $(T'/9)$. Only the latter option shown leads to a meaningful energy relationship, as it leads to an *antisymmetrical* auxiliary torque diagram sympathetic to the *antisymmetrical* actual torque diagram. Other-

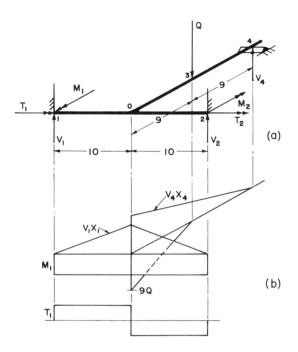

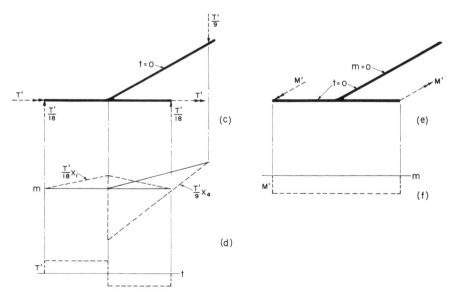

Figure 15.5 The tee, clamped at 1 and 2 and simply supported at 4, develops actual moment and torque diagrams shown. Two auxiliary preloads are required: (*c*) and (*e*).

wise, the auxiliary diagram is symmetrical, and the product integral is degenerative. In effect, the first option fails to incorporate any elasticity information pertaining to the leg 0–4.

From Fig. 15.5d we have

$$0 = \frac{T'}{9} \sum_0^{18} \frac{A\bar{x}_4}{EI} + \frac{T'}{18} 2 \sum_0^{10} \frac{A\bar{x}_1}{EI} + 2 \int_0^{10} \frac{T'T_1}{GJ}\, ds \qquad (15.8)$$

With auxiliary couples M' (Fig. 15.5e), only constant bending is induced in 1–2, with a correspondingly simple equation:

$$0 = \frac{M'}{EI} 2\sum_0^{10} A \qquad (15.9)$$

By writing statics equations and substituting numerical values, the fourth degree problem is reduced to three simultaneous linear equations:

$$\begin{cases} 23.33 V_4 - V_1 - 2.70 T_1 = 7.29Q \\ 0 \qquad + 18V_1 - T_1 \quad = 4.50Q \\ V_4 + 2V_1 + \quad 0 \quad = Q \end{cases} \qquad (15.10)$$

from which

$$\left(\frac{V_1}{Q}\right) = 0.293 \qquad \left(\frac{V_4}{Q}\right) = 0.414 \qquad \left(\frac{M_1}{Q}\right) = 1.465 \qquad \left(\frac{T_1}{Q}\right) = 0.774$$

In this sharing of constraint supports, we note the simple outboard support 4 carries less than half of Q. Also, checking for maximum bending moment,

$$\left(\frac{M_3}{Q}\right) = 3.726 \qquad \left(\frac{M_0}{Q}\right) = -1.548 \quad \text{(in the 0–4 direction)}$$

$$\left(\frac{M_0}{Q}\right) = 1.465 \quad \text{(in the 1–2 direction)}$$

Thus maximum bending stress is encountered at 0 in the 0–4 direction, but there is no torque here for combined stress purposes.

If the structure is unsymmetrical by virtue of unequal cross sections or unequal distances at the head of the tee, we have

$$V_1 \neq V_2 \qquad M_1 \neq M_2 \qquad T_1 \neq T_2$$

This solution is considerably more challenging, as we now have three additional variables. Also, the end at 4 could introduce an M_4 and T_4 if completely clamped. Such solutions will require additional auxiliary loading combinations and equations from statics. Although lengthy, the more general case can be analyzed from elasticity using fundamental strain energy procedures indicated.

15.6 *Solution by End-Displacement Restoration*

The fixed–fixed bent with unequal legs now illustrates the use of displacement equations for an indeterminate solution with several unknowns. Taking distributed loading (Fig. 15.6a), we completely decouple at 2. From Table 14.3a we have expressions for the resulting translational and rotational displacements z, θ_x, and θ_y under load. To determine the deflections due to loads at the end 2 for the bent, we use Table 14.1a; however, it is now important to relate the space orientation properly and to establish a sign convention. One approach to the signs is to completely reverse the previous signs (Table 15.2a); that is, positive results for reactions indicate a sense opposing the positive sense of Table 14.1, but conforming to the assumed reverse sense (Fig. 15.6a and b). Obviously, these directions can be unlikely, and the M_2 couple cannot be in the direction assumed in Fig. 15.6a.

We then proceed to quantify the coefficients and the deflections as indicated in Table 15.2b with $(a/b) = 0.75$ and $\lambda = 1.25$, canceling the *EI*

Figure 15.6 The fixed–fixed bent with distributed loading is decoupled at 2 to obtain the displacements produced for the restoration solution in the redundant phase. Auxiliary loading for the deflection z_1 is shown in (c).

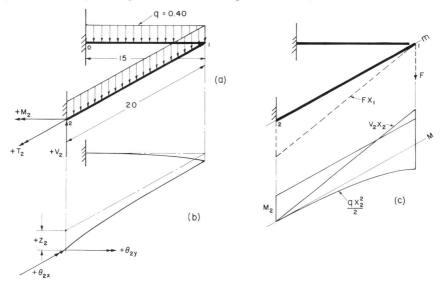

terms from all equations:

$$\left\{ \begin{array}{l} 11292V_2 + \quad 575M_2 + 112.5T_2 = 49530 \\ 112.5V_2 + \qquad 0 + \quad 40T_2 = \quad 1125 \\ 575V_2 + 38.75M_2 + \qquad 0 = \quad 2033 \end{array} \right. \tag{15.11}$$

which yields, for the total weight of 14,

$$V_2 = 6.63 \qquad M_2 = -45.92 \qquad T_2 = 9.48$$

Reactions at 0 follow from statics.

15.7 *Intermediate Deflection*

For the previous case, we now indicate the source of the Table 15.2b expression for z_1 at the corner, and this is the maximum. In this situation, knowing the reactions at 2, it is desirable to adopt a procedure directed from this end. Having used available displacement data in lieu of energy equations, we now return to auxiliary loading. Decoupling at 1 (Fig. 15.6c), we preload the cantilever 1–2 applying F at 1, and $m = -Fx_1$. Taking the assumed directions for V_2 and M_2 (Fig. 15.6a), the M diagrams are shown in Fig. 15.6c:

$$Fz_1 = \frac{1}{EI} \sum A\bar{x}_1$$

$$z_1 = \frac{1}{EI} \left[-\frac{V_2 b^3}{6} - \frac{M_2 b^2}{2} + \frac{qb^4}{24} \right] \tag{15.12}$$

This is the equation of Table 15.2b.

15.8 *Once-Redundant Arcuate Beam*

A fixed–fixed semicircular bar with a central out-of-plane load (Fig. 15.7a) is similar to the equivalent rectangular bent (Fig. 15.1a). Analytically, it is simpler, described geometrically by a single radius rather than the a and b dimensions. As in Sec. 15.1 the only redundancy remaining, after considerations of statics and symmetry, are the equal values of the torque reactions T_0.

For the solution we apply opposed T' couples to produce the auxiliary moment and torque distributions (Fig. 15.7c). Then, using θ_0 from the fixed

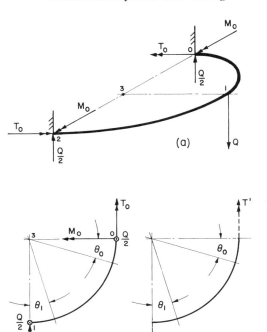

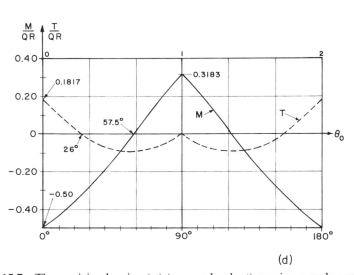

Figure 15.7 The semicircular ring (*a*) (once redundant) requires couple preload (*c*). Distributions of moment and torque are shown in (*d*).

end,

$$m = -T'\sin\theta \qquad t = +T'\cos\theta$$

$$M = -T_0\sin\theta - M_0\cos\theta + \frac{Q}{2}R\sin\theta \qquad (15.13)$$

$$T = +T_0\cos\theta - M_0\sin\theta + \frac{Q}{2}(1 - \cos\theta)$$

$$0 = 2\int_0^{\pi/2}\frac{mM}{EI}\,ds + 2\int_0^{\pi/2}\frac{tT}{GJ}\,ds$$

and after integration

$$T_0 = QR\left[\frac{0.1427(1 + \lambda)}{0.7854(1 + \lambda)}\right] = 0.1817QR \qquad (15.14)$$

We note by the cancellation that T_0 is independent of the $\lambda = EI/GJ$ parameter and will be the same regardless of the cross section or material of the beam.

15.9 Simplified Solution

In the previous analysis of the semicircular ring, opposed auxiliary T' couples were applied at 0 and 1 following the procedure for the bent (Fig. 15.2a). And this is completely logical; however, a shorter solution is possible if we recognize that *the internal torque at 1 is 0* (Fig. 15.7b). For this torque to exist, the tendency to rotate inwardly on one side of 1 must be countered by a tendency to rotate outwardly on the other side. Since for symmetry about the 1–3 axis, the elastic behavior in the two halves must be identical, this condition is impossible and the torque at this point is indeed 0.

Proceeding from 1 with θ_1 (Fig. 15.7b and c), and involving M_1,

$$m = -T'\cos\theta \qquad t = -T'\sin\theta$$

$$M = M_1\cos\theta - \frac{QR}{2}\sin\theta$$

$$T = M_1\sin\theta - \frac{QR}{2}(1 - \cos\theta) \qquad (15.15)$$

and we now have four components for integration rather than six [Eqs. (15.13)]. Summing the complementary energies to 0 and integrating from 0 to $\pi/2$,

$$M_1 = 0.3183QR \quad \text{and} \quad T_0 = 0.1817QR$$

with the final result for T_0 from statics.

Note the auxiliary loading (Fig. 15.7c) is identical in the two solutions, and often there is no choice as to how the preload is applied. The more expeditious choice is seen to lie with the choice of coordinate direction.

15.10   *Moment and Torque Distribution*

Returning to θ_0 and Eqs. (15.13), we now evaluate total distributed loading along the bar. Substituting for T_0

$$M(\theta_0) = QR[0.3183 \sin \theta - 0.50 \cos \theta] \qquad (15.16a)$$

$$T(\theta_0) = QR[-0.3183 \cos \theta - 0.50 \sin \theta + 0.50] \qquad (15.16b)$$

with these equations valid from 0 to $\pi/2$, and the resulting symmetrical functions are plotted (Fig. 15.7d). Although the abscissa is in terms of θ_0, it also represents the developed or rectified length of the beam of constant radius.

The moment curve is nearly straight, maximum at the supports, and 0 at 57.5°, or at the point of inflection of the elastic curve. Torque is also maximum at the supports, reversing sense from outward to inward. It is 0 at three points as shown, verifying symmetry and $T_1 = 0$.

Maximum stresses (bending, torsional, and combined) are located at the supports.

15.11   *Tabulated Results for Circular Segments*

Results for the fixed–fixed circular beam with various central angles are given in Tables 15.6 and 15.7. Table 15.6a includes the special case of $\psi = 90°$ (Sec. 15.8). In order to accommodate the variable angle, these expressions are developed utilizing deflection data from Chap. 14 tables. The coordinate θ is taken from the decoupling point at the center and superposition methods applied as in Sec. 15.2.

In Table 15.7a the situation is symmetrical with ($M_x/2$) transmitted as a torsional load to the two halves. Table 15.7b is similar with the bending load at 1 divided equally; however, the two halves behave antisymmetrically, as indicated by the end reactions.

15.12   *Transversely Loaded Complete Ring*

The planar ring (Table 15.8) is alternately loaded and supported on knife edges by a number of equally spaced out-of-plane forces (Table 15.8a). In the figure this is illustrated by four loads and four supports with $\psi = 45°$.

Table 15.6 The fixed–fixed symmetrical circular segment bar with out-of-plane loading is solved in terms of the tabulated displacement coefficients.

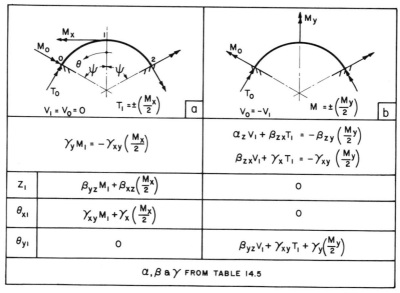

M_I		$\dfrac{\beta_{yz}}{\gamma_y}$	M_I	$\dfrac{\theta_y}{\gamma_y}$
$M(\theta)$	$\dfrac{Q}{2}$	$\left(\dfrac{\beta_{yz}}{\gamma_y}\right)\cos\theta - R\sin\theta$	$M(\theta)$	$\left[\left(\dfrac{\theta_y}{\gamma_y}\right)+qR^2\right]\cos\theta - qR^2$
$T(\theta)$		$\left(\dfrac{\beta_{yz}}{\gamma_y}\right)\sin\theta - R(1-\cos\theta)$	$T(\theta)$	$\left[\left(\dfrac{\theta_y}{\gamma_y}\right)+qR^2\right]\sin\theta - qR^2\theta$
Z_I		$\alpha_z - \left(\dfrac{\beta_{yz}^2}{\gamma_y}\right)$	Z_I	$z - \left(\dfrac{\beta_{yz}\theta_y}{\gamma_y}\right)$

α_z, β_{yz} & γ_y FROM TABLE 14.5 | Z & θ_y FROM TABLE 14.6

Table 15.7 Similar to Table 15.6, but with central couple loading about the x and y axes.

$V_I = V_0 = 0$ $T_I = \pm\left(\dfrac{M_x}{2}\right)$ [a]

$V_0 = -V_I$ $M = \pm\left(\dfrac{M_y}{2}\right)$ [b]

	(a)	(b)
	$\gamma_y M_I = -\gamma_{xy}\left(\dfrac{M_x}{2}\right)$	$\alpha_z V_I + \beta_{zx}T_I = -\beta_{zy}\left(\dfrac{M_y}{2}\right)$
		$\beta_{zx}V_I + \gamma_x T_I = -\gamma_{xy}\left(\dfrac{M_y}{2}\right)$
Z_I	$\beta_{yz}M_I + \beta_{xz}\left(\dfrac{M_x}{2}\right)$	0
θ_{xI}	$\gamma_{xy}M_I + \gamma_x\left(\dfrac{M_x}{2}\right)$	0
θ_{yI}	0	$\beta_{yz}V_I + \gamma_{xy}T_I + \gamma_y\left(\dfrac{M_y}{2}\right)$

α, β & γ FROM TABLE 14.5

	2	3	4	6	8	2	3	4	6	8
N	2	3	4	6	8	2	3	4	6	8
ψ	$\frac{\pi}{2}$	$\frac{\pi}{3}$	$\frac{\pi}{4}$	$\frac{\pi}{6}$	$\frac{\pi}{8}$	$\frac{\pi}{2}$	$\frac{\pi}{3}$	$\frac{\pi}{4}$	$\frac{\pi}{6}$	$\frac{\pi}{8}$
A	1	0.5774	0.4142	0.2679	0.1989	1.5708	1.2092	1.1107	1.0472	1.0262
B	0.2854	0.0604	0.0229	0.0063	0.0026	0.4483	0.0633	0.0180	0.0033	0.0010
C	0.0708	0.0066	0.0014	0.0004	—	0	-0.0610	-0.0383	-0.0152	-0.0073
M_I (QR)	0.5000	0.2887	0.2071	0.1340	M_I (qR^2)	0.5708	0.2092	0.1107	0.0472	0.0262
M_O (QR)	-0.5000	-0.2887	-0.2071	-0.1340	M_O	-1	-0.3954	-0.2146	-0.0931	-0.0519

$M(\theta)$	$\frac{QR}{2}$	$A\cos\theta - \sin\theta$			qR^2	$A\cos\theta - 1$
$T(\theta)$		$A\sin\theta - (1 - \cos\theta)$				$A\sin\theta - \theta$
Z_I	$\frac{QR^3}{EI}$	$B + \lambda C$			$\frac{qR^4}{EI}$	$B + \lambda C$

Table 15.8 The continuous planar ring with a discrete number of equally spaced transverse loads is supported by an equal number of knife edge supports.

For $N = 2$, one half of the ring resembles Fig. 15.7a, and there is zero slope at 0 and 2, with the other half of the ring supplying the moment reactions; however, there is a major difference between the two cases. By previous arguments relating to symmetry, there *can now be no torque at 0 or 2* over the knife edge supports provided by the mating ring half. In Table 15.8 there is zero torque at all support points and at all points midway between the support points. Thus with the ring permitted to roll at the supports, the redundant torque reactions when fixed are eliminated and the solution is *statically determinate*. Using the geometric relations for various discrete values of ψ, we find the moments and torques acting.

A ring having uniformly distributed loading with equally-spaced simple supports (Table 15.8b) is similarly determinate. This static solution involves centroidal locations of the loaded circular segments. Moments, torques, and deflections are given.

Technically, these cases, being determinate, should not be included in this chapter; however, since most readers would expect the continuous ring in bending and torsion to be of sufficient complexity to introduce redundancies, they are included and distinguished from the clamped cases.

15.13 *Torsionally Loaded Complete Ring*

Solutions for the ring with couple loading are given in Table 15.9. The symmetrical aspects of these cases are rather challenging, and we now illustrate by analyzing the Table 15.9b problem, which can be visualized as a type of torsional coupling element interposed between shafts aligned in the 1–3 direction (Fig. 15.8a). For the analysis it is irrelevant whether the ring rotates or is stationary, provided we neglect centrifugal loading that can be superimposed.

Separating into two halves for free-body purposes (Fig. 15.8b and c), we use symmetry to divide M_y equally relative to the transmission of the total torque. For the left half (Fig. 15.8b), 1 tends to move down due to $(M_y/2)$; therefore, we anticipate a shear V upward at 1. This must then be opposed by an equal shear at 3.

Next considering the T_1 torques, we assume these are in the *same* direction at 1 and 3. If in opposite directions, they cancel, but they must function to oppose the shear couple:

$$\sum M_{02} = 2VR - 2T_1 = 0 \qquad T_1 = VR \qquad (15.17)$$

Thus from statics we have partially solved the problem. Although our assumed sense of V has not as yet been verified by elastic energy, Eq. (15.17) is valid, with the two couples in opposition.

Table 15.9 The complete ring carries opposed couple loading with distribution functions valid for 180°. [See Sects. 15.13 and 15.14 for a discussion of (b).]

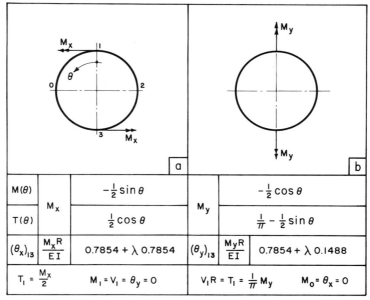

		a			b
$M(\theta)$		$-\frac{1}{2}\sin\theta$			$-\frac{1}{2}\cos\theta$
$T(\theta)$	M_x	$\frac{1}{2}\cos\theta$	M_y		$\frac{1}{\pi}-\frac{1}{2}\sin\theta$
$(\theta_x)_{13}$	$\frac{M_x R}{EI}$	$0.7854 + \lambda\,0.7854$	$(\theta_y)_{13}$	$\frac{M_y R}{EI}$	$0.7854 + \lambda\,0.1488$
$T_1 = \frac{M_x}{2}$		$M_1 = V_1 = \theta_y = 0$	$V_1 R = T_1 = \frac{1}{\pi}M_y$		$M_0 = \theta_x = 0$

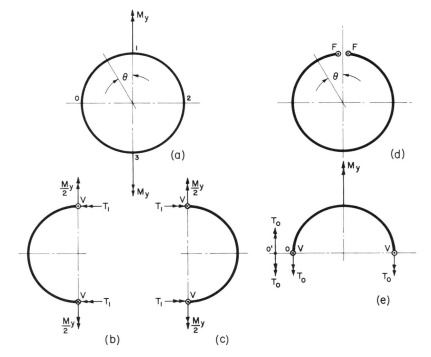

Figure 15.8 A continuous ring (Table 15.9*b*) with opposed torque loading about the 1–3 axis requires a single antisymmetrical force preload (*d*).

The mating half (Fig. 15.8*c*) reacts the first with respect to V and T_1; however, the M_y couple is shared. The two halves now typify antisymmetry, as a 180° rotation of Fig. 15.8*b* relative to Fig 15.8*c* will cause Fig. 15.8*b* to conform identically to Fig. 15.8*c*. Furthermore, comparing the ends of the halves at 1, or at 3, the sense of the external moments, torques, and shears are all opposed.

With antisymmetry present in the actual loading, we must beware of attempting a symmetrical auxiliary loading or the energy equation will be degenerative. Selecting the shear redundancy V to simulate, we apply opposed F shear forces at 1 (Fig. 15.8*d*). Writing the equations from $\theta = 0$ to $\theta = \pi/2$,

$$m = FR \sin \theta \qquad t = FR(1 - \cos \theta) \qquad (15.18a)$$

$$M = -\left(\frac{M_y}{2}\right)\cos \theta \qquad T = VR - \left(\frac{M_y}{2}\right)\sin \theta \qquad (15.18b)$$

With antisymmetry, signs for all actual and auxiliary terms are reversed for the second half, and the energy integrals are additive:

$$0 = 2\int_0^\pi \frac{mM}{EI}\,ds + 2\int_0^\pi \frac{tT}{GJ}\,ds$$

$$= \frac{1}{GJ}\big[(VR) - M_y\big]$$

$$VR = \frac{1}{\pi}M_y = 0.3183M_y \qquad (15.19)$$

Only the torsional terms exist, and the positive result confirms the assumed directions in Fig. 15.8b and 15.8c for V and T_1.

15.14   *Moment and Torque Distribution*

Substituting $\theta = \pi/2$ in Eqs. (15.18), we find at 0 on the horizontal diameter (Fig. 15.8e)

$$M_0 = 0 \qquad T_0 = -0.1817M_y$$

The zero moment is logical because, considering equilibrium of the 0–1 quadrant,

$$\sum M_{02} = T_1 - VR + M_0 = 0$$

$$T = VR \quad \text{and} \quad M_0 = 0$$

Similarly, from statics

$$\sum M_{13} = T_0 - VR - \left(\frac{M_y}{2}\right) = 0 \qquad T_0 = -0.1817M_y$$

The negative sign indicates a reversal of torsional sense between 1 and 0. T_1 tends to turn the bar outwardly, with the T_0 vector at 0 turning inwardly. As detailed at 0′ (Fig. 15.8e), this vector is up, but the external reactive torque is down in opposition. Thus while the load torque is *inward*, the balancing external torque is *outward* at 1, and there is no inconsistency. This reversal phenomenon does not occur as we proceed in a given direction from a selected free end of a structure.

15.15   *Symmetrically Disposed Prismatic Elements*

The circular bar (Fig. 15.8a) has a counterpart in the structure of Fig. 15.9, in which two stiff flanges carrying torsion about the central axis are connected by two straight bars. We then have a rectangular version of the

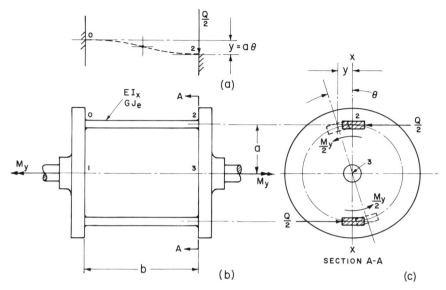

Figure 15.9 Torque is transmitted by two stiff flanges integrally connected by two elastic bars subjected to combined bending and torsion.

previous circle. Using a procedure similar to Sec. 15.14, we use an auxiliary loading as in Fig. 15.8d. There is only one redundancy and strain energy only in the connecting bars. The resulting solution becomes,

$$\frac{Va}{M_y} = \frac{1}{2} B \tag{15.20a}$$

$$\frac{T_1}{M_y} = \left(\frac{b}{4a}\right) B \tag{15.20b}$$

$$\theta_y = \frac{bM_y}{2EI}\left[\left(\frac{b}{a}\right)^2 \frac{B^2}{12} + \lambda(B-1)^2\right] \tag{15.20c}$$

where $B = \{\lambda/[\frac{1}{12}(b/a)^2 + \lambda]\}$
 θ_y = total relative angular deflection
 I = area moment area of the bar as flexed
 $\lambda = EI/GJ$ = relative stiffness parameter

and V and T_1 are developed within the volumes of the stiff flanges.

The indicated procedure involving energy integrals is not particularly difficult, but requires a degree of organization and time. In a simpler approach we visualize the elastic elements as prismatic springs subjected to a combination of bending and torsional loading.

15.16 *Combined Spring Derivation*

Given an angular displacement θ (Fig. 15.9c), the end 2 of the elastic element has a tangential displacement of $y = a\theta$. As shown in Fig. 15.9a with the ends constrained to planes, the slope is 0 at 2 and the beam is equivalent to two cantilevers. Each has a length of $b/2$ with an end load at the center point of $Q/2$, and

$$y = a\theta = \frac{2\left(\frac{Q}{2}\right)\left(\frac{b}{2}\right)^3}{3EI} = \frac{Qb^3}{24EI} \tag{15.21}$$

Torque carried by the two parallel elements in bending is

$$T_B = 2a\left(\frac{Q}{2}\right) = \left(\frac{24EIa^2}{b^3}\right)\theta \tag{15.22}$$

As seen in Fig. 15.9c, the elements are simultaneously twisted through an angle θ, and the corresponding torsional couple is

$$T_T = \left(\frac{2GJ}{b}\right)\theta \tag{15.23}$$

and the total torque transmitted is

$$M_y = \sum T = (T_B + T_T) = \left[\left(\frac{24EI}{b^3}\right)a^2 + 2\left(\frac{GJ}{b}\right)\right]\theta \tag{15.24a}$$

For a single element, or on a unit basis, the torque will be one half of this value, although it is unsymmetrical.

This stiffness factor can be interpreted as composed of two springs in parallel for an individual bar, even though physically there is only one elastic prism. The first term $(12EI/b^3)$ is the transverse spring constant of the double cantilever with the radius term a^2 converting to polar coordinates. The second term (GJ/b) is the conventional simple torsional spring constant.

On the basis of Eq. (15.24a) for two bars, we have established a fundamental relationship for quantifying the equivalent flexibility of any number of symmetrically disposed elements, and it is supplemental to the Table 13.1 relations:

$$\frac{T}{\theta} = N\left[\frac{12EIa^2}{b^3} + \frac{GJ}{b}\right] \tag{15.24b}$$

where N = the total number of elements.

Note that, unlike simple torsion in which stiffness is inversely proportional to length, there is now a length cubed factor deriving from the bending component.

These results are equivalent to Eq. (15.20b), but in simpler form.

Examples

15.1. A shaft has simple shear supports at each end but is completely fixed with respect to rotation at each end. It carries a rigid arm 1–2 supporting the load Q. Find the deflection of the load.

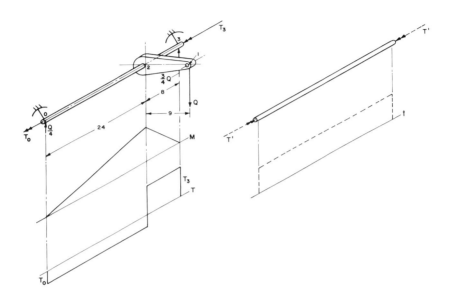

Solution: The structure is indeterminate in a torsional sense. We first introduce opposed auxiliary torques T'. There is actual bending in 0–3 but no auxiliary moment:

$$0 = \int_0^\ell \frac{tT}{GJ}\, ds = \frac{T'}{GJ}\left[-\int_0^{24} T_0\, ds + \int_0^8 T_3\, ds\right]$$

$$-24T_0 + 8T_3 = 0$$

And from statics, $T_0 + T_3 = 9Q$

$$T_0 = 2.25Q \qquad T_3 = 6.75Q$$

For the deflection solution,

$$\frac{Qz_1}{2} = \frac{1}{2EI}\left[\int_0^{24}\left(\frac{Q}{4}x\right)^2 dx + \int_0^8\left(\frac{3}{4}Qx\right)^2 dx\right]$$

$$+ \frac{1}{2GJ}\left[\int_0^{24}(2.25Q)^2\, dx + \int_0^8 (6.75Q)^2\, dx\right]$$

$$z_1 = Q\left[\frac{384}{EI} + \frac{486}{GJ}\right]$$

where the respective terms indicate the respective contributions of bending and twisting.

15.2. A rectangular planar frame of constant cross section has a distributed transverse load on one side and is simply supported at each corner. Find all moment and torque reactions:
(a) Using elastic energy
(b) By deflection superposition

Solution:
(a) From symmetry the beam 0–3 has equal restraining couples at each end, developed by the resistance of the 0-1-2-3 section to distortion, and this U-section carries only M_0 as moment and torques. Vertical supports at 0 and 3 each carry half of the distributed load. Actual moment in 0–3 is symmetrical but can be taken as continuous components from 0 to 3 as shown. To find M_0 we introduce opposed M' couples at 0 with auxiliary bending in 0–3 and 1–2 and auxiliary torques in the short legs:

$$0 = \frac{M'}{EI} \sum_0^3 A + \frac{M'}{EI} \sum_1^2 A + \frac{2M'}{GJ} \int_0^1 M_0 \, ds$$

$$= \left[18q \frac{(36)^2}{2} - 36M_0 - \frac{q(36)^3}{6} \right] - 36M_0 - 48\lambda M_0$$

Only the first term represents a positive product; however, M_0 is positive and is therefore in the assumed negative bending sense:

$$M_0 = \left[\frac{54q}{1 + \frac{2}{3}\lambda} \right]$$

(b) For the couple–slope behavior of the U-section at 0 and 3 we take $\frac{1}{2}$ the total angle from Table 14.1b:

$$\theta_0 = \frac{\theta y}{2} = \left(\frac{1}{2} \right) \frac{36M_0}{EI} \left[1 + 2\lambda \left(\frac{2}{3} \right) \right]$$

From Tables 4.1b and 5.3a for end slopes of the loaded beam 0–3 with end couples and distributed load,

$$\theta_0 = \frac{36}{3EI} + \frac{36}{6EI} M_0 + \frac{q(36)^3}{24EI}$$

Equating,

$$\frac{18}{EI} M_0 \left(1 + \frac{4}{3}\lambda \right) = \frac{1944q}{EI} - \frac{18M_0}{EI}$$

$$M_0 = \left[\frac{54q}{1 + \frac{2}{3}\lambda} \right]$$

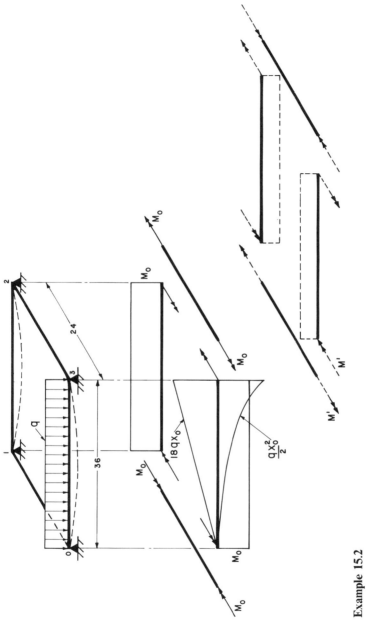

Example 15.2

15.3. A rectangular bar, considered rigid flexurally and torsionally, is supported by two fixed–fixed vertical cantilevers, for which $E/G = 2.50$ and

$$I_1 = 0.90 \qquad I_2 = 1.00$$
$$J_1 = 1.80 \qquad J_2 = 2.25$$

Load point 4 and attachment points 1 and 2 are assumed to lie in the same horizontal plane. Determine all reactions.

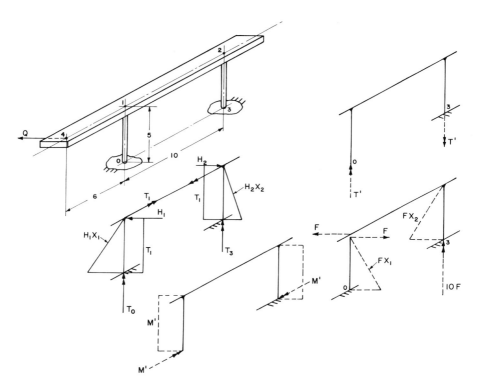

Solution: Testing by decoupling at 2, the Q load could be carried entirely by 0–1, with the bar having horizontal displacement, twist about 1–2, and rotation in the plane. With the double supports, there are *three* redundancies corresponding to the three coordinate displacements, and three auxiliary loadings are required as shown. In conjunction with the actual moments and torques we obtain,

$$0 = \frac{M'}{EI_1} \sum_0^1 A + \frac{M'}{EI_2} \sum_3^2 A \tag{a}$$

$$0 = \frac{T'}{GJ_1} \int_0^1 T_0 \, ds + \frac{T'}{GJ_2} \int_3^2 T_3 \, ds \tag{b}$$

$$0 = \frac{F}{EI_1} \sum_0^1 A\bar{x}_1 + \frac{F}{EI_2} \sum_3^2 A\bar{x}_2 + \frac{10F}{GJ_2} \int_3^2 T_3 \, ds \tag{c}$$

In (a) there is only bending energy and in (b) there is only torsional. With the bar stiff, the 1–2 span is not involved. After evaluation, these equations reduce to

$$2.5H_1 + 2.25H_2 - 1.90T_1 = 0 \qquad \text{(a)}$$
$$T_0 - 0.80T_3 = 0 \qquad \text{(b)}$$
$$-8.333H_1 - 7.5H_2 + 4.75T_1 + 10T_3 = 0 \qquad \text{(c)}$$

With three equations from elasticity and five unknowns, we add two from static equilibrium of the bar:

$$\sum F = Q - H_1 + H_2 = 0 \qquad \text{(d)}$$
$$\sum M_2 = T_0 + T_3 + 10H_1 - 16Q = 0 \qquad \text{(e)}$$

By substitution from (d) and (e) we have three final reduced equations:

H_1	T_1	T_3	Q
4.75	−1.90	0	2.25
10	0	1.8	16
−15.833	4.75	10	−7.50

And from the simultaneous solution,

$$\left(\frac{H_1}{Q}\right) = 1.53 \qquad \left(\frac{T_1}{Q}\right) = 2.63 \qquad \left(\frac{T_3}{Q}\right) = 0.42$$

$$\left(\frac{H_2}{Q}\right) = 0.53 \qquad \left(\frac{T_0}{Q}\right) = 0.33$$

Note the torques T_0 and T_3 are related through the simple ratio of the respective J values, as the bar exerts the same rotational angle on both as it rotates.

15.4. A symmetrical U-shaped bar of constant round section carries a pair of equal transverse loads. If $\lambda = 1.25$, solve the redundancy.

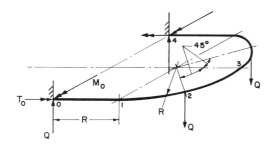

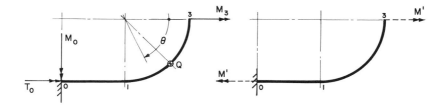

Solution: This problem resembles other symmetrical cases (Fig. 15.1) and we conclude that there is zero shear and zero torque at 3. It is then convenient to integrate from 3 toward 0 with only M_3 and Q present. In the circular quadrant, taking first M_3 and using Table 14.4,

$$\left. \begin{array}{l} M(\theta) = M_3 \cos \theta \\ T(\theta) = M_3 \sin \theta \end{array} \right]_0^{\pi/2}$$

Due to Q_2,

$$\left. \begin{array}{l} M(\theta) = -QR \sin(\theta - 45°) = -QR(0.707)(\sin \theta - \cos \theta) \\ T(\theta) = -QR[1 - \cos(\theta - 45°)] = -QR[1 - 0.707(\cos \theta + \sin \theta)] \end{array} \right]_{\pi/4}^{\pi/2}$$

In the straight section 0–1 we need only the actual torque, as $m = 0$:

$$T_0 = M_3 - 0.2929QR$$

Auxiliary loading in the curved section is obtained with a pair of internal couples M' at 3. From symmetry we can equate the auxiliary energy to 0 for the half shown:

$$m(\theta) = M' \cos \theta \qquad t(\theta) = M' \sin \theta$$

and $t = M'$ in the straight section. Canceling M' from all terms, the energy equation is

$$0 = \frac{M_3}{EI} \int_0^{\pi/2} \cos^2 \theta R \, d\theta + \frac{M_3}{GJ} \int_0^{\pi/2} \sin^2 \theta R \, d\theta$$

$$- \frac{0.707QR}{EI} \int_{\pi/4}^{\pi/2} (\sin \theta \cos \theta - \cos^2 \theta) R \, d\theta$$

$$+ \frac{QR}{GJ} \int_{\pi/4}^{\pi/2} (-\sin \theta + 0.707(\sin \theta \cos \theta + \sin^2 \theta) R \, d\theta$$

$$+ \frac{(M_3 - 0.293QR)}{GJ} \int_0^R dx$$

$$M_3 = \frac{0.5368}{3.0172} QR = 0.178QR$$

From statics

$$\sum M_0 = M_3 - 0.2929QR + T_0 = 0$$

$$T_0 = 0.115QR, \qquad M_0 = 1.707QR$$

15.5. Repeat Example 15.4, replacing the twin loads by the distributed gravity load of the bar.

Solution: For the distributed loading the procedure is similar, proceeding from the symmetrical point 3 and using Table 14.4*d*:

$$M(\theta) = M_3\cos\theta - qR^2(1 - \cos\theta)$$

$$T(\theta) = M_3\sin\theta - qR^2(\theta - \sin\theta)$$

$$T_0 = T_1 = M_3 - qR^2\left(\frac{\pi}{2} - 1\right)$$

With the previous auxiliary couples,

$$0 = \frac{M_3}{EI}\int_0^{\pi/2}\cos^2\theta R\,d\theta - \frac{qR^2}{EI}\int_0^{\pi/2}\cos\theta(1 - \cos\theta)R\,d\theta$$

$$+ \frac{M_3}{GJ}\int_0^{\pi/2}\sin^2\theta R\,d\theta - \frac{qR^2}{GJ}\int_0^{\pi/2}(\theta\sin - \sin^2\theta)R\,d\theta$$

$$+ \frac{(M_3 - 0.571qR^2)}{GJ}\int_0^R dx$$

$$M_3 = 0.397qR^2$$

16

Springs and Resilience

IN ALL PREVIOUS CHAPTERS, conservation of energy has been the single pervasive argument upon which all analytical relations have been developed. Energy as a concept has enabled us to quantify deflections and load distributions in a multitude of elastic structures and in many combinations of geometric elements; however, these procedures have consistently related to *energy equivalence*. At no time has it been necessary or meaningful to evaluate the *absolute strain energy levels* existing in a system, and these quantities are typically meager at best. Where complementary energy is used, it is proportional to the auxiliary load (say F). Since the magnitude of F has no bearing on the solution this energy can indeed be infinitesimal.

Alternately, there are instances, particularly in mechanisms, in which it is necessary to store a known quantity of energy in an elastic element and to retrieve it on a controlled basis, as in a clock spring. In other devices it may be necessary to absorb a given amount of energy when cushioning an impact. We must then consider the absolute force-deflection product numerically. Energy-storage capacities of various configurations and for various materials are studied in this context with respect to the total volume of spring material required.

Finally, we consider the possibility of structural loads producing stresses in excess of the elastic limit. Although plastic deformation of ductile materials violates our usual assumptions of linear load-deflection behavior and superposition, we will nevertheless demonstrate that certain aspects of energy conservation are applicable even under these circumstances. In fact, this investigation establishes the validity of our previously developed energy techniques even when the structures under analysis are in a state of complex residual stress due to forming processes.

16.1 Total Strain Energy in a Spring

A simple tensile bar (Fig. 16.1) illustrates a gradually applied axial load on an elastic member. More generally, the deformed element can be a helical spring or a beam with a load at a point, since it is unnecessary to specify the state of stress in the structure if the linear compliance C or the linear stiffness K are known. We need only equate the work done on the spring element to the total strain energy stored in the spring:

$$U_e = \frac{1}{2}Q\delta = U_s = \frac{1}{2}CQ^2 = \frac{1}{2}\frac{Q^2}{K} = \frac{1}{2}\frac{\delta^2}{C} = \frac{1}{2}K\delta^2 \qquad (16.1)$$

For present purposes, we will define the spring characteristic as the *rate* K, load per unit deflection, rather than C.

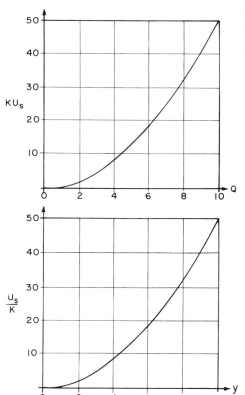

Figure 16.1 For a given spring rate K, strain energy is proportional to the square of both the applied load and the deflection produced.

In terms of the load, with a given K, the strain energy in a spring is proportional to the *square* of the gradually applied load. Similarly, the stored energy is also proportional to the *square* of the gradually produced displacement (Fig. 16.1).

We also conclude that for a given load, the total energy stored is *inversely proportional* to the spring rate. For a given displacement, the total energy stored is *directly proportional* to the spring rate.

Another characteristic, during the recovery of strain energy from a stressed spring, relates to the decrease in the spring force available for actuation purposes which decreases linearly from maximum at maximum deflection to 0 at zero deflection. This is analogous to utilizing energy from a compressed air reservoir with the pressure dropping, but with residual energy still available. Thus an increasing force during the storing cycle is associated with a decreasing force during the use, or return cycle.

16.2 *Dynamic Vertical Loading*

In all previous chapters, it has been assumed that the actual and auxiliary loads are gradually applied and that the structures are essentially in the static domain. Response to dynamic loading involves acceleration, and can be a very complex problem when considering distributed elastic and mass functions, but the effects of locally applied dynamic forces can often be interpreted using static procedures if the mass of the elastic member is negligible.

For instance, if a weight W is attached to a spring K (Fig. 16.2), and the weight is suddenly released from a position above the position of zero spring force, a vibratory motion ensues. We are primarily interested in the maximum resulting instantaneous downward load and deflection, and secondarily the peak stress and the peak value of the energy to be absorbed.

This system is subjected to two sources of energy input. One is the change in potential energy of the weight. The other is the initial strain energy related to the initial tension in the spring:

$$U_e = W(y_0 + y_1) + \frac{K}{2}y_0^2 \qquad (16.2)$$

where y_0 = initial upward (positive) displacement
  y_1 = maximum downward (positive) displacement

As the weight drops it develops velocity and kinetic energy, and the initial strain energy is imparted. When the weight stops at the instant of maximum travel, its contribution to the energy of the system is the product of W and the total vertical displacement, and its kinetic energy is then 0.

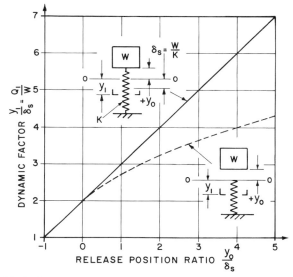

Figure 16.2 Variation in dynamic load factor with release position of a weight attached to a spring is linear (solid curve). Freely falling weight to a spring has lower factors (dashed).

Equating the external energy to the strain energy to which it has been completely converted,

$$U_e = U_s = \frac{K}{2} y_1^2 \qquad (16.3)$$

Combining Eqs. (16.2) and (16.3), the resulting quadratic equation reduces to

$$\left(\frac{y_1}{\delta_s}\right) = \left[2 + \frac{y_0}{\delta_s}\right] \qquad (16.4)$$

where $\delta_s = W/K$ = the final static equilibrium displacement of the weight on the spring.

Equation (16.4) is valid for both positive and negative values of y_0, and it describes a linear relationship between the dimensionless dynamic factor and the release position ratio (Fig. 16.2). With release from the zero position, the dynamic load effect due to the sudden release is 2, and the dynamic force is twice the actual weight. With release at $y_0/\delta_s = -1$, the weight is initially at its sympathetic or at rest position, and there is no subsequent motion. The dynamic factor is $+1$ and the maximum resulting spring force is W. At high values of y_0, there will be large dynamic factors corresponding to significant overshoot beyond δ_s. This factor, incidentally, is independent of K, except as it enters δ_s.

Required strain energy capacity of the spring is

$$U_s = \frac{1}{2}Q_1 y_1 = \frac{1}{2}Ky_1^2 = \frac{1}{2}\frac{Q_1^2}{K} \tag{16.5}$$

with this design parameter increasing as the square of the dynamic factor. From Eq. (16.4) the dimensionless expression for the required strain energy capacity is

$$\frac{U_e}{W\delta_s} = \frac{1}{2}\left[2 + \frac{y_0}{\delta_s}\right]^2 \tag{16.6}$$

16.3 The Freely Falling Weight

Given a weight freely falling a distance y_0 before contacting a spring (Fig. 16.2), we have input potential positional energy converted to kinetic during the fall, and finally to potential internal energy as strain in the spring:

$$U_e = W(y_0 + y_1) = U_s = \frac{K}{2}y_1^2 \tag{16.7}$$

which reduces to

$$\left(\frac{y_1}{\delta_s}\right) = 1 + \sqrt{1 + 2\left(\frac{y_0}{\delta_s}\right)} \tag{16.8}$$

This behavior is shown dashed in Fig. 16.2, and applies to all cases with positive y_0 where there is separation initially between the spring and the weight, with the peak condition occurring obviously during the contact phase. If below the contact point, it behaves exactly as in Sec. 16.2 because under contact conditions it does not matter whether the spring is attached or not. Thus we have the straight line between $y_0/\delta_s = -1$ and 0 for release during partial spring compression.

We conclude that the *attached* spring provides a *linear* system with the force varying continuously through the zero position in either direction. The transition from free-fall to constant contact represents a transitional system and *nonlinear* behavior. In both systems, we have neglected energy losses due to damping and have not been concerned with the post-maximum recovery phase.

On the basis of Eq. (16.8), the energy to be absorbed by the spring becomes

$$\frac{U_e}{W\delta_s} = 1 + \left(\frac{y_0}{\delta_s}\right) + \sqrt{1 + 2\frac{y_0}{\delta_s}} \tag{16.9}$$

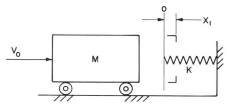

Figure 16.3 Horizontal impact of a mass with initial velocity compresses the spring a distance x_1.

16.4 *Dynamic Horizontal Loading*

If a mass with an initial horizontal velocity strikes a spring (Fig. 16.3), the energy exchange becomes a simple conversion of kinetic to strain at the peak deflection:

$$U_e = \tfrac{1}{2}MV_0^2 = U_s = \tfrac{1}{2}Kx_1^2$$

$$x_1 = \sqrt{\frac{M}{K}}\,V_0 \tag{16.10}$$

With squared terms in both energy expressions, the peak displacement is directly proportional to the initial velocity. The dynamic peak force is Kx_1. There is, in all cases of impact loading, a finite time required from initial release or initial contact to the instant of maximum displacement.

16.5 *Stresses in the Helical Spring*

A wire formed to a helical geometry becomes a *helical spring* (Fig. 16.4), and it can be loaded along its central axis in tension or compression, depending on the configuration of the ends. For simple compression, both ends are ground as planes perpendicular to the axis. For tension the ends are bent to form hooks or eyes. If the ends are extended along the central axis, and clamped to ground or to a load, the spring is able to carry either tension or compression but this is uncommon.

Analytically, we consider the tensile spring with round cross section (Fig. 16.4). From the free-body diagrams, we see that every cross section is subjected to a couple QR in any plane containing the central axis; however, there is a helix angle α and the orientation of the loads must coincide with the principal beam axes. The moment and torque components of QR are

$$M = QR \sin \alpha \tag{16.11a}$$

$$T = QR \cos \alpha \tag{16.11b}$$

where $\alpha = \tan^{-1}(p/2\pi R)$.

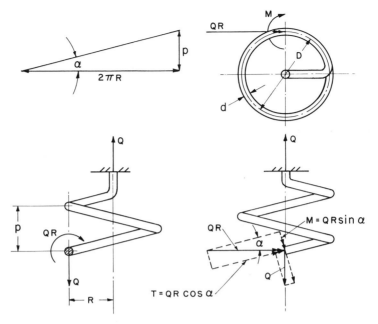

Figure 16.4 The helical spring in tension has continuously distributed torque and moment loading.

If α varies significantly during loading, an average value can be taken.

Similarly the tensile and shear force components on any section are (Fig. 16.4) $Q \sin \alpha$ and $Q \cos \alpha$, respectively. We neglect these effects.

Thus the spring is equivalent to a beam having the total length of the helix with constant loading along the length. Corresponding stresses in bending and torsional shear are

$$\sigma = \frac{QR}{I/c} \sin \alpha = \frac{16QD}{\pi d^3} \sin \alpha \qquad (16.12a)$$

$$\tau = \frac{QR}{J/r} \cos \alpha = \frac{8QD}{\pi d^3} \cos \alpha \qquad (16.12b)$$

Obviously, the shear stress is predominant and governs in design. These are nominal maximum stresses with tensile–compressive at the top and bottom fibers, and tangential shear at all points on the surface of the wire. In addition, we assume the spring has a slender geometry with a large index (R/r or D/d). If this ratio is less than about 6, there is an appreciable concentration of shear stress on the inner fibers of the helix. For present purposes, we are not concerned with this aspect of the problem.

Except for large wire sizes that may require hot forming, the majority of springs are cold wound from drawn and heat treated wire in an extremely hard condition. With successive drawing this wire tends to extremely high

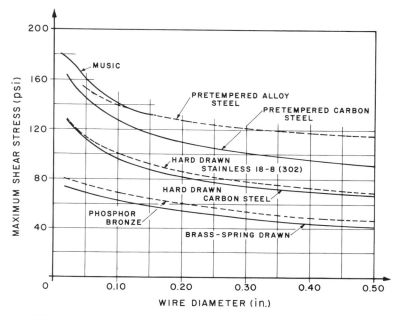

Figure 16.5 Recommended maximum stresses for helical springs show considerable decrease with increasing wire diameter, and are for compressive loading. For tensile springs the values should be reduced by a factor of 0.80.

ultimate stresses (Fig. 16.5), and these decrease as the wire diameter increases. Thus in Eq. (16.12a), the stress is not simply proportional to the cube of the wire diameter.

If the wire section is square or rectangular, we can use values for the maximum torsional shear stress from Table 13.1*d* or *e*, with $T = QR \cos \alpha$.

16.6 Deflection of the Helical Spring

With M and T values from Eq. (16.11), we equate the external energy due to the lowering of the load to the total strain energy distributed uniformly throughout the length. For round wire

$$\frac{Qy}{2} = \int_0^\ell \frac{M^2}{2EI}\, ds + \int_0^\ell \frac{T^2}{2GJ}\, ds$$

$$y = \left[\frac{QR^2}{EI}\sin^2\alpha + \frac{QR^2}{GJ}\cos^2\alpha\right]\left(\frac{2\pi RN}{\cos \alpha}\right)$$

$$= \frac{8QND^3}{Gd^4}C_y \tag{16.13}$$

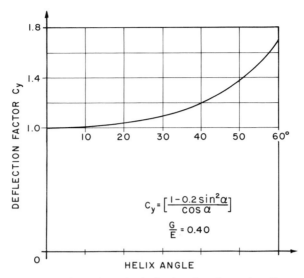

Figure 16.6   Increase in deflection of a helical spring due to bending components is a function of helix angle.

where N = total number of active coils

$$C_y = \left[\frac{1 + \left(2\dfrac{G}{E} - 1\right)\sin^2\alpha}{\cos \alpha}\right] = \left[\frac{1 - 0.2 \sin^2\alpha}{\cos \alpha}\right] \quad \text{if } \frac{G}{E} = 0.4$$

The C_y factor accounts for the effect of the helix angle and the bending, and it increases with α as shown in Fig. 16.6. For usual *close-coiled* springs with a small angle, the effect of this factor is negligible, but for *open-coiled*, the factor rises with increasing helix angle, and the deflection is correspondingly greater.

With axial loading there is a tendency for one end of the spring to rotate about the axis relative to the other. Considering the upper end grounded and the lower free to rotate (Fig. 16.7), the coupled angular deflection is determined by auxiliary M' loading:

$$M'\phi = \frac{M'}{EI}\int_0^\ell(-\cos \alpha)QR \sin \alpha \, ds + \frac{M'}{GJ}\int_0^\ell(\sin \alpha)QR \cos \alpha \, ds$$

$$\phi = \frac{16QD^2N}{Gd^4}\left[1 - 2\frac{G}{E}\right]\sin \alpha \qquad (16.14a)$$

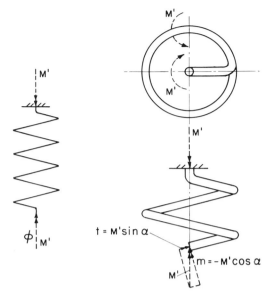

Figure 16.7 With axial loading, a spring tends to rotate about its axis as well as to elongate.

If $G/E = 0.40$,

$$\phi = \left(\frac{3.2QD^2N}{Gd^4} \right) \sin \alpha \qquad (16.14b)$$

Equations (16.14) indicate the absolute value of the angle of twist. The sense of ϕ is as assumed by M', and the helix tends to close. It also tends to close for a right-hand helix under tension. Reversal of Q to a compressive load will, however, obviously reverse the rotative effect.

If the ends are constrained angularly, the spring is redundant and the apparent axial stiffness is slightly increased.

16.7 *The Helical Spring as a Coupling*

A spring is sometimes used to transmit torque between two shafts, thereby providing flexibility in both a torsional and an alignment sense. With purely torque loading, the applied couple travels the length of the spring, and for usual helix angles the cross section of the wire is essentially in simple bending. This condition is indicated in Fig. 16.7, with an actual torque T replacing the auxiliary couple M'. Neglecting α, and for round wire,

$$\sigma = \frac{T}{I/c} = \frac{32T}{\pi d^3} \qquad (16.15)$$

For the rotational deflection about the central axis

$$\frac{T\phi}{2} = \int_0^\ell \frac{T^2}{2EI}\, ds$$

$$\phi = \frac{T\ell}{EI} = \frac{64NDT}{Ed^4} \tag{16.16}$$

With torque in one direction, the coil tends to open and vice versa.

16.8 *Resilience*

Resilience is a property of a structure relating to its capacity to store within itself elastic energy under the application of a load that can be utilized as mechanical energy when the load is removed. It is intrinsically a larger value than in usual structures and is quantified as $U_e = U_s$ (the magnitude of the possible energy interchange).

A uniform tensile link (Fig. 1.1) has constantly distributed stress, and the maximum energy that can be stored without causing permanent deformation of the bar corresponds to the yield stress σ_1 (Fig. 16.8). This condition also represents the maximum design value of both the elongation and the load Q_1:

$$U_s = \frac{1}{2} C Q_1^2 = \frac{1}{2}\left(\frac{\ell}{EA}\right) Q_1^2 \tag{16.17}$$

Figure 16.8 Idealized stress–strain curve for a ductile material as loaded to failure.

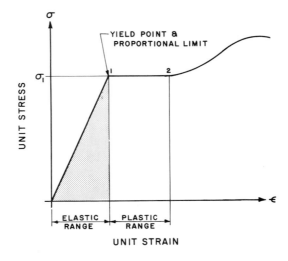

Reducing this to the strain energy per unit volume, we have the *tensile modulus of resilience* of the material:

$$u_r = \frac{U_s}{V} = \frac{\frac{1}{2}\left(\frac{\ell}{EA}\right)Q_1^2}{A\ell} = \frac{1}{2}\frac{\sigma_1^2}{E} \tag{16.18}$$

In Fig. 16.8, with stress plotted against strain for an elastic material, u_r is numerically equivalent to the shaded triangular area:

$$u_r = \frac{1}{2}\sigma_1\epsilon_1 = \frac{1}{2}\frac{\sigma_1^2}{E} \tag{16.19a}$$

If there is constant shear stress in a material, we have by analogy the *shear modulus of resilience*

$$u_r = \frac{1}{2}\frac{\tau_1^2}{G} \tag{16.19b}$$

of particular interest in torsional shear.

We note that the modulus of resilience increases as the *square* of the permissible stress and is inversely proportional to the modulus of elasticity. This indicates the importance of using spring materials capable of sustaining high stresses to minimize the size and weight of springs.

16.9   Specific Resilience

The tensile bar (Fig. 1.1) is not used as a spring because it is too stiff. Normally a spring has flexibility an order of magnitude greater than structural elements intended to effect rigidity and beam configurations are often used for this purpose. For instance, a cantilever beam spring (Fig. 16.9) of rectangular section has triangular moment distribution, and the governing stress is at the fixed end:

$$\sigma_1 = \frac{Q\ell}{I/c}$$

The energy stored in the entire cantilever is

$$U_s = \int_0^\ell \frac{M^2}{2EI}\,ds = \frac{Q^2\ell^3}{6EI}$$

Taking now the total energy relative to the total beam volume, we define the *specific resilience* as this ratio u_s:

$$u_s = \frac{U_s}{V} = \frac{\sigma_1^2\left(\frac{I}{c}\right)^2}{2}\left(\frac{\ell^3}{6EI}\right)\left(\frac{1}{bh\ell}\right)$$

$$= \frac{1}{18}\frac{\sigma_1^2}{E} = \frac{1}{9}\left[\frac{1}{2}\left(\frac{\sigma_1^2}{E}\right)\right] = \frac{1}{9}u_r = \zeta u_r \tag{16.20}$$

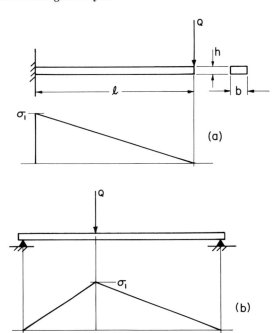

Figure 16.9 Triangular stress distributions result in the same specific resilience factors for (a) and (b).

where ζ is a dimensionless factor pertaining to the energy distribution in the cantilever of rectangular section with an end load.

With a factor of $\zeta = \frac{1}{9}$, the cantilever is only $\frac{1}{9}$, or approximately 10 percent as efficient as the tensile bar relative to energy storage. This dramatic decrease arises from the nonuniform nature of the stress distribution. In the cantilever there is linear stress variation vertically with zero stress at the neutral axis. Also the bending stress at the top and bottom fibers vary from 0 at the end to maximum at the wall. With strain energy a function of σ^2, most of the fibers in the beam are understressed. In fact, each of the linear distributions mentioned reduces ζ by a factor of three, yielding $\zeta = (\frac{1}{3})(\frac{1}{3}) = \frac{1}{9}$.

We now consider the generality of Eq. (16.20). It states that the storage capacity of a rectangular cantilever beam with end loading is governed only by the factors σ_1 and E for a given volume. The geometric arrangement is immaterial. Energy storage per unit volume *is independent of the absolute width, height and length of the beam*, and the total strain energy capacity is proportional to the total volume, regardless of how it is arranged:

$$U_s = u_s V = \left(\frac{1}{18}\frac{\sigma_1^2}{E}\right)(bh\ell) \qquad (16.21)$$

This relation enables us to predict early in the design process the total volume and weight of a cantilever spring required to absorb a given quantity of energy, independent of the final dimensional configuration of the beam.

16.10 *Specific Resilience Factors*

Obviously Eq. (16.20) is only restricted to the assumed conditions as these relate to:

1. A beam of rectangular cross section.
2. A continuous triangular bending moment.

Thus $\zeta = \frac{1}{9}$ for any simply supported rectangular beam with a concentrated load (Fig. 16.9*b*), or with the load outboard of the supports.

Using derivations similar to those for Eq. (16.20), we obtain ζ factors for a number of basic geometries and types of loading (Table 16.1). In all cases, the comparison is with the optimum for the uniform stress term u_r and indicates the relative volumetric efficiency of energy storage. The most ineffective geometry listed is the end-loaded cantilever, or its equivalent beam, of round cross section.

Several combinations of torsional shear are also shown. On an absolute energy basis, the moduli of resilience depend on σ_1^2/E relative to τ_1^2/G, usually favoring the former; however, ζ factors of $\frac{1}{2}$ are achievable in a helical spring and of 1 is approached in the thin-walled torque tube. As

Table 16.1 Specific resilience factors vary with the type of loading and the cross section for basic cases.

	LOADING	CROSS SECTION	ζ	u_r
a	TENSION–COMPRESSION	ANY	1	
b	CONSTANT MOMENT	RECTANGULAR	$\frac{1}{3}$	$\left(\dfrac{\sigma_1^2}{2E}\right)$
c		ROUND	$\frac{1}{4}$	
d	TRIANGULAR MOMENT	RECTANGULAR	$\frac{1}{9}$	
e		ROUND	$\frac{1}{12}$	
f	CONSTANT TORQUE	THIN-WALLED ROUND TUBE	1	$\left(\dfrac{\tau_1^2}{2G}\right)$
g		SOLID ROUND	$\frac{1}{2}$	
h		SOLID SQUARE	$\frac{3}{10}$	

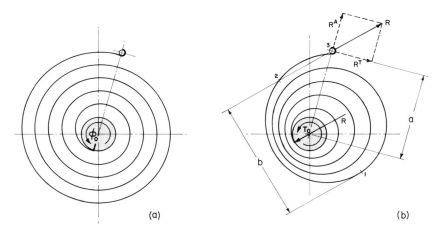

Figure 16.10 The power spring with a pinned extremity develops eccentrically displaced turns due to the force reaction.

these are much larger factors than obtainable with conventional beams, shear loading will tend to prevail in this competition. Actual stress values will provide specific comparisons; however, the power spring to be discussed shortly, is markedly superior to any other type of spring for energy storage.

Incidentally, the tapered beam (Table 4.9a) with $b_1 = 0$, will have $\zeta = \frac{1}{3}$, as in Table 16.1b. It has constant surface stress, but linear variation from the neutral axis. The approximate equivalent of the triangular flat beam is the laminated beam spring with graduated leaves.

16.11 *The Power Spring*

The spirally wound flat spring, or clock spring (Fig. 16.10) has been displaced by batteries and small motors in numerous applications. It is attractive, however, in several important respects, as power springs:

1. are highly effective in storing energy
2. can provide energy over long periods of time through escapement mechanisms
3. are completely reliable over long periods of time
4. can be reused indefinitely
5. are rugged and durable
6. cannot be overstressed
7. comply readily with torsional drive geometry
8. are simply manufactured

To form such a spring, it is only necessary to wrap pretempered flat spring stock tightly against a cylindrical mandrel, after which it can be immediately banded. On release from this closed condition the spring expands to an equilibrium spiral geometry (Fig. 16.10a). If pin-connected to ground as shown, subsequent tightening by rotation of the central arbor will return the same solid situation as when formed. It is then locked against overstressing and is a fully charged cylindrical energy container.

A power spring is thus subjected to bending loads induced by the shaft torque T_0. During winding the fixed anchor pin tends to distort the geometry (Fig. 16.10b), with the force reaction pulling the turns together eccentrically. As they eventually make contact, friction is created as the arbor rotates in this regime.

At the pin there is a single force reaction R that is opposed at the arbor bearings (Fig. 16.10b), and the torque reaction is (aR^T). Bending moment varies continuously throughout the beam. At 1 it is a maximum and equal to (bR), and at 2 it is 0. The device converts the shaft torque T_0 to a bending moment M, and a shaft rotation ϕ_0 to an angular displacement θ.

To analyze this type of geometry would involve integration of continuous functions from the geometry (Sec. 11.6). We hesitate to attempt this for several reasons:

1. The geometry varies during winding.
2. Friction effects are difficult to accommodate.
3. Our usual assumptions of linear stress–strain in a virgin beam are grossly violated by the yielding inflicted during forming.

Figure 16.11 The pinned power spring is nonlinear because of eccentricity and frictional effects.

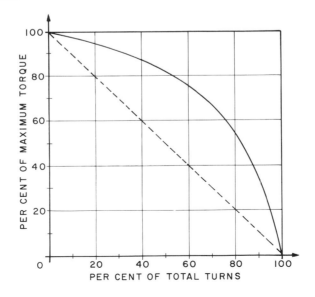

Although we could proceed further, we defer to experimental measurements to obtain torque–angle characteristics. The typical behavior resulting from such tests (Fig. 16.11) is nonlinear, as shown by the solid curve.

16.12   *Power Spring with Constant Couple*

Fortunately, we can achieve more insight into the mechanics and energy phenomena of power springs than previously indicated. In spite of the apparent complexity, we can invoke some simplifying but potentially realistic conditions. Thus we reverse the cowardly retreat of Sec. 16.11 into something of a counterattack on the problem. The discussion is rather extensive in order to explore the basic nature of plastic deformation and residual stress.

The most simple loading for a curved beam is the application of equal and opposite couples at each end. This result is essentially obtained by using a pair of opposed power springs in tandem (Fig. 16.12). The arbors are geared together to rotate in opposite directions at the same rate, and the extremities are clamped to each other by a stiff bar. This arrangement

Figure 16.12   Both concentricity and constant couple loading are approached by joining two units in series.

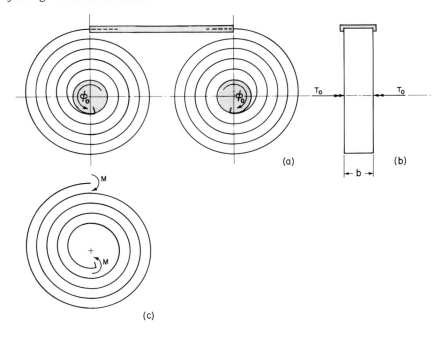

eliminates the fixed pin (Fig. 16.10). As the arbors wind oppositely the connecting bar moves downwardly with translation, and the coils tend to maintain a concentric rather than an eccentric relationship. Ideally, the constant couple $T_0 = M$ now travels completely from one arbor to the other through both springs and the bar.

This is a considerable improvement, both functionally and analytically. In addition to having constant couple loading, the contact friction is eliminated.

16.13   *Elastic–Plastic Behavior at Forming*

Consider the forming of the flat stock on a mandrel with the material at a constant radius of curvature R_0, and further assume that this is the exact curvature at which the extreme fiber stress has just been brought to the elastic limit σ_1 (Fig. 16.8) for the ductile material.

Although it has not been necessary to consider the radius of curvature in our energy procedures, we now apply the fundamental flexural relation

$$\frac{\sigma}{E} = \frac{y}{R} \tag{16.22}$$

where σ = tensile or compressive stress at a fiber
y = distance from the neutral axis to the fiber
R = radius of curvature at the neutral axis

As R decreases we arrive at the condition of imminent yielding when $\sigma = \sigma_1$. Denoting this radius as R_0,

$$y_0 = \frac{\sigma_1 R_0}{E} \quad \text{and} \quad R_0 = y_0 \frac{E}{\sigma_1} \tag{16.23}$$

To produce the necessary cold forming, we must reduce the radius further to R_1 (Fig. 16.13*b*).

In so doing, zones of plastically deformed material develop at both top and bottom surfaces that progress toward the neutral axis. We assume the stresses in these regions to remain substantially at the yield stress σ_1 as shown. This corresponds to the constant stress transition from 1 to 2 (Fig. 16.8). Metal in these regions suffers significant structural alteration, broadly termed *plastic deformation*, and when the stock is released it will no longer return to its straight condition.

However, the central core $2y_1$ has not suffered yielding, *and remains in its completely elastic state*. Thus we have a sandwich with the plastic layers enclosing the elastic layer. As R_1 approaches 0, the elastic core approaches 0, and in the extreme half of the material is at the tensile yield and half at the compressive yield stress, assumed equal.

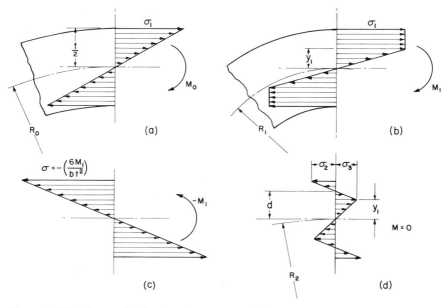

Figure 16.13 Stress distributions in a rectangular beam as its curvature is decreased to produce yielding. Residual stresses are shown in (d).

Knowing the stress pattern (Fig. 16.13b) we evaluate the forming moment M_1, which is also the maximum torque the spring can exert when in service. Summing moments for the two stress regions, we have for the elastic or triangular stress

$$\sigma_1 = \frac{M_e}{I/c} = \frac{6M_e}{b(2y_1)^2}$$

$$M_e = \frac{2}{3}\sigma_1 b y_1^2$$

For the external moment from the plastic rectangular sections, we have two areas and resultant forces for each.

$$M_p = \left[\sigma_1 b\left(\frac{t}{2} - y_1\right)\right]\left[2\left(\frac{\frac{t}{2} + y_1}{2}\right)\right] = \sigma_1 b\left(\frac{t^2}{4} - y_1^2\right)$$

with a total couple of

$$M_1 = \sigma_1 b\left(\frac{t^2}{4} - \frac{y_1^2}{3}\right) \tag{16.24}$$

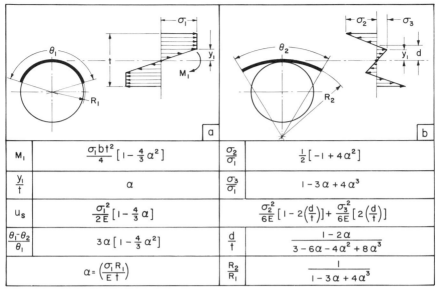

M_1	$\dfrac{\sigma_1 b t^2}{4}[1-\tfrac{4}{3}\alpha^2]$	$\dfrac{\sigma_2}{\sigma_1}$	$\tfrac{1}{2}[-1+4\alpha^2]$
$\dfrac{y_1}{t}$	α	$\dfrac{\sigma_3}{\sigma_1}$	$1-3\alpha+4\alpha^3$
u_s	$\dfrac{\sigma_1^2}{2E}[1-\tfrac{4}{3}\alpha]$		$\dfrac{\sigma_2^2}{6E}[1-2(\tfrac{d}{t})]+\dfrac{\sigma_3^2}{6E}[2(\tfrac{d}{t})]$
$\dfrac{\theta_1-\theta_2}{\theta_1}$	$3\alpha[1-\tfrac{4}{3}\alpha^2]$	$\dfrac{d}{t}$	$\dfrac{1-2\alpha}{3-6\alpha-4\alpha^2+8\alpha^3}$
	$\alpha=\left(\dfrac{\sigma_1 R_1}{E t}\right)$	$\dfrac{R_2}{R_1}$	$\dfrac{1}{1-3\alpha+4\alpha^3}$

Table 16.2 Formulas for calculating parameters of a rectangular beam yielded to a circular arc, as loaded (a) and at release after yielding (b).

In Table 16.2 we facilitate the use of the various relations by means of a dimensionless parameter

$$\alpha = \left(\frac{\sigma_1 R_1}{E t}\right) = \left(\frac{y_1}{t}\right) \tag{16.25}$$

and this factor cannot be greater than $\tfrac{1}{2}$. The simplified equation for M_1 is given in Table 16.2a.

16.14   *Internal Energy when Wound*

When formed initially, or subsequently fully wound, we can calculate the total strain energy in an arbitrary length of the spring ℓ, by summing the elastic and plastic volumes. From Table 16.1

$$U_{e1} = U_{s1} = 2b\ell y_1\left(\frac{\sigma_1^2}{6E}\right) + (t-2y_1)b\ell\left(\frac{\sigma_1^2}{2E}\right)$$

$$u_{s1} = \frac{U_{s1}}{b\ell t} = \frac{\sigma_1^2}{2E}\left[1 - \frac{4}{3}\left(\frac{y_1}{t}\right)\right] \tag{16.26}$$

As y_1 approaches 0 (Fig. 16.13b), the specific energy approaches the maximum possible value (u_r) (Table 16.1a). Then all fibers of the entire volume are at capacity, or at the yield stress; however, there must always be a finite elastic core, and R_1 cannot become zero. The maximum of half volume at maximum tension and the other half at maximum compression is the limiting situation.

In Eq. (16.26) having the final u_s, the product of u_s and the total volume of the spring determines the total elastic energy stored in the solidly wound spring. Even with plastic deformation, there is no escape or dissipation of the external work input. To apply to the entire spring, it is necessary that the radius R_1 be taken at a mean location in order to average the variation from the mandrel to the outer loop.

16.15 Conditions at Release

Upon release from the solidly wound state, the spring both rotates and expands to the free equilibrium condition. Taking the applied moment M_1 to 0 is equivalent to superimposing a moment of $-M_1$ at the free end. This $-M_1$ loading produces a *normal linear elastic stress pattern* (Fig. 16.13c) in spite of the previous plastic yielding, provided we do not exceed σ_1 in a reverse sense. The superimposed stress is determined by classical means, as shown, with the upper surface in compression.

Superimposing the stress patterns (Fig. 16.13b and c) algebraically, we arrive at a final antisymmetrical distribution of *residual stress* (Fig. 16.13d). There are now three neutral axes, two regions in tension, and two regions in compression. In addition, as the spring is in equilibrium in the curved free state, the counterclockwise couple from the outer sections must be equal and opposite to the clockwise couple from the stresses between the two external neutral axes. The several parameters describing the residual stress distribution follow from the geometry and are given in Table 16.2b.

We calculate the number of release, or usable service turns, per turn of the solidly wound spring using radius of curvature equations. For the free radius

$$\frac{1}{R_1} - \frac{1}{R_2} = \frac{M_1}{EI} - \frac{M_2}{EI} \tag{16.27}$$

where R_2 = released radius of curvature
$M_2 = 0$

Then for the same length of spring stock, $R_1\theta_1 = R_2\theta_2$ and $\theta_1 = 2\pi$, and

$$\frac{\theta_1 - \theta_2}{\theta_1} = 1 - \frac{R_1}{R_2} = \frac{M_1 R_1}{EI} \tag{16.28}$$

Multiplying this ratio by the total number of turns in the spring will indicate the working turns available in service from 0 to maximum torque.

16.16   *Residual and Operational Energy Range*

Finally, we have a case in which the residual energy does not return to 0. The remaining internal energy derives from the triangular stress factor (Table 16.1b). Energy in the top and bottom volumes are determined by σ_2, and in the rest of the volume by σ_3, in Table 16.2b. This residual energy resides permanently in the unwound spring and *is not recoverable*; therefore, the available operational energy is represented by the *difference between the strain energies in the wound and unwound states*.

$$u_s = u_{s1} - u_{s2} \qquad (16.29)$$

where these are given in Table 16.2.

If we base the strain energy cycle upon the maximum for the material, $u_r = \sigma_1^2 / 2E$ and

$$\text{efficiency} = \frac{u_{s1} - u_{s2}}{u_r} \qquad (16.30)$$

Depending upon how closely y_1 approaches 0 and the extent of the residual energy, calculated efficiencies can approach 80 percent. Since this pertains to the maximum energy that can be applied and recovered from a given material and a given volume, the power spring does remarkably well in this respect.

16.17   *Elastic–Plastic Behavior in Torsion*

A cylindrical element torqued beyond the elastic regime develops stress patterns similar to those in Fig. 16.13 for bending. The outer fibers will come to the yield stress in shear τ_1, and the annular volume so effected will progress in toward the center under additional torque. When released, the cylinder has suffered a permanent angular set. Relations similar to Table 16.2 for rectangular geometry could be derived.

The torsional bar is probably seldom used as a spring motor because of its stiffness and length. Since the helical spring is stressed torsionally, it is theoretically possible to yield these in the manner of Fig. 16.13. The helical spring must store energy when deflected, but again it is not often used specifically for this purpose. The problem is more apt to relate to possible fatigue failure, for which the residual stresses of interest are compressive due to cold working of all outer surfaces.

16.18   *Significance to Energy Methods*

This excursion into the elastic–plastic domain has particular implications for all other chapters that rely on strain energy and superposition. Most importantly, the application of $-M_1$ (Fig. 16.13c) to the overstressed condition (Fig. 16.13b) follows conventional behavior and linear stress pattern even though $-M_1$ in turn exceeds the yield stress limitation of M_0 (Fig. 16.13a).

We conclude that an element or structure *behaves completely elastically whether it is loaded from a zero stress or from a residual stress datum.*

As far as final stresses are concerned, these must include the residual stress contribution. This has been demonstrated in Fig. 16.13d, with the final stresses obtained by simple superposition of Fig. 16.13b and c.

The same conclusions follow for deflections, with those due to an actual loading being superimposed upon any residual deflection pattern. Actual deflections are determined by conventional means, disregarding any pre-existing stresses.

16.19   *Resilience in Bolts*

Sections 16.2–16.4 explained how an elastic element carries peak dynamic loads when subjected to impact energy. From Eq. (16.1) the peak force is

$$Q = \sqrt{2KU_e} \qquad (16.31)$$

For a given energy input, the force increases as $\sqrt{K}$. The larger the stiffness the larger the peak force, because of the lesser compliance. Prevention of excessive stress in the spring is another problem.

As an example, we take a conventional bolt (Fig. 16.14a) and subject it to tensile impact in absorbing input kinetic energy along its axis. Although stress patterns in the threads are quite complex, we assume simple tension and constant stress based upon the reduced threaded cross section. There are thus two different stress levels in the thread and in the body:

$$\sigma_t = \frac{Q}{A_t} \qquad \sigma_b = \frac{Q}{A_b} \qquad (16.32)$$

Summing resilience,

$$\begin{aligned}
U_e = U_s &= \frac{\sigma_t^2}{2E} V_t + \frac{\sigma_b^2}{2E} V_b \\
&= \frac{\sigma_t^2 V_t}{2E}\left[1 + \left(\frac{A_t}{A_b}\frac{\ell_b}{\ell_t}\right)\right] \qquad (16.33)
\end{aligned}$$

where the second term pertains to the energy absorbed in the body.

The area ratio varies for different diameters and thread types, but to illustrate we take a typical value of 0.7 to indicate the reduction at the threads. Substituting in Eq. (16.33), we have the lower curve (Fig. 16.14). For a given nut and material, the factor ($\sigma_t^2 V_t / 2E$) is a constant.

We conclude that it is advantageous to use as long a bolt length as possible, thereby increasing the volume containing the strain energy.

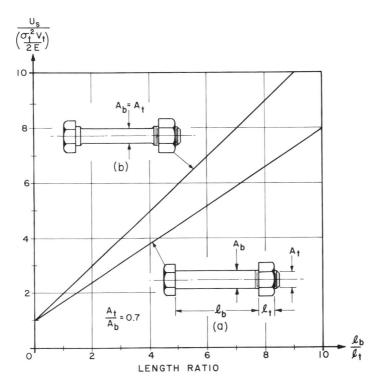

Figure 16.14 Impact resistance of a bolt is improved by designing for uniform tensile stress throughout the length to increase the energy storage capacity.

For high-strength, impact-resistant bolts the body diameter is sometimes reduced (Fig. 16.14b). This seems inconsistent with the previous statement favoring volume in the body; however, the net result is favorable. We succeed in raising the stress level in the body to the limiting stress in the thread. With $A_b = A_t$,

$$U_e = U_s = \frac{\sigma_t^2 V_t}{2E}\left[1 + \frac{\ell_b}{\ell_t}\right]$$ (16.34)

the higher curve results (Fig. 16.14).

The improvement, in turn, is more pronounced at greater bolt lengths. By removing material we have been able to absorb a larger quantity of impact energy at a given limiting tensile stress.

16.20 *Beam Design by Resilience*

Total strain energy in a simple cantilever beam with an end load (Fig. 16.9a) is equal to the external energy applied at the end. In terms of maximum bending stress, and from Table 16.1c,

$$U_e = \frac{Qy}{2} = U_s = \frac{1}{18}\left(\frac{\sigma^2}{E}\right)V \qquad (16.35)$$

where σ = maximum bending stress at the fixed end
  V = total volume of the prismatic beam

Solving for the several variables,

$$y = \frac{\sigma^2 V}{9EQ} \qquad (16.36a)$$

$$V = \frac{9EQy}{\sigma^2} \qquad (16.36b)$$

$$\sigma = \sqrt{\frac{9EQy}{V}} \qquad (16.36c)$$

$$Q = \frac{\sigma^2 V}{9Ey} \qquad (16.36d)$$

Since these expressions derive from specific resilience for triangular moment distribution, they apply to any rectangular beam, whether cantilever or simply supported with a concentrated load at any point between or beyond the supports; however, it is understood that y is the deflection at the load and σ is the maximum bending stress, wherever this occurs.

If the beam is of solid round section, the equations are similar, but the factor 9 is replaced by 12.

This procedure appears to be a rather obtuse approach to the design of beams, but Eqs. (16.36) do allow us to draw broad or limiting conditions for a particular type of beam. Given four specific design parameters, we can calculate the fifth dependent variable as indicated, and this overview of the beam design is made possible by the concept of resilience.

Examples

16.1. A weight of 14 lb, attached to a vertical spring with $K = 60$ lb/in. is released when the tensile force in the spring is 8 lb. Find:
(a) The maximum compressive deflection of the spring
(b) The peak dynamic force in the spring
(c) The peak strain energy in the spring
(d) The sources of (c)

Solution:
(a) From Eq. (16.4), $\delta_s = \frac{14}{60} = 0.233$, $y_0 = \frac{8}{14} = 0.571$

$$\frac{y_1}{\delta_s} = 2 + \frac{0.571}{0.233} = 4.45 \qquad y_1 = 1.038$$

(b) $Q_1 = Ky_1 = 60(1.308) = 62.3$ lb
(c) $U_s = \frac{1}{2}Ky_1^2 = 32.3$ in. lb
(d) W has fallen a total of $(0.571 + 1.038) = 161$ in.

$$U_e = Wh = 14(1.61) = 22.53 \text{ in. lb}$$

$$\text{initial strain energy input} = \frac{1}{2}(60)(0.571)^2 = 9.78 \text{ in. lb}$$

$$\text{total input energy} = 22.5 + 9.8 = 32.3 \text{ in. lb}$$

The results can be checked in Fig. 16.2 at $y_0/\delta_s = 2.45$. We check (c) using Eq. (16.6) at 32.2 in. lb.

16.2. The mass in Example 16.1 is not attached, but is released from the same height ($y_0 = 0.57$). Repeat the calculations.

Solution:
(a) Using Eq. (16.8)

$$\frac{y_1}{\delta_s} = 1 + \sqrt{1 + 2\frac{y_0}{\delta_s}} = 3.43 \qquad y_1 = 0.80 \text{ in.}$$

(b) $Q_1 = Ky_1 = 60(0.80) = 48$ lb
(c) $U_s = \frac{1}{2}Ky_1^2 = 19.2$ in. lb
(d) Input energy is entirely potential from the fall of W.

$$U_e = 14(0.80 + 0.57) = 19.2 \text{ in. lb}$$

Again we check the dynamic factor in Fig. 16.2 at the dashed curve, horizontally at 2.45, and agree with the 3.43 factor. We check (c) using Eq. (16.9) at 19.2 in. lb.

16.3. A mass of 25 kg moving horizontally at a velocity of 6 m/s strikes a helical spring, by means of which it is brought to a stop and rebounds. If the maximum shear stress in the spring is 60 kN/cm², determine:
(a) The required volume of the spring
(b) The required weight of the spring

Solution:
(a) Kinetic energy of the mass is $\frac{1}{2}(25)(6)^2 = 450$ N · m. From Table 16.1, and taking $G = 8000$ kN/cm²

$$u_r = \frac{1}{4}\frac{(60)^2}{8000} = 0.113 \text{ kN/cm}^2 \text{ or } 113 \text{ N} \cdot \text{cm/cm}^3$$

$$V = \frac{45,000}{113} = 400 \text{ cm}^3$$

(b) For a density of steel of 0.008 kg/cm³,

$$W = 400(0.008) = 3.2 \text{ kg}$$

Note these results are independent of K, the deflection, the peak force, and the actual spring geometry.

16.4. For the data of Example 16.3 specify a helical compression spring of round cross section with a pitch diameter of 9 cm.

Solution: From Eq. (16.10), neglecting α,

$$60,000 = \frac{8Q(9)}{\pi d^3}$$

A number of different combinations of Q and d satisfy the stress condition. Iteration is required to find a reasonable geometry:

With $\quad d = 1.0$ cm $\quad Q_1 = 2620$ N $\quad x_1 = \dfrac{45,000(2)}{2620} = 34.4$ cm

$\qquad\qquad$ 1.5 $\qquad\qquad$ 8840 $\qquad\qquad\qquad$ 10.2

$\qquad\qquad$ 1.8 $\qquad\qquad$ 15,280 $\qquad\qquad\qquad$ 5.9

The deflection of 5.9 cm seems reasonable for the 9 cm diameter. From Example 16.3

$$V = \frac{\pi d^2}{4}\ell = 400 \text{ cm}^3$$

$$\ell = 157.2 \text{ cm} \qquad N = 5.6 \text{ turns}$$

To verify this result based upon total resilience, we check the actual dynamic deflection, using Eq. (16.11):

$$x_1 = \frac{8(15,280)(5.56)(9)^3}{8(10)^6(1.8)^4} = 5.9 \text{ cm}$$

There are many combinations of d, D, and N that will satisfy the conditions of this problem, but the volume of all these possibilities will be the predicted 400 cm³.

16.5. A strip of aluminum alloy $\frac{1}{8}$ in. thick and $\frac{3}{4}$ in. wide has a yield stress of 40,000 lb/in.² It is deformed on a circular die to a constant radius.
 (a) What is the mean forming radius corresponding to the threshold of permanent set?
 (b) If the mean radius of the strip is taken to 9 in., what is the radius of the arc after springback?

Solution:
 (a) From Table 16.2a, we approach yielding at $\alpha = y_1/t = \frac{1}{2}$.

$$\alpha = \frac{\sigma_1 R_1}{Et} = \frac{(40,000) R_1}{(10)^7(0.125)} = \frac{1}{2} \qquad R_1 = 15.62 \text{ in.}$$

 (b) $$\alpha = \frac{(40,000)(9)}{(10)^7(0.125)} = 0.288$$

From Table 16.2*a*

$$\frac{\theta_1 - \theta_2}{\theta_1} = 3(0.288)\left[1 - \tfrac{4}{3}(0.288)^2\right] = 0.768$$

$$\frac{\theta_2}{\theta_1} = \frac{R_1}{R_2} = 0.232 \qquad R_2 = \frac{9}{0.232} = 38.8 \text{ in.}$$

Checking with Eq. (16.25)

$$M_1 = 104.2 \text{ in. lb} \qquad I = \frac{1}{12}\left(\frac{3}{4}\right)\left(\frac{1}{8}\right)^3 = 0.000122 \text{ in.}^4$$

$$\frac{1}{9} - \frac{1}{R_2} = \frac{104.2}{(10)^7(0.000122)} = 0.0854$$

$$R_2 = 38.8 \text{ in.}$$

16.6. Using Table 16.2*b* relations at $\alpha = 0.25$, verify that the summation of external moments from the residual stress pattern is 0 over the cross section.

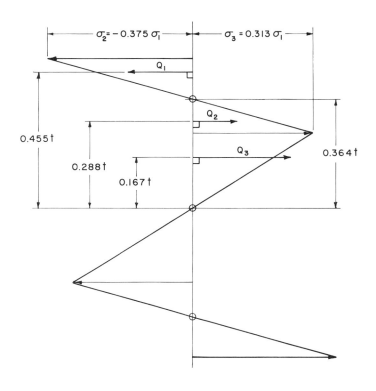

Solution: Since the pattern is antisymmetrical, it is only necessary to consider the stress-area moments about the central neutral axis. Taking the three triangular areas, the resultant forces are

$$Q_1 = \frac{(0.375)(0.1364)}{2}\,\sigma_1 bt = 0.0256\sigma_1 bt \quad (\text{compression})$$

$$Q_2 = \frac{(0.313)(0.1136)}{2}\,\sigma_1 bt = 0.0178\sigma_1 bt \quad (\text{tension})$$

$$Q_3 = \frac{(0.313)(0.250)}{2}\,\sigma_1 bt = 0.0391\sigma_1 bt \quad (\text{tension})$$

where the intermediate neutral axis is located by the geometry. Multiplying by the respective moment arms from the neutral axis,

$$M_{cg} = (0.0256)(0.455) - (0.178)(0.288) - (0.0391)(0.167)$$
$$= 0$$

The beam is in equilibrium under the action of the residual stress.

16.7. A rectangular steel beam carrying an overhung load Q must satisfy the requirements:

$$\text{maximum stress} = 12,000 \text{ lb/in.}^2 \qquad \text{volume} = 140 \text{ in.}^3$$

$$\text{deflection at } q = 0.90 \text{ in.} \qquad E = 30(10)^6 \text{ lb/in.}^2$$

Specify the cross section of the beam.

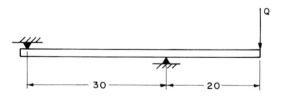

Solution: From Eq. (16.36d)

$$Q = \frac{(12,000)^2(140)}{9(30)(10)^6(0.090)} = 829.6 \text{ lb}$$

From Table 4.2a with α_{22} the deflection is,

$$y = \frac{(829.6)(30)^3}{3EI}\left(\frac{20}{30}\right)^2\left(\frac{5}{3}\right) = 0.90 \text{ in.}$$

$$I = 2.048 \text{ in.}^4$$

For the stress,

$$12,000 = \frac{M}{I/c} = \frac{(20)(829.6)(6)}{bh^2}$$

$$bh^2 = 8.296 \text{ in.}^3$$

For the area,

$$bh = \frac{V}{\ell} = \frac{140}{50} = 2.80 \text{ in.}^2$$

Solving for b and h

$$b = 0.945 \text{ in.} \qquad h = 2.96 \text{ in.}$$

We have satisfied all beams conditions, and determine $Q = 829.6$ lb, the unique load for this combination of parameters.

Problems

CHAPTER 1

1.1 A load of 8 kN is applied at the pin at 1. If the link a has a compliance of $3(10)^4$ cm/kN, determine:
  a. The change in the length of a
  b. The horizontal deflection x_1

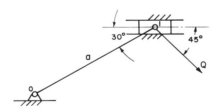

1.2 Three aluminum links all have a constant section of 0.75 in.^{2} and the length of c is 24 in. With $Q = 3,000$ lb calculate:
  a. The deflection δ_{02} in the direction of loading
  b. Vertical deflection y_1 relative to the 0–2 base

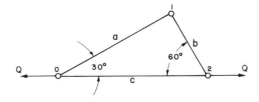

1.3 The steel links *a* and *b* are each 0.25 in.^{2} in cross section with $\ell_a = 7.40$, $\ell_b = 4.40$, and $\ell_{12} = 8.22$ in. Find:

a. Deflection in the direction of *Q*
b. The horizontal deflection x_1
c. The vertical deflection y_1
d. The resultant deflection of 1

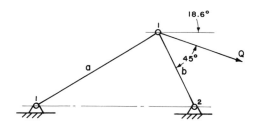

1.4 A vertical load *Q* acts on a linkage restrained by a spring *K*. Assuming small deflections, all deformations in the spring, and that the mechanism is shown in its mean position, determine:

a. The vertical compliance at 1
b. The displacement of 2 due to *Q*
c. The horizontal displacement x_1

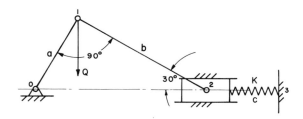

1.5 In Prob. 1.4 find the direction and magnitude of the maximum possible compliance with a load at 1.

1.6 Repeat Prob. 1.5 for the minimum compliance at 1.

1.7 A pinned planar linkage has a *Q* loading as shown on the axis of symmetry. If

$$C_a = C_f = C_c = C_h = 0.60(10)^{-6}$$

$$C_b = C_d = C_e = C_g = 0.80(10)^{-6}$$

calculate:

a. Total deflection Δ_{03}
b. Change in the width $\delta_d = \delta_e$

(Note: with no dimensional units given, these results can be finalized in either SI or British units—cm or in.)

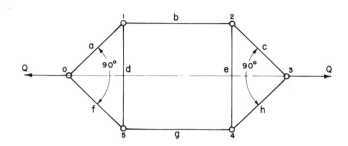

1.8 All links have a common AE factor. Taking a as reference, find:

a. y_2 **b.** y_1 **c.** x_2 **d.** x_1

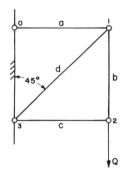

1.9 With all links having the same area and material, taking d as reference and $C_d = 0.70(10)^{-6}$, calculate:

a. x_2 **b.** x_1 **c.** y_2 **d.** y_1

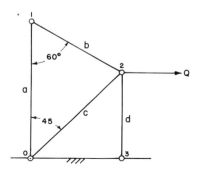

1.10 A crane-type structure carries a vertical load. Given lengths are

$$a = 145 \qquad b = 196 \qquad c = 380 \qquad d = 400$$
$$0\text{–}4 = 215 \qquad 3\text{–}4 = 130$$

For equal AE values find displacements
a. y_2 **b.** x_2

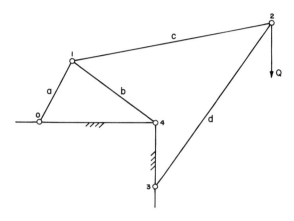

1.11 In Prob. 1.10 all links are rigid relative to a, which is a cable and relatively flexible. Again find the two displacement components of 2.

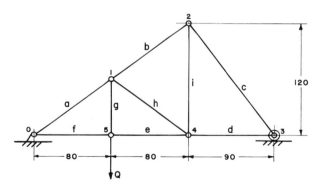

1.12 A steel pinned structure incorporates rods to carry all tensile loads and structural members for all compressive loads. As a result, the equal areas of the compressive elements are 10 times the equal areas of the tensile elements. With a as reference determine the following vertical deflections:

a. y_5 **b.** y_4 **c.** y_1 **d.** y_2

CHAPTER 2

2.1 The pinned truss has similar AE values for all links. Length of b is 12, and this is the reference link for both stiffness and the redundant solution. Calculate the load in all links.

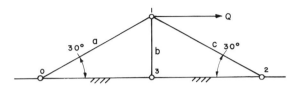

2.2 Repeat Prob. 2.1, but with c rigid relative to the other links.

2.3 Compliance terms for the truss are

$$C_a = C_e = 0.35(10)^{-6} \qquad C_c = C_d = 0.54(10)^{-6} \qquad C_b = 0.45(10)^{-6}$$

Determine:
a. Load in each link
b. Deflection in the direction of Q

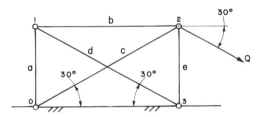

2.4 The linkage carries a single load and has a roller support at 2. Lengths are

$$a = 10.0 \qquad b = 12.0 \qquad c = 14.8 \qquad d = 12.6 \qquad 0\text{–}3 = 6.63$$

What percentage of Q is carried by the lower (c–d) branch at 2 if the bars have a common cross section?

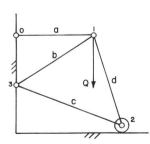

2.5 In Prob. 2.4 determine the horizontal deflection at 2.

(***Hint:*** By proper decoupling this solution only involves c.)

2.6 If all bars have the same AE factor, take c as the reference link and determine:

a. y_2 **b.** x_2

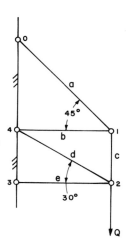

2.7 A symmetrical pinned structure has the following dimensions:

$$A_a = A_c = A_d = A_f = 0.90 \qquad A_b = A_e = 1.20$$
$$A_h = A_i = 0.80 \qquad A_g = 0.40$$
$$\ell_b = \ell_e = 70 \qquad \ell_h = \ell_i = 20$$

Calculate:

a. The total deflection Δ_{03}

b. The relative stress factor q_i/A_i for all links

(***Hint:*** For T loading decouple to only g. Use b and e as reference links for F loading and a for reference compliance.)

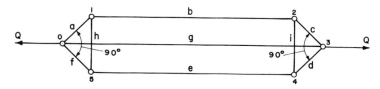

2.8 The truss has the following ℓ/AE factors:

$$C_a = C_c = 0.7(10)^{-6} \qquad C_b = C_d = 0.9(10)^{-6} \qquad C_e = 0.3(10)^{-6}$$

Solve for loads in all links.

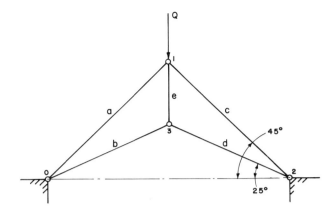

2.9 In Prob. 2.8 *e* has a length of 27 and is short 0.015 of its correct length, requiring forced assembly. Determine the resulting preload in all links.

2.10 In Prob. 2.8 *e* is enlarged in section to become an effectively rigid link. What load is developed in *e* by Q?

2.11 For the symmetrical pinned frame, given: $C_c = 0.8(10)^{-6}$

$$C_a = C_e = 1.5(10)^{-6} \qquad C_b = C_d = C_f = C_g = 1.1(10)^{-6}$$

Determine:
a. Loads in all links
b. Deflection y_4

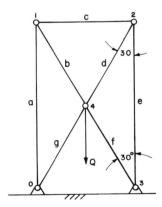

2.12 All connected links have the same AE term, and *a* is continuous from 1 to 4. Dimensionally, (1–2) and (2–3) are of equal length. The

extended link (2–4) is one half as long as (1–2). Calculate:
a. The load in *a*
b. Resultant (total) displacement of 4

(***Hint:*** Use radial and transverse components of 4.)

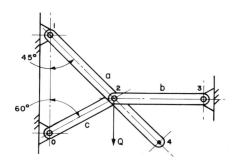

2.13 A rigid triangular link is pivoted at 0 and supported by three springs acting tangentially, with

$$K_a = 70 \qquad K_b = 50 \qquad K_c = 35$$

If the springs behave linearly with no preload, find:
a. The forces in the springs developed by *Q*
b. The axial deflection of each spring
Verify that the deflections are consistent with the respective radii.

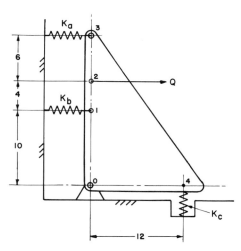

CHAPTER 3

3.1 A rectangular cantilever is 0.75 wide and 0.25 deep. End loads on a rigid cross piece are

$$Q_x = 800 \qquad Q_y = 3$$

Find:
a. The maximum tensile stress
b. The maximum compressive stress
c. The transverse displacement
d. The axial displacement

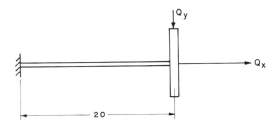

3.2 The beam of Prob. 3.1 has horizontal loads of

$$Q_1 = 32 \qquad Q_2 = 40$$

Find:
a. The maximum bending stress
b. The maximum direct stress
c. The angular deflection at the end
d. The axial deflection at the end

3.3 The beam is considered rigid in the larger section, with constant I in the smaller. Determine the vertical displacement at Q.

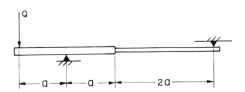

3.4 A uniform beam is loaded by a couple M_1 as shown. Find the slope to the elastic curve at the couple.

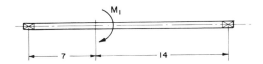

3.5 Two uniform cantilever beams are connected by a helical spring. Free length of the spring is 4.5 and $K_b = 85$. If

$$(EI)_a = 2.0(10)^6 \qquad (EI)_c = 0.8(10)^6$$

what is the force in the spring?

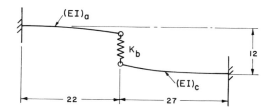

3.6 In Prob. 3.3 the supports are not rigid, but have equal compliances of $1.5(10)^{-6}$. What additional deflection occurs at Q?

3.7 The symmetrical beam carries equal overhung loads. Find y_Q.

3.8 Repeat Prob. 3.7 if the overhung sections are rigid.

3.9 The symmetrical beam has equal overhung couples at each end. What slopes are produced at the ends?

3.10 Repeat Prob. 3.9 for rigid overhung sections.

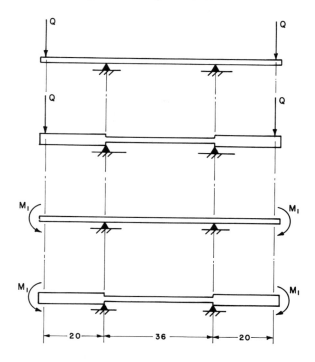

3.11 An extended exciter rotor has effectively rigid support from the main rotor at 0, and there is a fixed simple support provided by an outboard bearing at 3. Determine the bending stress at 0 for a misalignment of 0.050 at 3 during rotation.

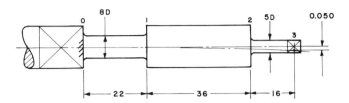

3.12 A rigid pivoted bracket rests on a uniform beam at 3. Find the vertical deflection y_2 at the load.

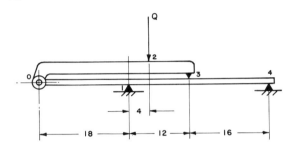

CHAPTER 4

4.1 The simple beam carries a single load Q_1. Find the deflection y_2:
  a. Using basic energy methods
  b. Using influence coefficients (Table 4.1*b*)
  c. Using equations (Table 4.6)
  d. From curves (Figure 4.11)
  e. Using volumetric integration [Eq. (4.8)]

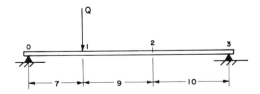

4.2 A simple beam has a concentrated load Q at $\frac{1}{4}$ of the total length. For the beam determine the following displacements:
  a. y_1 **b.** y_2 **c.** θ_0

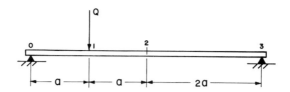

4.3 An overhung load is applied to a beam having two different stiffness sections. Derive expressions for the deflections:

    a. y_0 **b.** θ_1 **c.** θ_2

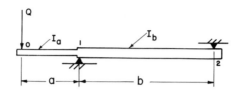

4.4 A rectangular steel beam 21 in. long has a section 1 in. wide and $\frac{1}{2}$ in. deep. It is supported at 2 by a steel rod $\frac{1}{4}$ in. in diameter and 48 in. long. Find the vertical deflections:

    a. y_1 **b.** y_2

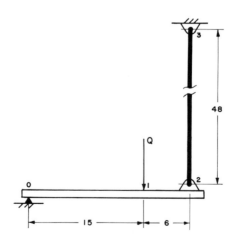

4.5 A uniform shaft in two bearings carries a rigid hub connected by a press fit at 3. With a transverse load on the hub at 2, find the deflections:

    a. y_2 **b.** y_3

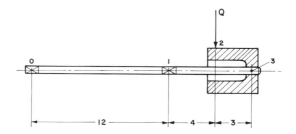

4.6 A constant beam suspends a rigid rectangular weight at the ends. With all deflection due to flexure of the beam, determine, for a suspended weight of 340 ($E = 30(10)^6$ and $I = 0.55$):
  a. Deflection of the centroid 5
  b. Resultant slope of the bar
  c. Relative deflection y_{25}

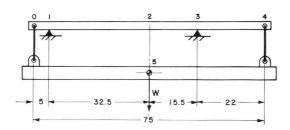

4.7 Repeat Prob. 4.6 with compliances in the vertical links of $C_0 = 0.9(10)^{-4}$ and $C_4 = 1.2(10)^{-4}$.

4.8 What value of M_2 for the beam of Prob. 4.1 will result in zero net vertical deflection at 1 with Q_1 and M_2 both acting?

4.9 A tapered conical cantilever (Table 4.9c) has an end load Q and $d_1 = 2.5$ and $d_0 = 4.0$. Find θ_1 at the end if $\ell = 18$.

CHAPTER 5

5.1 The simply supported beam has a uniformly distributed load over its length. Determine:
  a. y_1 **b.** θ_0

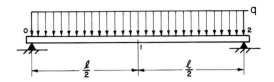

5.2 The beam has uniform loading, but unsymmetrically spaced supports Using basic energy methods find:

    a. y_0 **b.** y_2 **c.** θ_4

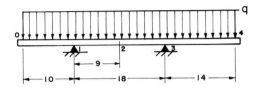

5.3 Repeat Prob. 5.2 using tabulated functions for verification.

5.4 For the beam with partial uniform loading find the deflection at 1, with the beam having constant *EI*.

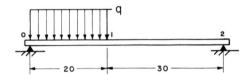

5.5 Repeat Prob. 5.4 assuming the section 0–1 to be rigid and flexure only in 1–2.

5.6 Repeat Prob. 5.4 assuming the section 1–2 to be rigid and flexure only in 0–1.

5.7 The constant cantilever has equal spans of reversed uniform loading. Find:

    a. y_1 **b.** θ_1

(**Hint:** In the span 1–2 construct the *M* diagram from 2 to 1 with $V_2 = 0$.)

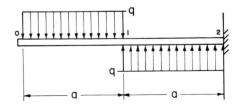

5.8 The distributed overhung load increases linearly to a maximum at 1. Calculate the displacements
  a. y_0 **b.** θ_2

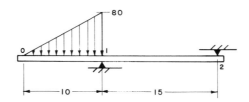

5.9 The uniform beam on simple supports has a distributed load in the 0–2 interval. Determine:
  a. y_0 **b.** θ_3 **c.** y_2

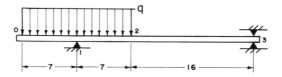

5.10 With two different opposed distributed loadings, find the value of q_1 that will result in zero vertical deflection of 2.

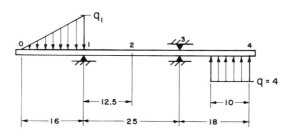

5.11 Under the loading conditions of Prob. 5.10, find the slopes:
  a. θ_1 **b.** θ_3 **c.** θ_2

5.12 Deflection at 1 is limited to 0.10. What is the minimum EI for the beam?

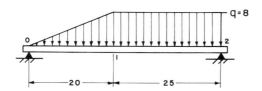

5.13 For the triangularly distributed loading calculate, for the free end
   a. y_4 **b.** θ_4

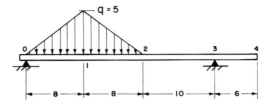

CHAPTER 6

6.1 A cantilever with an auxiliary support carries an end couple M_2. Solve
   for:
   a. V_1 **b.** V_0 **c.** M_0

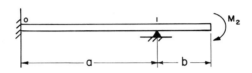

6.2 A shaft with three simple bearing supports has a single load Q. Find
   all support reactions.

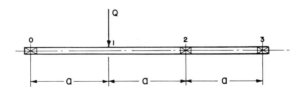

6.3 Repeat Prob. 6.2 altering the 2–3 span so that this shaft section is
   twice that in the 0–2 span ($I_{23} = 2I_{02}$).

6.4 An end-loaded cantilever has an additional spring support at 1.
   Determine the ratio of EI of the beam to K of the spring required for
   the spring to assume one-half of Q.

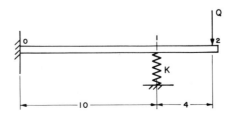

6.5 A fixed–fixed beam has symmetrical distributed loading. Calculate:

    a. V_0 **b.** M_0 **c.** y_2

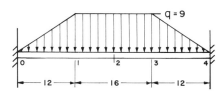

6.6 Two constant horizontal beams with $EI = 180(10)^6$ are connected by three links $[C_a = 5(10)^{-6}$ and $C_b = 3(10)^{-6}]$. Find:

    a. Loads in the three links

    b. Net downward deflection of the twin loads Q

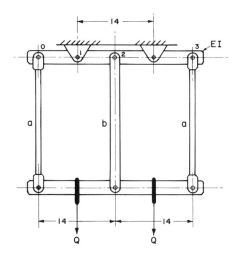

6.7 A symmetrical constant beam $[EI = 0.30(10)^6]$ carries equal end loads and is supported by two springs and a center contact. Before loading, the springs have no preload and the beam just touches the center support. Find the values of K for which there is no reaction at 2.

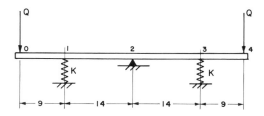

6.8   A fixed–fixed beam carries a concentrated load Q. Determine:
   a. All shear and moment reactions
   b. The deflection at 1

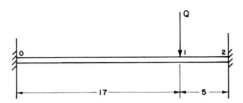

6.9   Similar to Prob. 6.8, but with a partial distributed load. Find all end reactions.

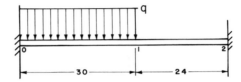

6.10   A cantilevered beam with two outboard supports has distributed loading over its length. Calculate all four external reactions.

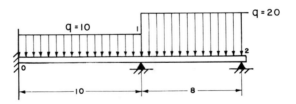

CHAPTER 7

7.1   Rework Prob. 6.1 using the three-moment method to verify moment reactions only.

7.2   Same for Prob. 6.2.

7.3   Same for Prob. 6.5.

7.4   Same for Prob. 6.8.

7.5   Same for Prob. 6.9.

7.6   Same for Prob. 6.10.

7.7 A fixed–fixed beam has a partial uniform load. Find both end moments and end shears at 0 and 3.

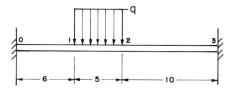

7.8 The fixed–fixed beam has three concentrated loads. Determine all end reactions using the three-moment approach, accounting for all three loads simultaneously.

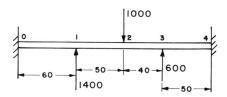

7.9 The rotor in three bearing supports carries a uniformly distributed load q over the entire length. Solve for
a. Internal bending moments at 1 and 2
b. Slope of the tangent to the elastic curve at 2

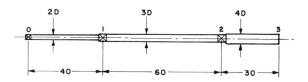

7.10 A fixed–fixed beam has a single load and an additional simple support. Determine the moments and shear reactions at 0, 2, and 3.

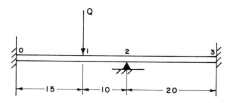

7.11 The cantilevered beam has two auxiliary supports. Find all external reactions at the supports.

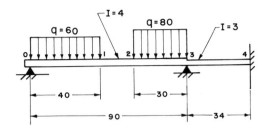

CHAPTER 8

8.1 A cantilevered bent has an end load Q. Find the spatial direction of the resultant deflection θ.

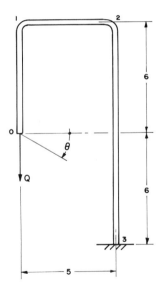

8.2 Instead of a constant section beam in Prob. 8.1, there are three different stiffnesses in the three sections:

$$I_{01} = 0.9 \qquad I_{12} = 1.4 \qquad I_{23} = 1.2$$

Repeat, to find again the angle θ.

8.3 The constant section planar bent has a horizontal end load. Determine the magnitude and direction of

    a. x_4 using the M^2 integration

    b. x_4 using auxiliary loading

    c. y_4

    d. Δ_{24} (the relative deflection)

    e. θ_4

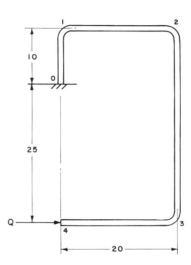

8.4 The bent with a uniformly distributed load is pinned at 0 and supported horizontally at 3. For the midpoint 1 calculate:

    a. y_1 **b.** θ_1

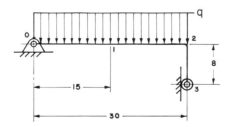

8.5 A rectangular weldment has a total length of 75 in. and weighs 140 lb. For the distributed weight loading, find:

    a. y_3 **b.** θ_3

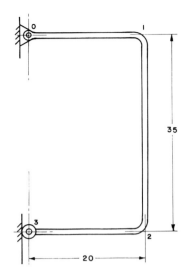

8.6 A beam in the shape of an equilateral triangle has separational loading at the apex. Find the expression for:

a. The relative deflection x_{01}

b. The relative vertical deflection y_{13}

c. y_{13} including both bending and direct stress energy

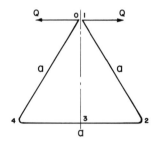

8.7 A pinned–supported vee beam has a horizontal load Q. Determine

a. The horizontal deflection x_2

b. The maximum bending moment in the beam

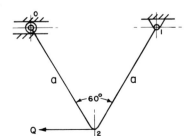

8.8 A long open rectangular tank filled with a liquid of density ρ is filled to a height of 33, and rests upon a rigid planar surface. At what horizontal dimension $2a$ will there be negligible contact pressure on the supporting plane between 2 and 4?

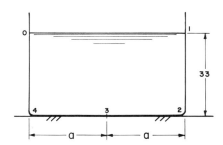

8.9 The bent pinned–supported beam has a concentrated load Q at 1. Find:

a. y_1 **b.** x_3

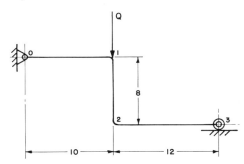

8.10 A cantilevered bent beam has an end load. Determine:

a. y_4 **b.** θ_4

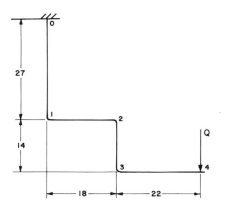

CHAPTER 9

9.1 A bent carries a uniformly distributed load on the horizontal span, and is pin-connected at the base. Determine:

   a. H_0 **b.** M_3 **c.** M_2

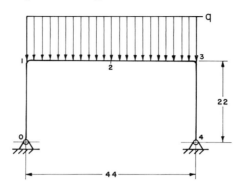

9.2 Repeat Prob. 9.1, but with the vertical legs 0–1 and 3–4 rigid in bending.

9.3 The cantilevered bent has horizontal constraint at 2. Find:

   a. H_2 **b.** M_1 **c.** M_0 **d.** y_2

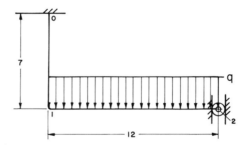

9.4 The continuous square bent is loaded at the corners 0 and 2 by equal but opposed couples. Calculate:

   a. The bending moment at 1

   b. The total relative rotation between the two applied couples

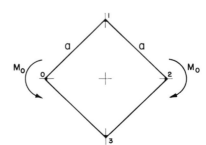

9.5 For the triangularly distributed loading, find:
   a. Bending moments at the corners
   b. Horizontal displacement x_2

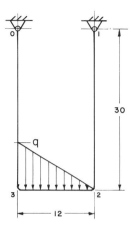

9.6 For a unit strip of an internally pressurized triangular vessel of wall thickness t, calculate:
   a. The maximum bending stress
   b. The direct tensile stress

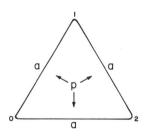

9.7 A rectangular tube has internal pressure. Determine the bending moment at all four corners.

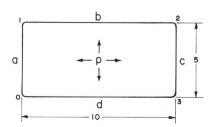

9.8 Repeat Prob. 9.7, but with each side having a different thickness, such that

$$I_a = 4 \qquad I_b = 3 \qquad I_c = 2 \qquad I_d = 1$$

Solve for the bending moment at all four corners.

(**Hint:** The Three-Moment Theorem is not applicable.)

9.9 A continuous square bent is loaded by equal and opposed couples at two corners. Find all internal reactions at all corners.

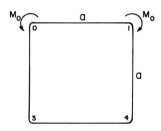

9.10 The assembly rotates at 1000 rpm with the tubes 0–1 welded to the rigid rotating disks at each end. If the tubes weigh 0.09 lb/in. and the radii shown are in inches, find:
a. M_0 **b.** M_1

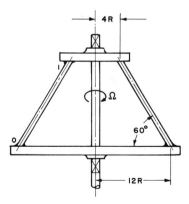

9.11 An axially loaded symmetrical structure consists of a rectangular bent of constant section [$EI = 5(10)^6$ and two auxiliary struts ($C_b = 4(10)^{-6}$]. Neglecting direct stress effects in the beam sections, determine the tensile load in the struts using elastic energy methods.

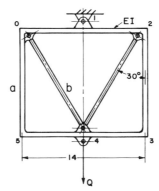

9.12 Repeat Prob. 9.11, but solve using equivalent displacements at 4 and Table 4.1*b* for beam deflection.

CHAPTER 10

10.1 Write the equation for the bending moment in the circular beam as a function of θ and α.

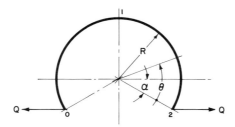

10.2 The cantilevered beam is loaded by its own weight q. Determine the equations for the bending moment:
a. In 2–1 in terms of x
b. In 1–0 in terms of θ

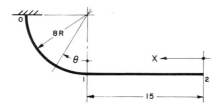

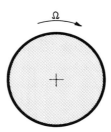

10.11 Repeat Prob. 10.10 if the central half of the shell is empty ($R_0 = R/2$).

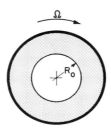

CHAPTER 11

11.1 For the partial circular beam in Prob. 10.1, find the expression for the relative deflection at the end x_{02}.

11.2 For the cantilevered beam of Prob. 10.2, use tabulated factors to find:
    a. y_1 **b.** y_2

11.3 In Prob. 10.3 determine the horizontal separation at the diameter x_{14}.

11.4 An arcuate cantilever has a horizontal load Q. Find:
    a. x_2 **b.** y_2 **c.** x_1 **d.** y_1

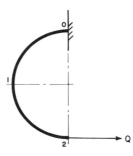

11.5 Repeat Prob. 11.4 for the distributed weight loading.

11.6 The split cylindrical container is half full of a liquid. Calculate:
a. The horizontal separation x_{32}
b. The absolute slope θ_2, with the vertical centerline as a fixed reference

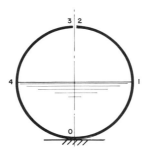

11.7 A symmetrical curved beam has twin loads Q. Using energy methods find:
a. y_2 b. θ_2

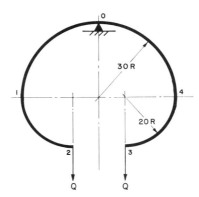

11.8 Verify the Prob. 11.7 results using superposition and tabulated factors.

CHAPTER 12

12.1 The semicircular cantilever beam has a horizontal end support. Find:

    a. V_2 **b.** M_0 **c.** x_2

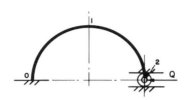

12.2 The complete ring has different diametral loadings as shown. Calculate, by superposition of tabulated deflections:

    a. M_0 **b.** M_1 **c.** y_{02} **d.** x_{13}

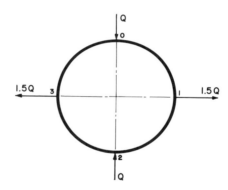

12.3 For the concentrated load on the ring, use energy methods to find

    a. M_0 **b.** H_3 **c.** M_1

(*Hint:* Two auxiliary loadings are required.)

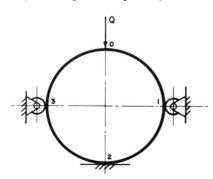

12.4 Verify the Prob. 12.3 results from Table 12.1.

12.5 Verify the Prob. 12.3 results from Table 12.11*b*.

12.6 A long pressurized cylindrical shell is rigidly supported radially on four sides. Determine the contact loading at the sides, using superposition of radial deflections.

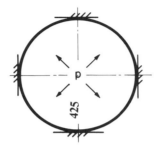

12.7 An antisymmetrical ring has two opposed quadrants that are large enough to be considered rigid. Find:
a. M_0 **b.** H_0 **c.** y_{02}

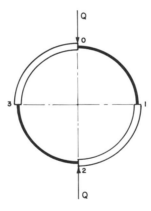

CHAPTER 13

13.1 A square shaft in torsion is to be replaced by a round shaft of the same length and material. Find the (D/b) ratio:
a. For the same deflection
b. For the same maximum stress

13.2 Repeat Prob. 13.1 for a rectangular shaft with $b/h = 3$ replacing the square shaft.

13.3 A hollow rectangular tube is 3 in. × 2 in. outside dimensions with $\frac{1}{8}$ in. walls. What angular displacement is developed in a length of 14 ft at a maximum shear stress of 15,000 lb/in.^{2} if $G = 4(10)^6$ lb/in.^{2}?

13.4 By what percentage are the shear stress and the torsional deflection changed by replacing a solid circular shaft to a hollow circular section with $(d/D) = \frac{2}{3}$ having the same outside diameter?

13.5 The stepped diameter shaft carries a torque. Determine the total angular deflection.

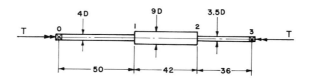

13.6 In Prob. 13.5 we consider the center section essentially rigid. What percentage error results in the angle of twist?

13.7 Gear meshes 1–2 and 3–4 provide speed reductions of 3.5 and 4.5, respectively. All shaft diameters are equal. Find the total angle of twist θ_{05} due to shaft deformation.

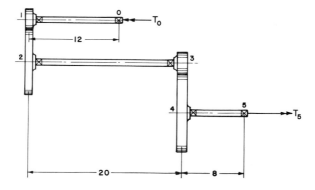

13.8 The shaft has an intermediate torque T_2 and both ends are grounded. Find T_0 and T_4.

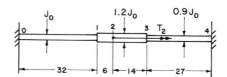

13.9 A cantilevered round beam carries Q on an offset. Find the total strain energy in the beam, including shear.

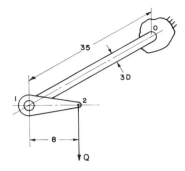

13.10 In Prob. 13.9 what is the vertical deflection of Q due only to transverse shear effects in the beam?

CHAPTER 14

14.1 The cantilevered bar carries an end torque T_1, with this vector in the xy plane, but at an angle to the bar axis. Find:
    a. z_1 **b.** θ_x **c.** θ_y **d.** y_1

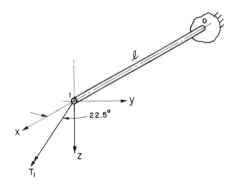

14.2 For the Prob. 13.9 system calculate y_2 due to combined bending and torsion.

14.3 A fixed–fixed beam of rectangular section has a central load at an angle to the vertical plane. Find the magnitude and direction of the resultant deflection Δ_1.

    (***Hint:*** For this beam $y = \frac{1}{192} Q\ell^3 / EI$.)

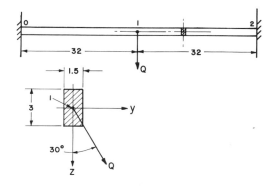

14.4 Repeat Prob. 14.3 with the load in the same direction, but translated from the centroid of the beam section as shown.

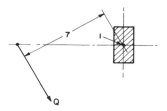

14.5 The bent has a uniform section, except that the two upper arms are rigid. Determine
  a. The absolute slope produced by Q at 6
  b. The relative slope between the ends θ_{06}

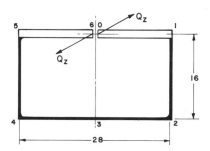

14.6 Verify Prob. 14.5 results using superposition and Table 14.1a.

14.7 The planar bent is loaded by an end couple. With the bent in the horizontal plane, find the vertical deflection z_0.

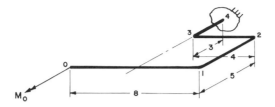

14.8 The Prob. 10.2 cantilever is located in the horizontal plane and subjected to out-of-plane weight loading. Determine:
   a. z_1 **b.** z_2 **c.** M_0 **d.** T_0

14.9 For the opposed couple loading at the ends, determine the total linear relative out-of-plane displacement that results.

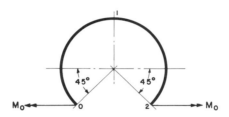

CHAPTER 15

Note: In order to achieve numerical values in the following problems, the ratio $EI/GJ = \lambda$ can be taken as 1.25.

15.1 A bent planar cantilever has an auxiliary simple vertical support at 2. Find the external reactions
   a. V_0 **b.** V_2 **c.** M_0 **d.** T_0

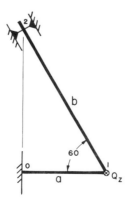

15.2 A rectangular bent is similar to Table 15.5b, except that at 3 the end has a simple vertical support rather than complete fixation. Determine:

    a. V_3 **b.** V_0 **c.** z_1

15.3 A fixed–fixed circular bar in the horizontal plane is subject to its own weight. Find:

    a. M_0 **b.** T_0

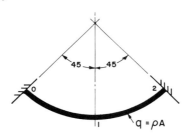

15.4 A cantilevered complete circular bar is constrained vertically at 2 and loaded by an end torque T_4. Calculate the vertical reaction at 2.

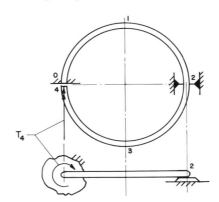

15.5 A fixed–fixed bent has a transverse load at 1. Using elastic energy find:

    a. V_0 **b.** M_0 **c.** T_0 **d.** z_1

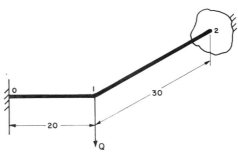

15.6 Repeat Prob. 15.5 using superposition and tabulated values.

15.7 The fixed–fixed bent has a transverse load at the two-thirds point of its outboard length. Solve the structure for
    a. V_4 **b.** M_4 **c.** T_4

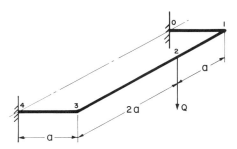

15.8 A complete closed circular ring carries two Q loads, and is supported by a radially connected cantilevered beam completely fixed at the center. Stiffness factors EI and GJ are the same for the ring and the cantilever. By elastic energy methods, find:
    a. M_3 **b.** T_3 **c.** z_3

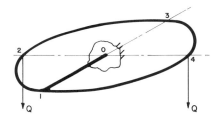

15.9 Repeat Prob. 15.8 using superposition of deflections from the various components.

15.10 A square frame in the horizontal plane is subjected to its own weight, and has a single fixed support at 0. Find:
    a. M_3 **b.** M_0 **c.** T_{12}

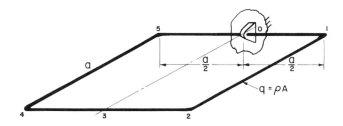

15.11 Verify the results of Prob. 15.10 using superposition.

CHAPTER 16

16.1 A mass of 3.6 kg is attached to a vertical helical spring and the spring stretched 20 cm upward before release. What value of K should apply to the spring if the maximum compressive force transmitted by the spring to its support is to be 450 N?

16.2 A steel helical spring is twisted 30° by a static torque about its central axis, with

$$D = 2 \text{ in.} \qquad d = 0.1875 \text{ in.} \qquad N = 16 \text{ turns}$$

Determine:
a. The strain energy stored in the spring
b. The maximum stress in the spring

16.3 A helical tension spring has an index (D/d) of 8 and a pitch diameter of 3 in. The initial pitch of the coils is 5 in. and there are 10 active turns. Calculate:
a. The spring rate as a close-coiled spring
b. Repeat as an open-coiled spring
c. The twist in degrees with a load of 200 lb applied at the lower end

16.4 A torque tube with an outside diameter twice the inner is to absorb 25 ft lb of kinetic energy torsionally during impact. If the shear stress is limited to 18,000 lb/in.^{2}, what weight of steel is required?

16.5 A mass of 30 kg drops from a height of 3 cm to the center of a simply supported steel beam, 1.50 m long. The beam width is 5 cm and the depth is 2 cm. Neglecting the mass of the beam, find:
a. The maximum beam deflection
b. The maximum beam stress

16.6 A tapered steel beam (Table 4.9b) has dimensions of:

$$h_1 = 1.5 \text{ in.} \qquad h_0 = 3 \text{ in.} \qquad b = 0.90 \text{ in.} \qquad \ell = 24 \text{ in.}$$

What is the total strain energy stored in the cantilever with an end load of 1500 lb?

16.7 Several different configurations are shown for a rectangular beam on simple supports. Determine the specific resilience factor ζ for each case based on the limiting stress.

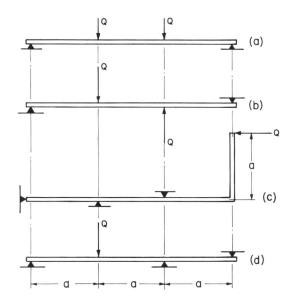

16.8 A split circular beam of rectangular cross section is loaded by separating forces as shown. Calculate the specific resilience factor ζ for this beam.

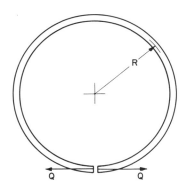

16.9 Determine the effect of stock thickness in a power spring with couple loading (Fig. 16.12c) at a tightly wound mean radius of 0.75 in. for one turn of a 1.0 in. wide strip with stresses as follows:

Thickness	Yield Stress
0.020	267,000 lb/in.^{2}
0.030	245,000
0.040	227,000
0.050	215,000
0.060	205,000

Tabulate the following quantities with respect to thickness:
a. The maximum moment required to form the spring
b. The radius at complete release
c. The usable, or working angle, as a percentage of the reference turn
d. The stress in the outer surface at zero load
e. Strain energy stored per cubic in. when wound
f. Same when released
g. Maximum possible energy storage per cubic in. for the steel
h. Actual usable energy in a working cycle, per cubic in.
i. Theoretical energy efficiency of the spring

16.10 For the data of Prob. 16.9 at $t = 0.060$ in., plot a curve of the forming couple M_1 versus the mean forming radius, from $R = 0$ to $R = 10$ in.

16.11 A standard $\frac{3}{4}$ in. bolt has 10 threads per in. and a tensile stress area in the threads of 0.334 in.^{2}. Maximum tensile yield stress is 40,000 lb/in.^{2}, and the length from nut to head is 3 in. Assuming the threaded region at maximum stress for a length of 0.40 in., determine:
a. Total preload strain energy due to tightening to the indicated stress
b. Same with the body relieved to the threaded area (Fig. 16.14b).

16.12 A cantilever beam has a round section and an end load of 300 N. Maximum stress is 9 kN/cm^2 and the end deflection is 1.10 cm. Find the dimensions and volume of the beam.

Answers to Problems

1.1 a. 0.002 cm b. 0.0023 cm   **1.2** a. 0.0096 in. b. 0.0042 in.

1.3 a. $0.855(10)^{-6}Q$ b. $0.876(10)^{-6}Q$ c. $0.075(10)^{-6}Q$

1.4 b. $0.433(Q_1/K_c)$ c. $0.325(Q_1/K_c)$

1.5 a. $-30°$ **1.6** a. $+60°$

1.8 a. $4.83C_aQ$ b. $3.83C_aQ$

1.9 a. $2.68(10)^{-6}Q$ c. $-0.70(10)^{-6}Q$

1.10 a. $8.38C_aQ$ b. $4.23C_aQ$

1.12 a. $25C_aQ$ b. $15.34C_aQ$ d. $9.34C_aQ$

2.1 $q_a = 0.577$ $q_b = 0$ $q_c = -0.577$

2.3 a. $q_b = -0.389$ b. $0.57(10)^{-6}Q$

2.4 66.5 percent   **2.6** a. $1.033C_cQ$ b. $1.096C_cQ$

2.8 $q_e = -0.20$ **2.10** $q_e = -0.217$

2.12 a. $q_a = 0.89$ **b.** $1.57C_aQ$ **2.13 a.** $q_a = 0.515$

3.1 a. $65,700$ **b.** $-57,100$ **c.** $\dfrac{1.02(10)^6}{E}$

3.2 a. $65,500$ **d.** $\dfrac{0.0077(10)^6}{E}$ **3.4** $\dfrac{2.333M_1}{EI}$

3.6 $2.833(10)^{-6}Q$ **3.7** $\dfrac{9870Q}{EI}$

3.9 $\dfrac{38M_1}{EI}$ **3.11** $\pm 1.52(10)^{-4}E$

3.12 $\dfrac{235.8Q}{EI}$

4.1 $\dfrac{236.5}{EI}$ **4.2 a.** $\dfrac{3}{4}\dfrac{Qa^3}{EI}$

4.4 a. 0.0342 in.   **4.5 a.** $94.3\dfrac{Q}{EI}$

4.6 a. 0.0695 **b.** 0.0695 **4.7 a.** 0.0874

4.9 $24.74\dfrac{Q}{E}$

5.2 a. $\dfrac{4760q}{EI}$ **b.** $\dfrac{1630q}{EI}$ **c.** $\dfrac{952q}{EI}$

5.4 $\dfrac{28,000q}{EI}$ **5.7 a.** $\dfrac{11}{24}\dfrac{qa^4}{EI}$

5.9 b. $\dfrac{4.35q}{EI}$ **5.10** 12.19

5.11 a. $\dfrac{2170}{EI}$ **b.** $-\dfrac{1085}{EI}$ **5.13 b.** $\dfrac{1183}{EI}$

6.1 a. $\dfrac{3}{2}\dfrac{M_2}{a}$ **c.** $\dfrac{1}{2}M_2$ **6.2** $\dfrac{3}{8}Q, \dfrac{7}{8}Q, \dfrac{1}{4}Q$

6.4 733.3

6.6 a. $V_a = 0.434Q$ **b.** $3.30(10)^{-6}Q$

6.7 340 or greater   **6.8 b.** $18.8\dfrac{Q}{EI}$

6.10 $V_1 = 151.3$ $V_2 = 64.6$ $M_0 = 64.2$

7.3 $M_0 = 1016$

7.4 $M_0 = 0.878Q$ $M_2 = 2.986Q$

7.5 $M_0 = 186.1q$

7.7 $M_0 = 14.67q$ $V_0 = 3.192q$

7.9 **a.** $M_2 = 450q$ **b.** $\dfrac{2057q}{EI_3}$

7.11 $V_0 = 1890$ $V_3 = 4400$ $M_4 = -16{,}900$

8.1 $77.6°$ **8.2** $79.2°$

8.3 **b.** $47{,}900\dfrac{Q}{EI}$ **8.4** **a.** $0.138(10)^6\dfrac{q}{EI}$

8.5 **a.** $\dfrac{1.35(10)^6}{EI}$ **8.7** $\dfrac{1}{8}\dfrac{Qa^3}{EI}$

8.9 **a.** $456\dfrac{Q}{EI}$ **b.** $45\dfrac{Q}{EI}$

9.1 **a.** $5.50q$ **c.** $121q$ **9.2** **a.** $7.33q$

9.4 **a.** $\dfrac{1}{4}M_0$ **b.** $\dfrac{1}{4}\dfrac{M_0 a}{EI}$ **9.5** **a.** $2.25q$

9.6 **b.** $0.289q\left(\dfrac{a}{t}\right)$ **9.8** $M_0 = 8.69p$ $M_1 = 5.32p$

9.10 **a.** 416 in. lb   **9.11** $0.334Q$

10.3 7220ρ **10.6** **a.** $0.234qR^2$

10.7 $22{,}100$ lb/in.^{2} **10.8** 53.7 in.

10.11 $\dfrac{3}{8}\dfrac{\rho R^2 \Omega^2}{gt}$

11.2 **a.** $\dfrac{13{,}260}{EI}$ **b.** $\dfrac{54{,}000}{EI}$ **11.3** $\dfrac{2.36(10)^6\rho}{EI}$

11.4 **c.** $1.7854\dfrac{QR^3}{EI}$ **d.** $1.0708\dfrac{QR^3}{EI}$

11.6 **a.** $0.822\dfrac{\rho R^5}{EI}$ **b.** $0.215\dfrac{\rho R^4}{EI}$

11.8 **a.** $14{,}200\dfrac{Q}{EI}$

12.1 **b.** $0.722\dfrac{QR^3}{EI}$

12.3 **a.** $M_0 = 0.1515QR$ $H = 0.918Q$

12.6 $13.7\left(\dfrac{t^2 p}{R}\right)$ **12.7** **c.** $0.0744\dfrac{QR^3}{EI}$

13.1 **a.** 1.095 **b.** 1.019 **13.3** $31.8°$

13.5 $4.5\dfrac{T}{G}$ **13.8** $T_4 = 0.47T_2$

13.10 $5.50\dfrac{Q}{G}$

14.1 **a.** $0.192\dfrac{T_1\ell^2}{EI}$ **14.3** $880\left(\dfrac{Q}{E}\right)$ at $66.6°$

14.5 **a.** $\dfrac{Q}{EI}(98 + \lambda 2.29)$ **b.** 0 **14.7** $\dfrac{M_0}{EI}(56 + \lambda 52)$

14.8 $\dfrac{q}{EI}(37{,}470 + \lambda 26{,}460)$

15.1 **a.** $1.025Q$ **d.** $0.043Qa$ **15.2** **a.** $-0.75q$ **c.** $2450\dfrac{q}{EI}$

15.4 $\dfrac{T_4}{R}\left[\dfrac{1+\lambda}{1+3\lambda}\right]$ **15.7** **a.** $0.333Q$ **b.** $0.354Qa$

15.8 **a.** $0.293QR$ **b.** 0 **c.** $5.21\dfrac{QR^3}{EI}$

15.10 **a.** $0.0139qa^2$ **b.** $0.764qa^2$ **c.** $0.111qa^2$

16.1 18.97 N/cm

16.3 **a.** 109.9 lb/in. **b.** 101.6 lb/in. **c.** $46.3°$

16.5 **a.** 1.71 cm **b.** 18.1 kN/cm^2

16.7 **a.** 0.185 **16.8** $\dfrac{1}{8}\dfrac{\sigma^2}{2E}$

16.11 **a.** 23.8 in. lb **b.** 30.3 in. lb **16.12** $V = 978$ cm^2

Comments on Elastic
Energy Solutions

- All systems are linear, with stress and deflection directly proportional to loading. Thus superposition applies.
- A system must be in static equilibrium under the action of auxiliary loading.
- Equations developed from strain energy based upon elastic behavior are always supplemental to the basic equations of static equilibrium.
- In an indeterminate system, as many equations are required from elasticity as there are redundancies.
- In deflection solutions, the direction is assumed by the direction of the auxiliary loading. A negative result indicates the actual deflection to have opposite sense due to the negative external work.
- In redundant solutions, the sense of a reaction is assumed for purposes of actual loading. A negative result indicates the true reaction to be of an opposite sense. This conclusion is completely independent of the assumed sense of the auxiliary loading.
- Auxiliary loads simulate and are closely related to corresponding redundant constraints and must be so chosen.
- Sign convention for the sense of moments or torques can be established arbitrarily but must then be observed consistently when relating actual to auxiliary loading.

- With axial or bending loads, the sense of the stress produced establishes the corresponding algebraic sign.

- Redundant forces or couples are best assumed to have positive sense, simplifying the sign situation.

- Mistakes in sign are probably the most common source of error in the computational process and should be carefully checked.

- When solving for deflection in an indeterminate system, as many constraints as necessary can be removed to reduce the system to static determinacy for purposes of preloading. There are no restrictions regarding which are selected.

- When developing mM integrals for a beam element, we can proceed from one end for m and the opposite end for M.

- Shear supports can take load in either a positive or negative sense unless otherwise specified.

- If a *symmetrical* actual loading exists in a symmetrical structure and the auxiliary loading is *antisymmetrical*, the resulting complementary energy summation is degenerate and will not yield a valid equation.

- To solve redundant systems by displacement restoration at reactions:

  Maintaining actual loading, decouple by removing as many constraints as necessary to reduce to static determinacy.

  Determine the corresponding deflection at each constraint in the direction of each constraint.

  With the *actual loading removed* apply loads independently to simulate each restraint and find the corresponding load-deflection relationship.

  To write the equations, assume directions of all loads and reactions in accordance with the positive convention for the tabulated deflection factors.

  Sum all loading and reactive displacements to 0.

  Resulting sense of the reactions will be in a direction corresponding to the positive tabulated sign convention if positive and vice versa.

Bibliography

1. Blake, A., "Design of Curved Members for Machines," Kreiger, Huntington, New York, 1979.
2. _____, "Practical Stress Analysis in Engineering Design," Marcel Dekker, Inc., New York, 1982.
3. Griffel, W., "Handbook of Formulas for Stress and Strain," Frederick Ungar, New York, 1966.
4. Roark, R. J., "Formulas for Stress and Strain," 5th ed., McGraw-Hill, New York, 1975.
5. Spotts, M. F., "Design of Machine Elements," 2nd ed., Prentice-Hall, Englewood Cliffs, N.J., 1953.
6. Timoshenko, S., "Strength of Materials," 2nd ed., Van Nostrand, New York, 1940.
7. Ugaral, A. C., and S. K. Fenster, "Advanced Strength and Applied Elasticity," Elsevier, New York, 1981.
8. Van den Broek, J. A., "Elastic Energy Theory," Wiley, New York, 1942.

Appendix

The following tables augment those included in the various chapters. There are solutions for several redundant beam conditions (Tables A.1 and A.2), with single loads. Additionally Tables A.3 through A.5 provide relations for combinations of two equal loads, force or couple.

433

Table A.1. Reactions and deflections of the fixed-supported beam with intermediate and outboard concentrated loads. The loads are either a transverse force or a couple.

	a		b		c		d	
$\alpha=\left(\dfrac{a}{\ell}\right)$	Q	M	Q	M	Q	M		
V_0	$\dfrac{1}{2}[2-3\alpha^2+\alpha^3]$	$\dfrac{M}{\ell}$	$\dfrac{3}{2}\alpha[2-\alpha]$	Q	$\dfrac{3}{2}\alpha$	$\dfrac{M}{\ell}$	$\dfrac{3}{2}$	
V_2	$\dfrac{\alpha^2}{2}[3-\alpha]$	M	$\dfrac{1}{2}[2-6\alpha+3\alpha^2]$	$Q\ell$	$[1+\tfrac{3}{2}\alpha]$	M	$-\dfrac{1}{2}$	
M_0	$\dfrac{\alpha}{2}[2-3\alpha+\alpha^2]$	$M\ell^2/EI$	$\dfrac{\alpha^2}{4}[2-6\alpha+5\alpha^2-\alpha^3]$	$\dfrac{Q\ell^3}{EI}$	$\dfrac{\alpha}{2}$	$\dfrac{M\ell^2}{EI}$	$\dfrac{\alpha}{4}[1+2\alpha]$	
y_1	$\dfrac{\alpha^3}{12}[12-\alpha(3-\alpha)^2]$	$M\ell/EI$	$\dfrac{\alpha}{4}[4-3\alpha(2-\alpha)^2]$	$\dfrac{Q\ell^2}{EI}$	$\dfrac{\alpha^2}{12}[3+4\alpha]$	$\dfrac{M\ell}{EI}$	$\dfrac{1}{4}[1+4\alpha]$	
θ_1	$\dfrac{\alpha^2}{4}[2-6\alpha+5\alpha^2-\alpha^3]$		$\dfrac{\alpha}{4}[-2+3\alpha]$		$\dfrac{\alpha}{4}$		$-\dfrac{1}{4}$	
θ_2	$\dfrac{\alpha^2}{4}[-1+\alpha]$				$\dfrac{\alpha}{4}[1+2\alpha]$			

		a		b
V_0	Q	$[1-3\alpha^2+2\alpha^3]$	$\dfrac{M}{\ell}$	$6\alpha[1-\alpha]$
V_2		$\alpha^2[3-2\alpha]$		
M_0		$\alpha(1-\alpha)^2$		$[1-4\alpha+3\alpha^2]$
M_1	$Q\ell$	$2\alpha^2(1-\alpha)^2$	M	$[1-4\alpha+9\alpha^2-6\alpha^3]$
M_2		$\alpha^2[1+\alpha]$		$\alpha[2-3\alpha]$
y_1	$\dfrac{Q\ell^3}{EI}$	$\dfrac{\alpha^3}{3}(1-\alpha)^3$	$\dfrac{M\ell^2}{EI}$	$\dfrac{\alpha^2}{2}[1-4\alpha+5\alpha^2-2\alpha^3]$
θ_1	$\dfrac{Q\ell^2}{EI}$	$\dfrac{\alpha^2}{2}[1-4\alpha+5\alpha^2-2\alpha^3]$	$\dfrac{M\ell}{EI}$	$\alpha[1-4\alpha+6\alpha^2-3\alpha^3]$

Table A.2. Similar to Table A.1, but for the fixed–fixed beam with intermediate loads.

Table A.3. The symmetrical simply supported constant beam carries twin loads, inboard or outboard of the supports.

		a	b		c	d
y_0		$-b\theta_1$	$y_1-b\theta_1$		$-b\theta_1$	$y_1-b\theta_1$
y_1		0	$\alpha^2[\tfrac{1}{2}+\tfrac{\alpha}{3}]$		0	$\dfrac{\alpha}{2}[1+\alpha]$
y_2	$\dfrac{Q\ell^3}{EI}$	$\dfrac{\alpha^2}{6}[3-4\alpha]$	0	$\dfrac{M\ell^2}{EI}$	$\dfrac{\alpha}{2}[1-2\alpha]$	0
y_3		$\dfrac{\alpha}{6}[\tfrac{3}{4}-\alpha^2]$	$-\dfrac{\alpha}{8}$		$\dfrac{1}{2}[\tfrac{1}{4}-\alpha^2]$	$-\dfrac{1}{8}$
θ_0		$\dfrac{\alpha}{2}[1-\alpha]$	$-\dfrac{\alpha}{2}[1+\alpha]$			$-[\tfrac{1}{2}+\alpha]$
θ_1	$\dfrac{Q\ell^2}{EI}$			$\dfrac{M\ell}{EI}$	$\dfrac{1}{2}[1-2\alpha]$	
θ_2		$\dfrac{\alpha}{2}[1-2\alpha]$	$-\dfrac{\alpha}{2}$			$-\dfrac{1}{2}$

Table A.4. Similar to Table A.3, but for antisymmetrical loading.

		a		b		c		d
y_0		$-b\theta_1$		$y_1-b\theta_1$		$-b\theta_1$		$y_1-b\theta$
y_1		0		$\frac{\alpha^2}{6}[1+2\alpha]$		0		$\frac{\alpha}{6}[1+3\alpha]$
y_2	$\frac{Q\ell^3}{EI}$	$\frac{\alpha^2}{6}[1-4\alpha+4\alpha^2]$		0	$\frac{M\ell^2}{EI}$	$\frac{\alpha}{6}[1-6\alpha+8\alpha^2]$		0
y_3		0		0		0		0
θ_0 θ_1		$\frac{\alpha}{6}[1-3\alpha+2\alpha^2]$		$-\frac{\alpha}{6}[1+3\alpha]$		$\frac{1}{6}[1-6\alpha+6\alpha^2]$		$-\frac{1}{6}[1+6\alpha]$
θ_2	$\frac{Q\ell^2}{EI}$	$\frac{\alpha}{6}[1-6\alpha+8\alpha^2]$		$-\frac{\alpha}{6}$	$\frac{M\ell}{EI}$	$\frac{1}{6}[1-6\alpha+12\alpha^2]$		$-\frac{1}{6}$
θ_3		$\frac{\alpha}{12}[-1+4\alpha^2]$		$\frac{\alpha}{12}$		$\frac{1}{12}[-1+12\alpha^2]$		$\frac{1}{12}$

Table A.5. The fixed–fixed beam is subjected to twin loads, with both the symmetrical and antisymmetrical cases shown. In the latter (*b*) note that the total length is 2ℓ.

$+y$ $+\theta$ $\alpha=\left(\dfrac{a}{\ell}\right)$		a		b	c	d
	Q		**M**			
V_0	Q	1	$\dfrac{M}{\ell}$	0	SEE TABLE A-1a	SEE TABLE A-1b
M_0	$Q\ell$	$\alpha[1-\alpha]$	M	$[1-2\alpha]$		
M_2		$-\alpha^2$	M	-2α		
y_1	$\dfrac{Q\ell^3}{EI}$	$\dfrac{\alpha^3}{6}\left[2-3\alpha\right]$	$\dfrac{M\ell^2}{EI}$	$\dfrac{\alpha^2}{2}\left[1-2\alpha\right]$		
y_2	$\dfrac{Q\ell^2}{EI}$	$\dfrac{\alpha^2}{12}\left[3-8\alpha+6\alpha^2\right]$	$\dfrac{M\ell}{EI}$	$\dfrac{\alpha}{4}\left[1-2\alpha\right]$		
θ_1	$\dfrac{Q\ell^2}{EI}$	$\dfrac{\alpha^2}{2}\left[1-2\alpha\right]$	$\dfrac{M\ell}{EI}$	$\alpha\left[1-2\alpha\right]$		

Index

Note: Italicized page numbers indicate tables.